Micrometeorology

Thomas Foken

Micrometeorology

Edited for English by Carmen J. Nappo

 Springer

Prof. Dr. Thomas Foken
University of Bayreuth
Laboratory of Micrometeorology
Universitaetsstrasse 30
95440 Bayreuth
Germany
Email: thomas.foken@uni-bayreuth.de
http://www.bayceer.uni-bayreuth.de/mm/

ISBN: 978-3-540-74665-2   e-ISBN: 978-3-540-74666-9

Library of Congress Control Number: 2008929337

Original German edition published by Springer-Verlag, Berlin Heidelberg, 2006

Cover illustration: Measuring system consisting on a sonic anemometer (CSAT3, Campbell Sci. Inc.) and an IR gas analyzer (LI-7500, LI-COR Inc.) for estimating the fluxes of sensible and latent heat and carbon dioxide using the eddy-covariance method as explained in Chapter 4.1 (Photograph: Foken)

Typesetting: Camera-ready by the Author

Cover design: WMXDesign GmbH

Printed on acid-free paper

9 8 7 6 5 4 3 2 1

springer.com

Preface

After the successful issue of two editions of the German book *Applied Meteorology – Micrometeorological basic* I am happy that the Springer publishing house has agreed to publish an English edition for a probably much larger community of readers. The present edition is the translation of the second German edition of 2006 with only small corrections and changes. It is named only *Micrometeorology* because this title is more appropriate to the context of the book. I am extremely happy that I found with Carmen Nappo a scientist, who has edited my first translation into the English language in such a way that keeps alive the style of a German or European book and also makes it easily readable.

It was not my aim to transfer the book into a style where the German and Russian backgrounds of my teachers cannot be seen. On the other hand, I hope that the reader will find some references of interest. These are mainly references to German standards or historical sources. The book is addressed to graduate students, scientists, practical workers, and those who need knowledge of micrometeorology for applied or ecological studies. The main parts are written as a textbook, but also included are references to historical sources and recent research even though the final solutions are still under discussion. Throughout the book, the reader will find practical hints, especially in the chapters on experimental techniques where several such hints are given. The appendix should give the reader an overview of many important parameters and equations, which are not easily found in other books.

Now I hope that the number of small printing errors converges to zero, but I would be very happy if the reader would inform me about such problems or necessary clarifications as well as parts that should be added or completed in a hopefully further edition.

I am extremely thankful for the wonderful cooperation with Carmen Nappo in the translation of the book and to Ute who gave me the opportunity and encouragement to do this time consuming work.

Bayreuth and Knoxville, Spring 2008

Thomas Foken

Preface of the 1st German Edition

Even though the beginning of modern micrometeorology was started 60–80 years ago in the German speaking countries, the division of meteorological phenomena according their scales in time and space is generally not used in Germany. Perhaps because the classic book by R. Geiger (1927) *The climate near the ground* focused on a phenomenological description of the processes at the ground surface, the word *micro* became associated with very small-scale processes. The development of micrometeorology combined with progress in turbulence theory started in the 1940s in the former Soviet Union, and continued e.g. in Australia and the USA. In these countries, *micrometeorology* is a well-established part of meteorology of processes near the ground surface occurring on scales some decameters in height and some kilometers in horizontal extension. In the Russian-speaking countries, *experimental meteorology* is more common than micrometeorology, but not connected to meteorological scales, and instead is used for all experiments. Because the area of investigation of micrometeorology is nearly identical with the area of human activity, it is obvious that applied meteorology and micrometeorology are connected; however, the latter is more theoretical and orientated toward basic research. That both fields are connected can be seen in the papers in the *Journal of Applied Meteorology*. In Germany, applied meteorology is well established in the environment-related conferences METTOOLS of the German Meteorological Society, but the necessary basics are found only with difficulty mostly in the English literature. While Flemming (1991) gave a generally intelligible introduction to parts of meteorology which are relevant for applied meteorology, the present book gives the basics of micrometeorology. It is therefore understandable that a book in German will provide not only quick and efficient access to information, but also encourage the use of the German scientific language. Furthermore, this book seeks to give the practical user directly applicable calculating and measuring methods and also provides the basics for further research.

This book has a long history. Over nearly 30 years spent mainly in experimental research in micrometeorology, it was always fascinating to the author, how measuring methods and the applications of measuring devices are directly dependent on the state of the atmospheric turbulence and many influencing phenomena. This connection was first discussed in the book by Dobson et al. (1980). The present book is a modest attempt along these lines. It is nearly a didactically impossible task to combine always divided areas of science and yet show their interactions. Contrary to the classical micrometeorological approach of looking only at uniform ground surfaces covered with low vegetation, this book applies these concepts to heterogeneous terrain covered with tall vegetation. References are given to very recent research, but these results may change with future research. The progressive application of micrometeorological basics in ecology (Campbell and Norman 1998) makes this step necessary.

The sources of this book were lectures given in *Experimental Meteorology* at the Humboldt-University in Berlin, in *Micrometeorology* at the University of

Potsdam, and since 1997 similar courses at the University of Bayreuth. The book would not be possible to write without my German and Russian teachers, my colleagues and co-workers, and master and PhD students who supported me in many cases. Many publishing houses and companies very kindly supported the book with pictures or allowed their reproduction. Mr. Engelbrecht has drawn some new pictures. Special thanks are given to Prof. Dr. H. P. Schmid for his critical review of the German manuscript, and to Ute for her understanding support of the work and for finding weaknesses in the language of the manuscript.

Bayreuth, October 2002

Thomas Foken

Contents

Contents   XI

Symbols

Symbols that are used only in single equations are not included in this list, but are explained in the text.

a	albedo	
a	absolute humidity	kg m^{-3}
a	scalar (general)	*
a_G	molecular heat conductance coefficient in soil	W m^{-1} K^{-1}
a_T	molecular heat conductance coefficient in air	W m^{-1} K^{-1}
A	rain fall	mm
A	Austausch coefficient	*
b	accuracy (bias)	*
b	constant used for REA measurements	
b_{st}	species-dependent constant according to Jarvis	W m^{-2}
B	sublayer Stanton number	
Bo	Bowen ratio	
C_D	drag coefficient	
C_E	Dalton number	
C_G	volumetric heat capacity	W s m^{-3} K^{-1}
C_H	Stanton number	
$C_n^{\,2}$	refraction structure-function parameter	m$^{-2/3}$
$C_T^{\,2}$	temperature structure-function parameter	K m$^{-2/3}$
$C_{x,y}$, Co	cospectrum (general)	*
c	sound speed	m s^{-1}
c	concentration (general)	*
c	comparability	*
c_p	specific heat at constant pressure	J kg^{-1} K^{-1}
c_v	specific heat at constant volume	J kg^{-1} K^{-1}
c_x	structure constant (general)	
d	displacement height	m
d	measuring path	m
D	molecular diffusion constant (general)	*
D	structure function (general)	*
Da_k	Kolmogorov-Damköhler number	
Da_t	turbulent Damköhler number	
DOY	day of the year: Jan. 1 = 1	
e	basis of the natural logarithm	
e	water vapor pressure	hPa
e'	fluctuation of the water vapor pressure	hPa
E	water vapor pressure at saturation	hPa
E	power spectra (general)	*
E_a	ventilation term	hPa m s^{-1}
EQT	equation of time	h

Eu	Euler number	
f	function (general)	
f	frequency	s^{-1}
f	Coriolis parameter	s^{-1}
f	footprint function	*
f_g	cut frequency	s^{-1}
f_N	Nyquist frequency	s^{-1}
F	flux (general)	*
F	power spectra (general)	*
Fi	inverse Froude number	
Fi_o	inverse external Froude number	
F_w	ventilation flow	$kg\ m^{-1}\ s^{-1}$
g	acceleration due to gravity	$m\ s^{-2}$
h	height of a volume element	m
h	wave height	m
H	water depth	m
I	long-wave radiation	$W\ m^{-2}$
$I{\downarrow}$	down-welling long-wave radiation	$W\ m^{-2}$
$I{\uparrow}$	up-welling long-wave radiation	$W\ m^{-2}$
I^*	long-wave net radiation	$W\ m^{-2}$
k	wave number	m^{-1}
k	absorption coefficient	m^{-1}
k	reaction rate	*
K	turbulent diffusion coefficient (general)	$m^{-2}\ s^{-1}$
K_E	turbulent diffusion coefficient of latent heat	$m^{-2}\ s^{-1}$
K_H	turbulent diffusion coefficient of sensible heat	$m^{-2}\ s^{-1}$
K_m	turbulent diffusion coefficient of momentum	$m^{-2}\ s^{-1}$
$K{\downarrow}$	down-welling short-wave radiation (at surface) global radiation	$W\ m^{-2}$
$K{\downarrow}_{extr}$	extra terrestrial radiation	$W\ m^{-2}$
$K{\uparrow}$	reflected short-wave radiation (at the ground surface)	$W\ m^{-2}$
l	mixing length	m
L	Obukhov length	m
L	characteristic length	m
L	distance constant	m
L_h	horizontal characteristic length	m
L_s	shearing parameter	m
L_z	vertical characteristic length	m
LAI	leaf area index	$m^2\ m^{-2}$
m	mixing ratio	$kg\ kg^{-1}$
n	dimensionless frequency	
N	precipitation	mm
N	Brunt-Väisälä frequency	Hz

N	cloud cover	
N	dissipation rate (general)	*
Nu	Nusselt number	
Og	ogive function	*
p	air pressure	hPa
p_0	air pressure at sea level	hPa
p'	pressure fluctuation	hPa
PAR	photosynthetically active radiation	μmol m^{-2} s^{-1}
Pr	Prandtl number	
Pr_t	turbulent Prandtl number	
q	specific humidity	kg kg^{-1}
q_c	specific concentration	*
q_a	specific humidity near the ground	kg kg^{-1}
q_e	conversion factor from specific humidity into water vapor pressure	kg kg^{-1} hPa^{-1}
q_s, q_{sat}	specific humidity by saturation	kg kg^{-1}
q_*	scale of the mixing ratio	kg kg^{-1}
Q	source density (general)	*
Q_c	dry deposition	kg m^{-2} s^{-1}
Q_E	latent heat flux	W m^{-2}
Q_E	latent heat flux expressed as a water column	mm
Q_G	ground heat flux	W m^{-2}
Q_H	sensible heat flux	W m^{-2}
Q_{HB}	buoyancy flux	W m^{-2}
Q^*_s	net radiation	W m^{-2}
Q_η	source density of the η parameter	*
r	correlation coefficient	
r_a, r_t	turbulent atmospheric resistance	s m^{-1}
r_c	canopy resistance	s m^{-1}
r_g	total resistance	s m^{-1}
r_{mt}	molecular-turbulent resistance	s m^{-1}
r_{st}	stomatal resistance	s m^{-1}
r_{si}	stomatal resistance of a single leaf	s m^{-1}
R	resistance	Ω
R	relative humidity	%
R_G	relative humidity near the surface	%
R_L	gas constant of dry air	J kg^{-1} K^{-1}
R_s	relative humidity near the surface	%
R_W	gas constant of water vapor	J kg^{-1} K^{-1}
R	universal gas constant	mol kg^{-1}K^{-1}
Re	Reynolds number	
Re_s	roughness Reynolds number	
Rf	flux Richardson number	
Ri	Richardson number, gradient Richardson number	

Ri_B	bulk Richardson number	
Ri_c	critical Richardson number	
Ro	Rossby number	
s	precision	*
s_c	temperature dependence of specific humidity at saturation	$kg\ kg^{-1}\ K^{-1}$
S	power spectra (general)	*
S	solar constant	$W\ m^{-2}$
Sc	Schmidt number	
Sc_t	turbulent Schmidt number	
Sd	duration of sunshine	h
Sd_0	astronomical maximal possible sunshine duration	h
Sf	radiation error	K
t	time	s
t	temperature	°C
t'	wet-bulb temperature	°C
T	transfer function	
T	temperature, temperature difference	K
T'	fluctuation of the temperature	K
T_*	temperature scale	K
T^+	dimensionless temperature	
T_K	transmission coefficient	
T_0	surface temperature	K
T_p	wavelet coefficient	*
T_s	sonic temperature	K
T_v	virtual temperature	K
u	wind speed	$m\ s^{-1}$
u	longitudinal component of the wind velocity	$m\ s^{-1}$
u_g	horizontal component of the geostrophic wind velocity	$m\ s^{-1}$
u_{10}	wind velocity at 10 m height	$m\ s^{-1}$
u'	fluctuation of the longitudinal component of the wind velocity	$m\ s^{-1}$
u_*	friction velocity	$m\ s^{-1}$
v	lateral component of the wind velocity	$m\ s^{-1}$
v_g	lateral component of the geostrophic wind velocity	$m\ s^{-1}$
v'	fluctuation of the lateral component of the wind velocity	$m\ s^{-1}$
v_D	deposition velocity	$m\ s^{-1}$
V	characteristic velocity	$m\ s^{-1}$
w	vertical component of the wind velocity	$m\ s^{-1}$

w'	fluctuation of the vertical component of the wind velocity	m s^{-1}
w_*	Deardorff (convective) velocity	m s^{-1}
w_0	deadband for REA method	m s^{-1}
x	fetch	m
x	horizontal direction (length)	m
x, X	measuring variable (general)	*
y	horizontal direction (length, perpendicular to x)	m
y, Y	measuring variable (general)	*
Z	geopotential height	m
z	height (general, geometric)	m
z_i	mixed-layer height	m
z_m	measuring height	m
z_R	reference height	m
z_0	roughness parameter, roughness height	m
z_{0eff}	effective roughness height	m
z_{0E}	roughness height for water vapor pressure	m
z_{0T}	roughness height for temperature	m
z'	height (aerodynamic)	m
z^+	dimensionless height	
z_*	height of the roughness sublayer	m
α	angle of inflow	°
α_{pt}	Priestley-Taylor coefficient	
α_0	ratio of the exchange coefficients of sensible heat to momentum	
α_{0E}	ratio of the exchange coefficients of latent heat to momentum	
β	Kolmogorov constant (general)	
γ	psychrometeric constant	hPa K^{-1}
γ	factor in O'KEYPS-Formel	
Γ	profile coefficient	m s^{-1}
Γ_d	dry adiabatic temperature gradient	K m^{-1}
δ	depth of the internal boundary layer	m
δ	depth of the molecular-turbulent (viscous) sublayer	m
δ_{ij}	Kronecker symbol	
δ_w	thickness of the mixing layer	m
δ_T	thickness of the thermal internal boundary layer	m
δ_T	thickness of the molecular temperature boundary layer	m
δ_T^+	dimensionless thickness of the molecular temperature boundary layer	m
Δc	concentration difference	*
Δe	water vapor pressure difference	hPa

ΔP	pressure difference	hPa
ΔT	temperature difference	K
Δu	wind velocity difference	m s^{-1}
Δz	height difference	m
ΔQ_S	energy source or sink	W m^{-2}
ΔS_W	water source or sink	mm
Δ_z	characteristic vertical gradient	*
ε	energy dissipation	m^2 s^{-3}
ε_{IR}	infrared emissivity	
ε_{ijk}	Levi-Civita epsilon tensor	
ζ	dimensionless height z/L	
η	measurement variable	*
θ	potential temperature	K
θ_v	virtual potential temperature	K
κ	von-Kármán constant	
λ	heat of evaporation for water	J kg^{-1}
Λ	local Obukhov length	m
Λ_u	Eulerian turbulent length scale for the horizontal wind	m
Λ_x	ramp structure-parameter	m
μ	dynamic viscosity	kg m^{-1} s^{-1}
v	kinematic viscosity	m^2 s^{-1}
v_T	thermal diffusion coefficient	m^2 s^{-1}
ξ	scalar (general)	*
ξ	time delay	s
ρ	air density	kg m^{-3}
ρ'	air density fluctuation	kg m^{-3}
ρ_c	partial density	kg^2 kg^{-1} m^{-3}
σ_{cH}	cloud cover for high cloud	
σ_{cL}	cloud cover for low clouds	
σ_{cM}	cloud cover for middle high clouds	
σ_c	standard deviation of the concentration	*
σ_{SB}	Stefan-Boltzmann constant	W m^{-2} K^{-4}
σ_u	standard deviation of the longitudinal wind component	m s^{-1}
σ_v	standard deviation of the lateral wind component	m s^{-1}
σ_w	standard deviation of the vertical wind component	m s^{-1}
σ_T	standard deviation of the temperature	K
σ_θ	standard deviation of the potential temperature	K
σ_φ	standard deviation of the wind direction	°
τ	dew point temperature	K

τ	shear stress	$\mathrm{kg\ m^{-1}\ s^{-2}}$
τ	time constant	s
Φ	geopotential	$\mathrm{m^2\ s^2}$
φ	geographical latitude	°
φ_m	universal function for momentum exchange	
φ_H	universal function for sensible heat flux	
φ_E	universal function for latent heat flux	
φ_T	universal function for temperature structure function parameter	
φ_ε	universal function for energy dissipation	
φ_*	correction function for the roughness sublayer	
χ	scalar (general)	*
ψ	integral of the universal function	
Ψ	inclination of the sun	°
ω	circular frequency	
Ω	angular velocity of the rotation of the Earth	$\mathrm{s^{-1}}$

Indices:

a	air
w	water

Remark:
    * dimension according to the use of the parameter

1 General Basics

This introductory chapter gives the basics for this book, and terms such as micro-meteorology, atmospheric boundary layer, and meteorological scales are defined and presented in relation to the subject matter of this book. Besides an historical outline, the energy and water balance equations at the Earth's surface and the transport processes are discussed. The micrometeorological basics are given, which will be advanced in the following theoretical and experimental chapters.

1.1 Micrometeorology

Meteorology is one of the oldest sciences in the world. It can be traced back to Aristotle (384–322 BCE), who wrote the four volumes of the book *Meteorology*. In ancient times, appearances in the air were called meteors. In the first half of the 20th century, the upper soil layers were considered part of meteorology (Hann and Süring 1939). Today meteorology is understood in a very general sense to be the science of the atmosphere (Dutton 2002; Glickman 2000; Kraus 2004), and includes also the mean states (climatology). Sometimes the definition of meteorology is very narrow, and only related to the physics of the atmosphere or weather prediction. To understand atmospheric processes, many other sciences such as physics, chemistry, biology and all geosciences are necessary, and it is not easy to find the boundaries of these disciplines. In a pragmatic way, meteorology is related only to processes that take place *in-situ* in the atmosphere, while other sciences can investigate processes and reactions in the laboratory. This underlines the specific character of meteorology, *i.e.*, a science that investigates an open system with a great number of atmospheric influences operating at all times but with changing intensity. Meteorology is subdivided into branches (Glickman 2000; Houghton 1985; Hupfer and Kuttler 2005; Kraus 2004). The main branches are theoretical meteorology, observational meteorology, and applied meteorology. Applied meteorology includes weather prediction and climatology. Climatology must be seen in a much wider geosciences context. The subdivision is continued up to special areas of investigation such as maritime meteorology.

The applications of time and space scales became popular over the last 50 years, and subdivisions into macro-, meso- and micrometeorology were developed. Micrometeorology is not restricted to particular processes, but to the time and space scales of these processes (see Chap. 1.2). The significance of micrometeorology is in this limitation. The living environment of mankind is the main object of micrometeorological research. This is the atmospheric surface layer, the lowest 5–10% of the atmospheric boundary layer, which ranges in depth from about 0.5 to 2 km.

The surface layer is one of the main energy-exchange layers of the atmosphere, and accordingly the transformations of solar energy into other forms of energy are a main subject of micrometeorology. Furthermore, the surface layer is a source of friction, which causes a dramatic modification of the wind field and the exchange processes between the Earth's surface and the free troposphere. Due to the coupling of time and space scales in the atmosphere, the relevant time scale of micrometeorology is less than the diurnal cycle. A recent definition of micrometeorology is (Glickman 2000):

Micrometeorology is a part of meteorology that deals with observations and processes in the smaller scales of time and space, approximately smaller than 1 km and one day. Micrometeorological processes are limited to shallow layers with frictional influence (slightly larger phenomena such as convective thermals are not part of micrometeorology). Therefore, the subject of micrometeorology is the bottom of the atmospheric boundary layer, namely, the surface layer. Exchange processes of energy, gases, *etc.*, between the atmosphere and the underlying surface (water, soil, plants) are important topics. Microclimatology describes the time-averaged (long-term) micrometeorological processes while the micrometeorologist is interested in their fluctuations.

If one examines the areas of investigation of applied meteorology (Fig. 1.1) it will be apparent that the main topics are related to microscale processes. Therefore, we see that micrometeorology gives the theoretical, experimental, and climatological basis for most of the applied parts of meteorology, which are related to the surface layer and the atmospheric boundary layer. Also, recent definitions such as environmental meteorology are related to micrometeorology. Applied meteorology often includes weather prediction and the study of air pollution.

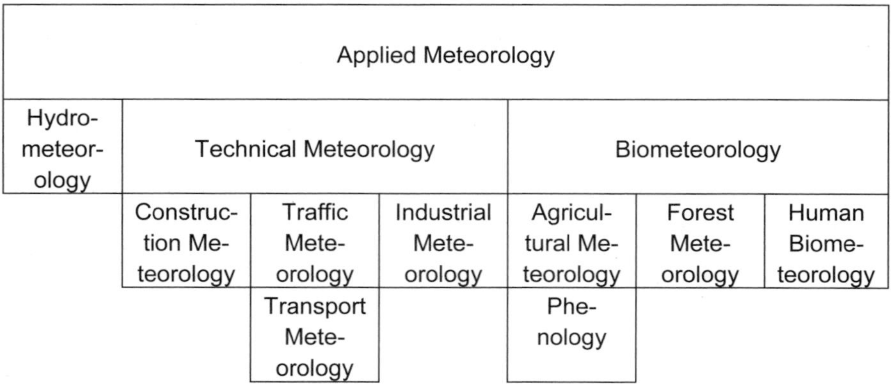

Applied Meteorology							
Hydro-meteor-ology	Technical Meteorology			Biometeorology			
	Construc-tion Me-teorology	Traffic Mete-orology	Industrial Mete-orology	Agricul-tural Me-teorology	Forest Mete-orology	Human Biome-teorology
		Transport Mete-orology		Phe-nology		

Fig. 1.1. Classification of applied meteorology

The basics of micrometeorology come from hydrodynamics. The following historical remarks are based on papers by Lumley and Yaglom (2001) and Foken (2006a). The origin may be dated to the year 1895, when Reynolds (1894) defined the averaging of turbulent processes, and described the equation of turbulent energy. Further steps were the mixing length approach by Taylor (1915) and Prandtl (1925), and the consideration of buoyancy effects by Richardson (1920). The every-day used term "turbulence element" is based on Barkov (1914), who found these in wind observations analyzed during a long stay in winter in the Antarctic ice shield. The actual term *micrometeorology* is dated to the determination of energy and matter exchange and the formulation of the *Austausch coefficient* by Schmidt (1925) in Vienna. At the same time in Munich, Geiger (1927) summarized microclimatological works in his famous book *The Climate near the Ground*, which is still in print (Geiger *et al.* 1995). The experimental and climatological application of these investigations of turbulent exchange processes were done mainly by Albrecht (1940) in Potsdam, who also wrote parts of the classical textbook by Kleinschmidt (1935) on meteorological instruments. In Leipzig, Lettau (1939) investigated the atmospheric boundary layer and continued his investigation after the Second World War in the U.S.A. (Lettau and Davidson 1957). With the end of the Second World War, an era of more than 20 years of famous German-speaking micrometeorological scientists ended, but the word *Austausch coefficient* was maintained from that period.

The origin of modern turbulence research in micrometeorology was in the 1940s in Russia. Following the fundamental studies on isotropic turbulence and the turbulence spectra by Kármán and Howardt (1938) and Taylor (1938), Kolmogorov (1941a, b) gave theoretical reasons for the turbulence spectra. In 1943, Obukhov (published in 1946), found a scaling parameter that connects all near-surface turbulence processes. This paper (Obukhov 1971) was so important because of its relevance to micrometeorology, it was once again published by Businger and Yaglom (1971). A similar paper was published by Lettau (1949), but was not applied because it used a different scaling. Using what became known as *similarity theory*, Monin and Obukhov (1954) created the basics of the modern stability-dependent determination of the exchange process. At the same time, a direct method to measure turbulent fluxes was developed (Montgomery 1948; Obukhov 1951; Swinbank 1951), which has become known as the *eddy-covariance* method. This method became truly established only after the development of the sonic anemometer, for which the basic equations were given by Schotland (1955). Following the development of a sonic thermometer by Barrett and Suomi (1949), a vertical sonic anemometer with 1m path length (Suomi 1957) was used in the O'Neill experiment in 1953 (Lettau and Davidson 1957). The design of today's anemometers was first developed by Bovscheverov and Voronov (1960), and later improved by Kaimal and Businger (1963) and Mitsuta (1966). The first anemometers used the phase shift between the transmitted and the received signal. The latter anemometers measured the time difference of the the running time along the path in both directions between the transmitted and received signal, and recent anemometers measure the time directly (Hanafusa *et al.* 1982). Following the early work by Sheppard (1947), the surface stress was directly measured with a drag plate in Australia (Bradley 1968b), and the sensible and latent heat fluxes were measured with highly sensitive modified classical sensors (Dyer *et al.* 1967).

These findings formed the basis for many famous experiments (see Appendix A5). Among them were the many prominent Australian experiments for studying turbulent exchange processes (Garratt and Hicks 1990). These were the so-called intercomparison experiments for turbulence sensors (Dyer *et al.* 1982; Miyake *et al.* 1971; Tsvang *et al.* 1973, 1985). Present day investigations are based mainly on the *KANSAS 1968* experiment (Izumi 1971; Kaimal and Wyngaard 1990). That experiment became the basis for the formulation of the universal functions (Businger *et al.* 1971) and the turbulence energy equation (Wyngaard *et al.* 1971a), which were based on an obscure earlier paper by Obukhov (1960). Twenty years after the issue of the first textbook on micrometeorology by Sutton (1953), an important state-of-the-art summary of turbulent exchange between the lower atmosphere and the surface was given in 1973 at the *Workshop on Micrometeorology* (Haugen 1973).

After some criticism of the experimental design of the KANSAS experiment by Wieringa (1980), and the reply by Wyngaard *et al.* (1982), who recommended a repetition of the experiment, several micrometeorological experiments were conducted, including investigations of fundamental micrometeorological issues, for example the Swedish experiments at Lövsta (Högström 1990). Finally, the corrected universal functions by Högström (1988) comprise our most current knowledge.

At the end of the 1980s, the step to micrometeorological experiments in heterogeneous terrain became possible. At about the same time, similar experiments were conducted in the U.S.A. (FIFE, Sellers *et al.* 1988), in France (HAPEX, André *et al.* 1990) and in Russia (KUREX, Tsvang *et al.* 1991). These experiments were to become the bases of many further experiments (see Appendix A5).

During the last 30 years, there were many of updates and increases in precision in the experimental and theoretical fields, but significant new results in problems such as heterogeneous surfaces and stable stratifications are still missing.

1.2 Atmospheric Scales

Contrary to other geophysical processes, meteorological processes have a clear time-space scaling (Beniston 1998). The reason for this is the spectral organization of atmospheric processes, where relevant wavelengths (extensions) are related to distinct durations in time (frequencies). The largest wavelengths are those of atmospheric circulation systems with 3–6 days duration and an extension of some thousands of kilometers (Rossby waves). The daily cycle is a prominent frequency. The time interval from minutes to seconds is characterized by exchange processes of energy and matter in the range of the so-called micro-turbulence, a main topic of micrometeorological investigations (see Chap. 1.4.3). The principle of classification was formulated by Orlanski (1975), see Fig. 1.2a. While atmospheric processes are strictly organized, the hydrological processes in soils and plant ecosystems have for the same time scales significantly smaller space scales. For example, in the case of coupling hydrologic with atmospheric models one must develop strategies to transfer hydrologic processes to the larger atmospheric space scale (Fig. 1.2b).

Different scale definitions in different scientific disciplines create problems in interdisciplinary communications. In addition, in climatology textbooks different scale concepts are discussed which are not easily comparable with the very clear meteorological scale concepts (see Chap. 7.1).

Because weather phenomena are classified according to space-time scales, the space-time scales of climate and weather prediction models must be classified in a similar way. For example, large-scale circulation models are assigned to the macro-β range. The classical weather forecast was formally related to the meso-α range, but with today's high-resolution models they are related to the meso-β,γ scale. Micrometeorology is related to all micro-scales and also partly to the meso-γ scale.

This scaling principle is basic for measurements of atmospheric processes. For instance, to measure the spatial extension of a small low-pressure system, high-resolution (small scale) measurements are necessary. The same is true for the movement of systems. The frequency of these measurements must be related to the velocity of the pressure system. This is valid for all scales, and phenomena can only be observed if the measurements are made on smaller space and time scales. Therefore, the sampling theorem (see Chap. 6.1.2) is of prime importance for all meteorological measurements and scales.

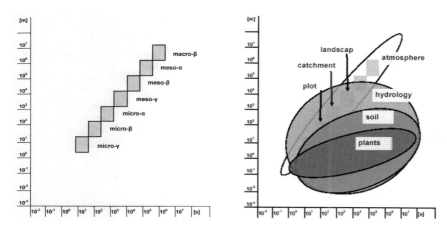

Fig. 1.2. (a) Scales of atmospheric processes according to Orlanski (1975); **(b)** Atmospheric scales added by typical scales in hydrology (Blöschl and Sivapalan 1995), in soils (Vogel and Roth 2003) and in plant ecosystems (Schoonmaker 1998)

1.3 Atmospheric Boundary Layer

The atmospheric boundary layer is the lowest part of the troposphere near the ground where the friction stress decreases with height. The wind velocity decreases significantly from its geostrophic value above the boundary layer to the wind near the surface, and the wind direction changes counter-clockwise up to 30–45° in the Northern hemisphere. In addition, thermal properties influence the boundary layer. Frequently the synonym *planetary boundary layer* is used in theoretical meteorology, where the general regularities of the boundary layers of planetary atmospheres are investigated. Above the boundary layer, lays a mostly statically-stable layer (inversion) with intermittent turbulence. The exchange processes between the atmospheric boundary layer and the troposphere take place in the entrainment zone. The thickness of this layer is approximately 10% of the atmospheric boundary layer, which has a thickness of about 1–2 km over land and 0.5 km over the oceans. For strong stable stratification, its thickness can be about 10 m or less.

The daily cycle is highly variable (Stull 1988), see Fig. 1.3. After sunrise, the atmosphere is warmed by the turbulent heat flux from the ground surface, and the inversion layer formed during the night brakes up. The new layer is very turbulent, well mixed (mixed layer), and bounded above by an entrainment zone. Shortly before sunset, the stable (night time) boundary layer develops near the ground. This stable layer has the character of a surface inversion and is only some 100 m deep. Above this layer, the mixed layer of the day is now much less turbulent. It is called the residual-layer, and is capped by a free (capping) inversion – the upper border of the boundary layer (Seibert *et al.* 2000). After sunrise, the developing mixed layer quickly destroys the stable boundary layer and the residual layer. On cloudy days and in winter, when the solar radiation and the energy transport to the surface are weak, the mixed layer will not disturb the residual layer, and the boundary layer is generally stratified. On days with strong solar radiation, the boundary layer structure will be destroyed by convective cells, which develop some 10 m above the ground. These cells have relatively small upwind cells relatively high vertical wind speeds, and typically develop over large areas with uniform surface heating. This is according to model studies over areas which are larger than 200–500 m^2 (Shen and Leclerc 1995).

In the upper part of the atmospheric boundary layer (upper layer or Ekman layer), the changes of the wind direction take place. The lowest 10% is called the surface or Prandtl layer (Fig. 1.4). Its height is approximately 20–50 m in the case of unstable stratification and a few meters for stable stratification. It is also called the *constant flux layer* because of the assumption of constant fluxes with height within the layer. This offers the possibility to estimate, for example, the fluxes of sensible and latent heat in this layer while in the upper layers flux divergences dominate. The atmospheric boundary layer is turbulent to a large degree, and only within a few millimeters above the surface do the molecular exchange processes dominate. Because the turbulent processes are about 10^5 fold more effective than molecular processes and because of the assumption of a constant flux, the linear vertical gradients of properties very near the surface must be very large. For example, temperature gradients up to 10^3 Km^{-1} have been measured (Fig. 1.5). Between this molecular boundary layer (term used for scalars) or laminar boundary layer (term used for the flow field) and the turbulent layer, a viscous sublayer (buffer layer) exists with mixed exchange conditions and a thickness of about 1 cm. According to the similarity theory of Monin and Obukhov (1954), a layer with a thickness of approximately 1 m (dynamical sublayer) is not influenced by the atmospheric stability – this layer is nearly neutral all of the time.

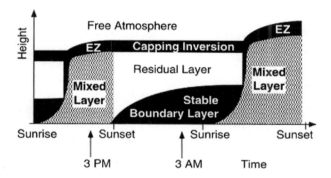

Fig. 1.3. Daily cycle of the structure of the atmospheric boundary layer (Stull 2000), EZ: Entrainment zone

height in m	name		exchange		stability
1000	upper layer (Ekman-layer)		turbulent	no const. flux	influence of stability
20	turbulent layer	surface layer (Prandtl-layer)		flux constant with height	
1	dynamical sublayer				no influence of stability
0.01	viscous sublayer		molecular/ turbulent		
0.001	laminar boundary layer		molecular		

Fig. 1.4. Structure of the atmospheric boundary layer

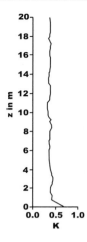

Fig. 1.5. Vertical temperature profile above a water surface with a molecular boundary layer, which has an linear temperature gradient (Foken *et al.* 1978)

All processes in the atmospheric boundary layer, mainly in the micrometeorological range near the ground surface (nearly neutral stratification), can be compared easily with measurements made in the laboratory (wind tunnels and water channels). Thus, the research of the hydrodynamics of boundary layers at a wall, for example by Prandtl, is applicable to atmospheric boundary layer research (Monin and Yaglom 1973, 1975; Oertel 2004; Schlichting and Gersten 2003). As will be shown in the following chapters, knowledge of micrometeorological is based to a large extent on hydrodynamic investigations. In the wind tunnel, many processes can be studied more easily than in nature. But, the reproduction of atmospheric processes in the wind tunnel also means a transformation of all the similarity numbers (see Chap. 2.1.2). Therefore, non-neutral processes can be studied in the laboratory only if extremely large temperature or density gradients can be realized in the fluid channels.

1.4 Energy Balance at the Earth's Surface

The earth's surface is the main energy transfer area for atmospheric processes (Fig. 1.6). It is heated by the shortwave down-welling irradiation from the sun ($K\downarrow$), and only a part of this radiation is reflected back ($K\uparrow$). Furthermore, the surface absorbs longwave down-welling radiation due to longwave emission by clouds, particles and gases ($I\downarrow$). The absorbed energy is emitted only partly into the atmosphere as longwave up-welling radiation ($I\uparrow$). In the total balance, the earth's surface receives more radiation energy than is lost, *i.e.* the net radiation at the ground surface is positive ($-Q_s^*$, see Chap. 1.4.1). The surplus of supplied energy will be transported back to the atmosphere due to two turbulent energy fluxes (see Chap. 1.4.3), the sensible heat flux (Q_H) and the latent heat flux (Q_E, evaporation). Furthermore, energy is transported into the soil due to the ground heat flux (Q_G) (see Chap. 1.4.2) and will be stored by plants, buildings, *etc.* (ΔQ_S). The sensible heat flux is responsible for heating the atmosphere from the surface up to some 100 m during the day, except for days with strong convection. The energy balance at the earth's surface according to the law of energy conservation (see also Chap. 3.7) is:

$$-Q_s^* = Q_H + Q_E + Q_G + \Delta Q_s \qquad (1.1)$$

The following convention will be applied:

> Radiation and energy fluxes are positive if they transport energy away from the earth's surface (into the atmosphere or into the ground), otherwise they are negative.

The advantage of this convention is that the turbulent fluxes and the ground heat flux are positive at noon. This convention is not used in a uniform way in the literature, *e.g.* in macro-scale meteorology the opposite sign is used. Often, all upward directed fluxes are assumed as positive (Stull 1988). In this case, the ground heat flux has the opposite sign of that given above.

The components of the energy balance shown in schematic form in Fig. 1.7 are for a cloudless day with unlimited irradiation. Changing cloudiness can produce typical variations of the terms of the energy balance. The same variations occur in the case of some micrometeorological processes, which will be discussed in the following chapters. The most important variances are a positive latent heat flux after sunset and a negative sensible heat flux that begins in the early afternoon (oasis effect). A negative sensible heat flux (evaporation) is identical with dewfall. The long-term mean values of the earth's energy balance are given in Supplement 1.1. Even though the values of the radiation fluxes are high (Supplement 1.2), the energy balance with a value of 102 Wm^{-2} is relatively low. The 1 K increase of the global mean temperature in the last century due to anthropogenic green house gas emissions corresponds to an additional radiation energy of 2 Wm^{-2}. Therefore, changes in the radiation and energy fluxes, *e.g.* due to changes in land use, can be a significant manipulation of the climate system.

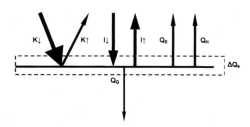

Fig. 1.6. Schematic diagram of the radiation and energy fluxes at the earth's surface. The net radiation according to Eq. (1.1) is the sum of the shortwave (K) and longwave radiation fluxes (I). In addition to the turbulent fluxes (Q_H and Q_E), the energy storage ΔQ_s in the air, in the plants, and in the soil are given.

Supplement 1.1.[*]   Components of the energy and radiation balance

This table gives the climatological time-average values of the components of the energy and radiation balance in W m^{-2} for the whole earth (Kiehl and Trenberth 1997):

reference level	K↓	K↑	I↓	I↑	Q_E	Q_H
upper atmosphere	−342	107	0	235	0	0
ground surface	−198	30	−324	390	78	24

Supplement 1.2.   Energy and radiation fluxes in meteorology

Energy and radiation fluxes in meteorology are given in terms of densities. While the unit of energy is the Joule (J) and for power is Watt (W= J s^{-1}), the unit for the energy flux density is W m^{-2}. It appears that the energy flux density has *apparently* no relation to time, but the exact unit is J s^{-1} m^{-2}. To determine the energy that one square meter gets during one hour, multiply the energy flux density by 3600 s. Energy fluxes expressed as J m^{-2} are unusual in meteorology except daily sums which are given in MJ m^{-2} (Mega-Joules per square meter), and sometimes kWh (kilo-Watt-hours).

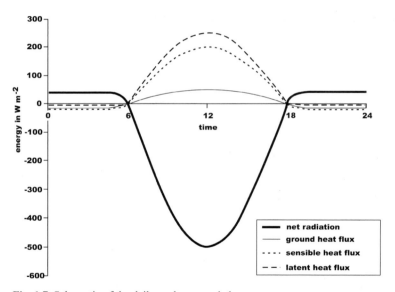

Fig. 1.7. Schematic of the daily cycle energy balance

[*] Supplements are short summaries from textbooks in meteorology and other sciences. These are included for an enhanced understanding of this book or for comparisons. For details, see the relevant textbooks.

1.4.1 Net Radiation at the Earth's Surface

The radiation in the atmosphere is divided into shortwave (solar) radiation and long-wave (heat) radiation. Longwave radiation has wavelengths > 3 µm (Supplement 1.3). The net radiation at the ground surface is given by:

$$Q_s^* = K\uparrow + K\downarrow + I\uparrow + I\downarrow \tag{1.2}$$

From Eq. (1.2) we see that the net radiation is the sum of the shortwave down-welling radiation mainly from the sun (global radiation), the longwave down-welling infrared (heat) radiation emitted by clouds, aerosols, and gases, the short-wave up-welling reflected (solar) radiation, and the longwave up-welling infrared (heat) radiation. The shortwave radiation can be divided into the diffuse radiation from the sky and direct solar radiation.

Supplement 1.1 gives the climatological-averages of the magnitudes of the components of the energy and radiation balance equation. These values are based on recent measurements of the mean solar incoming shortwave radiation at the upper boundary of the atmosphere, i.e. $S = -1368$ Wm^{-2} (Glickman 2000; Houghton et al. 2001) in energetic units or -1.125 Kms^{-1} in kinematic units. For conversion between these units, see Chapter 2.3.1. Figure 1.8 shows the typical daily cycle of the components of the net radiation. The ratio of reflected to the incoming shortwave radiation is called the albedo:

$$a = -\frac{K\uparrow}{K\downarrow} \tag{1.3}$$

In Table 1.1 the albedos for different surfaces are given.

Supplement 1.3.   Spectral classification of the radiation

Spectral classification of short and long wave radiation (Guderian 2000; Liou 1992)

notation	wave length in µm	remarks
ultraviolet radiation		
UV-C-range	0.100–0.280	does not penetrate the atmosphere
UV-B-range	0.280–0.315	does partly penetrate the atmosphere
UV-A-range	0.315–0.400	penetrates the atmosphere
visible radiation		
S-A-range	0.400–0.520	violet to green
S-B-range	0.520–0.620	green to red
S-C-range	0.620–0.760	red
infrared radiation		
IR-A-range	0.76–1.4	near infrared
IR-B-range	1.4–3.0	
IR-C-range	3.0–24	mean infrared

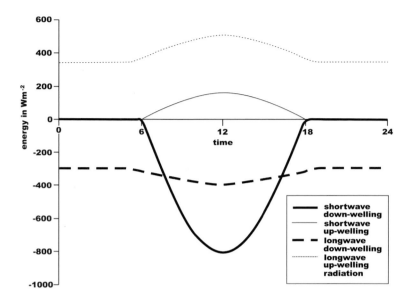

Fig. 1.8. Schematic of the diurnal cycle of the components of the net radiation

Table 1.1. Albedo of different surfaces (Geiger *et al.* 1995)

surface	albedo
clean snow	0.75–0.98
grey soil, dry	0.25–0.30
grey soil, wet	0.10–0.12
white sand	0.34–0.40
wheat	0.10–0.25
grass	0.18–0.20
oaks	0.18
pine	0.14
water, rough, solar angle 90°	0.13
water, rough, solar angle 30°	0.024

The longwave radiation fluxes will be determined according to the Stefan-Boltzmann law:

$$I = \varepsilon_{IR}\, \sigma_{SB}\, T^4 \tag{1.4}$$

where the infrared emissivities for different surfaces are given in Table 1.2, and $\sigma_{SB} = 5.67 \cdot 10^{-8}$ W m^{-2} K^{-4} is the Stefan-Boltzmann constant.

Table 1.2. Infrared emissivity of different surfaces (Geiger *et al.* 1995)

surface	emissivity
water	0.960
fresh snow	0.986
coniferous needles	0.971
dry fine sand	0.949
wet fine sand	0.962
thick green grass	0.986

In general, the up-welling longwave radiation is greater than the down-welling longwave radiation, because of the earth's surface is warmer than clouds and aerosols. It is only in the case of fog, that up-welling and down-welling radiation are equal. The down-welling radiation may be greater if clouds appear in a previously clear sky and the ground surface has cooled. For clear sky without clouds and dry air, the radiation temperature is approximately −55 °C.

Meteorological data are usually measured in UTC (Universal Time Coordinated) or in local time; however, for radiation measurements the mean or, even better, the true local time is desirable. Then, the sun is at its zenith at 12:00 true local time. Appendix A4 gives the necessary calculations and astronomical relations to determine true local time.

Measurements of global radiation are often available in meteorological networks, but other radiation components may be missing. Parameterizations of these missing components using available measurements can be helpful. However, it should be noted that such parameterizations are often based on climatological mean values, and are valid only for the places where they were developed. Their use for short-time measurements is invalid.

The possibility to parameterize radiation fluxes using cloud observations was proposed by Burridge and Gadd (1974), see Stull (1988). For shortwave radiation fluxes, the transmissivity of the atmosphere is used:

$$T_K = (0.6 + 0.2 \sin \Psi)(1 - 0.4\,\sigma_{c_H})(1 - 0.7\,\sigma_{c_M})(1 - 0.4\,\sigma_{c_L}) \quad (1.5)$$

For solar elevation angles $\Psi = 90°$, the transmission coefficient, T_K, ranges from 0.8 to 0.086. σ_c is the cloud cover (0.0–1.0) of high clouds, c_H, of middle clouds, c_M, and c_L of low clouds, (see Supplement 1.4). Notice that in meteorology the cloud cover is given in octas. (An octa is a fraction equal to one-eighth of the sky.) The incoming shortwave radiation can be calculated using the solar constant S:

$$K\downarrow = \begin{pmatrix} S\,T_K \sin \Psi, \Psi \geq 0 \\ 0 \qquad\qquad , \Psi \leq 0 \end{pmatrix} \quad (1.6)$$

Supplement 1.4   Cloud genera

The classification of clouds is made according their genera and typical height. In the middle latitudes, high clouds are at 5–13 km a.g.l.; middle high clouds at 2–7 km a.g.l., and low clouds at 0–2 km a.g.l. The cloud heights in Polar Regions are lower while in tropical regions clouds heights can be up to 18 km.

cloud genera	height	description
cirrus (Ci)	high	white, fibrously ice cloud
cirrocumulus (Cc)	high	small ice cloud, small fleecy cloud
cirrostratus(Cs)	high	white stratified ice cloud, halo
altocumulus (Ac)	mean	large fleecy cloud
altostratus (As)	mean	white-grey stratified cloud, ring
nimbostratus (Ns)	low	dark rain/snow cloud
stratocumulus (Sc)	low	grey un-uniform stratified cloud
stratus (St)	low	grey uniform stratified cloud
cumulus (Cu)	low*–mean	cumulus cloud
cumulonimbus (Cb)	low*–high	thundercloud, anvil cloud

* in parameterizations classified as low clouds.

The reflected shortwave radiation can be calculated according to Eq. (1.3) by using typical values of the albedo of the underlying surface (Table 1.1). The longwave radiation balance can be parameterized using the cloud cover:

$$I^* = I\uparrow + I\downarrow = \left(0.08 K\,ms^{-1}\right)\left(1 - 0.1\,\sigma_{c_H} - 0.3\,\sigma_{c_M} - 0.6\,\sigma_{c_L}\right) \qquad (1.7)$$

If the surface temperature is given, then the longwave up-welling radiation can be calculated using the Stefan-Boltzmann law Eq. (1.4). The longwave down-welling radiation can be calculated using Eqs. (1.4) and (1.7).

More often, parameterizations use the duration of sunshine because it was measured for a long time in agricultural networks. Note that time series of sunshine durations are often inhomogeneous because the older Campbell-Stokes sunshine autograph was replaced by electronic methods. These parameterizations are based on climatological cloud structures and can be used only for monthly and annual averaged values, and in the region where the parameterization was developed. The parameterization is based on the well-known Ångström equation with sunshine duration, Sd:

$$K\downarrow = K\downarrow_{extr}\left[a + b\left(Sd/Sd_0\right)\right], \qquad (1.8)$$

where constants a and b depend on the place of determination. For the German low lands, the constants are, for example, $a \sim 0.19$ and $b \sim 0.55$ (Wendling *et al.* 1997). The mean daily extraterrestrial incoming irradiation at the upper edge of the atmosphere can be calculated in Wm^{-2} according to:

$$K \downarrow_{extr} = 28.36 \left[9.9 + 7.08 \, \varsigma + 0.18 (\varsigma - 1)(\varphi - 51°) \right] \qquad (1.9)$$

This equation is given in a form such that for a geographical latitude of $\varphi = 51°$ no correction for the latitude is necessary. The theoretical sunshine duration is the time between sunrise and sunset and expressed in hours

$$Sd_0 = 12.3 + 4.3 \varsigma + 0.167 \varsigma (\varphi - 51), \qquad (1.10)$$

with (DOY= day of the year)

$$\varsigma = \sin \left[DOY (2\pi/365) - 1.39 \right]. \qquad (1.11)$$

Because direct measurements of the radiation components are now available, parameterizations should only be used for calculations with historical data.

1.4.2 Ground Heat Flux and Ground Heat Storage

The ground surface (including plants and urban areas) is heated during the day by the incoming shortwave radiation. During the night, the surface cools due to longwave up-welling radiation, and is cooler than the air and the deeper soil layers. High gradients of temperature are observed in layers only a few millimeters thick (see Chap. 1.3). The energy surplus according to the net radiation is compensated by the turbulent sensible and latent heat fluxes and the mainly molecular ground heat flux. For the generation of latent heat flux, an energy surplus at the ground surface is necessary, and water must be transported through the capillaries and pores of the soil. Energy for the evaporation can also be provided by the soil heat flux in the upper soil layer.

In meteorology, the soil and the ground heat fluxes are often described in a very simple way, e.g. the large differences in the scales in the atmosphere and the soil are often not taken into account. The heterogeneity of soil properties in the scale of 10^{-3}–10^{-2} m is ignored, and the soil is assumed to be nearly homogeneous for the given meteorological scale. For more detailed investigations, soil physics textbooks must be consulted. In the following, conductive heat fluxes in the soil and latent heat fluxes in large pores are ignored.

The ground heat flux, Q_G, is based mainly on molecular heat transfer and is proportional to the temperature gradient times the thermal molecular conductivity a_G (Table 1.3):

$$Q_G = a_G \frac{\partial T}{\partial z} \qquad (1.12)$$

This molecular heat transfer is so weak that during the day only the upper decimeters are heated. When considering the annual cycle of ground temperature, maximum temperature is at the surface during the summer, but 10–15 m below the surface during winter (Lehmann and Kalb 1993). On a summer day, the ground heat flux is about 50–100 Wm^{-2}. A simple but not reliable calculation (Liebethal and Foken 2007) is: $Q_G = -0.1 \, Q_s^*$ or $Q_G = 0.3 \, Q_H$ (Stull 1988).

Table 1.3. Thermal molecular conductivity a_G, volumetric heat capacity C_G, and molecular thermal diffusivity v_T for different soil and ground properties (Stull 1988)

ground surface	a_G in W m^{-1} K^{-1}	C_G in 10^6 W s m^{-3} K^{-1}	v_T in 10^{-6} m^2 s^{-1}
rocks (granite)	2.73	2.13	1.28
moist sand (40 %)	2.51	2.76	0.91
dry sand	0.30	1.24	0.24
sandy clay (15%)	0.92	2.42	0.38
swamp (90 % water)	0.89	3.89	0.23
old snow	0.34	0.84	0.40
new snow	0.02	0.21	0.10

The determination of the ground heat flux according to Eq. (1.12) is not practicable because the temperature profile must be extrapolated to the surface to determine the partial derivative there. This can be uncertain because of the high temperature gradients near the surface (Fig. 1.9) and the difficulties in the determining thermal heat conductivity.

However, the ground heat flux at the surface can be estimated as the sum of the soil heat flux measured at some depth using soil heat flux-plates (see Chapter 6.2.6) and the heat storage in the layer between the surface and the plate:

$$Q_G(0) = Q_G(-z) + \int_{-z}^{0} \frac{\partial}{\partial t} C_G(z) T(z) \, dz \qquad (1.13)$$

An optimal design for ground heat flux measurements was developed by Liebethal *et al.* (2005) using sensitivity analysis. According to their method, the soil heat-flux plate should be buried rather deeply (10–20cm) with some temperature measurements made above it to calculate the heat storage. A similar accuracy can be achieved if a temperature profile is used to calculate both the soil heat flux according to Eq. (1.12) at a certain depth and the heat storage between this depth and the surface.

The volumetric heat capacity $C_G = a_G / v_T$ (v_T is the molecular thermal diffusivity, Table 1.3), can be assumed constant with depth in the case of uniform soil moisture. A general measurement and calculation guide for the integral of the change of the soil temperature with time is not available. Most institutes have their own schemes, not all of which are optimal solutions. The simplest method is the measurement with an integrating temperature sensor of the mean temperature of the soil layer between the surface and the heat-flux plate. For the ground heat flux near the surface, it follows that:

$$Q_G(0) = Q_G(-z) + \frac{C_G |\Delta z| \left[\overline{T(t_2)} - \overline{T(t_1)} \right]}{t_2 - t_1} \qquad (1.14)$$

The change of the soil temperature with the time can be determined from the vertical gradient of the soil heat flux and using of Eq. (1.12):

$$\frac{\partial T}{\partial t} = \frac{1}{C_G}\frac{\partial Q_G}{\partial z} = \nu_T \frac{\partial^2 T}{\partial z^2} \tag{1.15}$$

The daily cycle of the soil temperature is to first-order a sine function (Fig. 1.10). Therefore, the surface temperature T_s can be calculated depending on a temperature T_M, which is not affected by the daily cycle, *i.e.*

$$T_s = T_M + A_s \sin\left[\left(\frac{2\pi}{P}\right)(t - t_M)\right], \tag{1.16}$$

where A_s is the amplitude and P is the period of the wave of the surface temperature and t_M is the time required for $T_s = T_M$ (Arya 2001).

Multiple layers are used for the modeling of the ground heat flux. Because during the daily cycle only the upper soil layers are heated (Fig. 1.10), the two-layer model (force-restore method) developed by Blackadar (1976) is widely used. The ground heat flux can be calculated from two components, *i.e.*, from the change of the temperature of the thin surface layer due to radiation and from the slow wave of the temperature difference between the surface layer and a deeper layer. The equation for the ground heat flux is (Stull 1988):

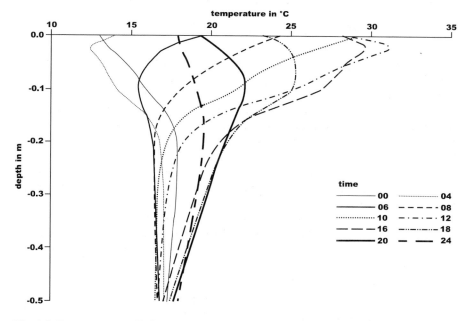

Fig. 1.9. Temperature profile in the upper soil layer on June 05, 1998, measured by the University of Bayreuth during the LITFASS-98 experiment (*bare soil*) at the boundary layer measuring field site of the Meteorological Observatory Lindenberg (high clouds from 12:00 to 14:00)

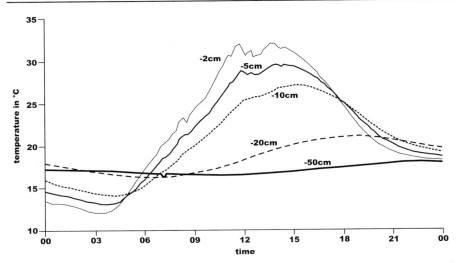

Fig. 1.10. Daily cycle of soil temperatures at different depth, measured by the University of Bayreuth during the LITFASS-98 experiment (*bare soil*) at the boundary layer measuring field site of the Meteorological Observatory Lindenberg (high clouds from 12:00 to 14:00)

$$Q_G = z_G\, C_G\, \frac{\partial T_G}{\partial t} + \left(2\pi\, \frac{z_G\, C_G}{P}\right)\left(T_G - T_M\right) \qquad (1.17)$$

Here is T_G the temperature of the upper soil layer, T_M is the temperature of the deeper soil layer, P is the time of the day, and z_G is the thickness of the surface layer. According to Blackadar (1976), during the day $2\pi/P$ is $3\cdot10^{-4}$ s^{-1}, and at night is $1\cdot10^{-4}$ s^{-1} (day: $T_a < T_G$, night: $T_a > T_{G}$, T_a: air temperature). The thickness of the upper soil layer depends on the depth of the daily temperature wave:

$$z_G = \sqrt{\frac{\nu_T\, P}{4\pi}} \qquad (1.18)$$

This can be calculated from Eq. (1.15) using Eq. (1.16), see Arya (2001). The force-restore method has the best results in comparison to other parameterization methods (Liebethal and Foken 2007).

1.4.3 Turbulent Fluxes

Contrary to the molecular heat exchange in the soil, heat exchange in the air due to turbulence is much more effective. This is because turbulent exchange occurs over scales of motions ranging from millimeters to kilometers. Turbulent elements can be thought of as air parcels with largely uniform thermodynamic characteristics. Small-scale turbulence elements join to form larger ones and so on. The largest eddies are atmospheric pressure systems. The heated turbulent elements transport their energy by their random motion. This process applies also for trace gases such as water vapor and kinetic energy. The larger turbulent elements receive their energy from the mean motion, and deliver the energy by a cascade process to smaller elements (Fig. 1.11). Small turbulent elements disappear by releasing energy (energy dissipation). On average, the transformation of kinetic energy into heat is about 2 Wm^{-2}, which is very small and not considered in the energy balance equation. The reason for this very effective exchange process, which is about a factor of 10^5 greater than the molecular exchange, is the atmospheric turbulence.

Atmospheric turbulence is a specialty of the atmospheric motion consisting in the fact that air volume (much larger than molecules: turbulent elements, turbulent eddies) achieve irregular and stochastic motions around a mean state. They are of different order with characteristic extensions and lifetimes ranging from centimeters and seconds to thousands of kilometers and days.

The characteristic distribution of turbulent elements (turbulent eddies) takes place according their size and is represented by the turbulence spectrum:

The turbulence spectrum is a plot of the energy distribution of turbulent elements (turbulent eddies) according their wavelength or frequency. Depending on the frequency, the distribution is classified as macro-, meso- or micro turbulence.

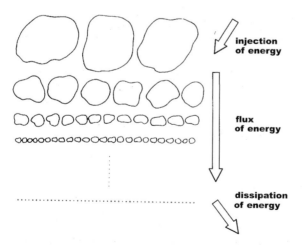

Fig. 1.11. Cascade process of turbulent elements (Frisch 1995), please note that the diminishing of the sizes of the elements is continuous.

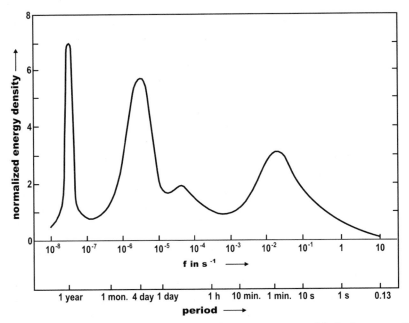

Fig. 1.12. Schematic plot of the turbulence spectra (Roedel 2000, modified); the range of frequencies > 10^{-3} Hz is called micro-turbulence

The division of atmospheric turbulence occurs over three time ranges, *i.e.* changes of high and low pressure systems within 3–6 days; the daily cycle of meteorological elements, and the transport of energy, momentum, and trace gases at frequencies ranging from 0.0001 to 10 Hz (Fig. 1.12). The transport of energy and trace gases is the main issue of micrometeorology.

Of special importance, is the inertial sub-range, which is characterized by isotropic turbulence and a constant decrease of the energy density with increasing frequency. In the range from about 0.01 to 5 Hz, no dominant direction exists for the motion of turbulent elements. The decrease of energy by the decay of larger turbulent elements into smaller ones takes place in a well-defined way according to Kolmogorov's *–5/3-law* (Kolmogorov 1941a). This law predicts that the energy density decreases by five decades when the frequency increases by three decades. The inertial sub-range merges into the dissipation range at the *Kolmogorov's microscale*. The shape of the turbulence spectra depends on the meteorological element, the thermal stratification, the height above the ground surface, and the wind velocity (see Chap. 2.5).

A typical property of turbulence, especially in the inertial sub-range, is that turbulent elements change little as they move with the mean flow. Thus, at two neighboring downwind measuring points the same turbulent structure can be observed at different times. This means that a high autocorrelation of the turbulent fluctuations exists. This is called *frozen turbulence* according to Taylor (1938).

Furthermore, the length scales of turbulent elements increase with the height above the ground surface. In an analogous way, the smallest turbulent elements with the highest frequencies are found near the ground surface. From these findings, it is reasonable to plot the turbulence spectra not as a function of frequency f, but as a dimensionless frequency, n, normalized with the wind speed u and the height z:

$$n = f\,\frac{z}{u} \tag{1.19}$$

(In the English literature, one often sees n as the frequency and f as the dimensionless frequency.)

Turbulent elements can be easily seen, e.g., in plots of temperature time series with high time resolution. In Fig. 1.13, we see short-period disturbances with different intensities superposed onto longer-period (~ 60 s) disturbances. We also see that even significantly larger structures are present.

The calculation of the heat fluxes (sensible and latent) caused by turbulent elements is analogous to Eq. (1.12) using the vertical gradients of temperature T and specific humidity q (see Supplement 2.1), respectively. However, the molecular transfer coefficient must be replaced by the turbulent diffusion coefficient. The sensible heat flux, Q_H, describes the turbulent transport of heat from and to the earth's surface. The latent heat flux, Q_E, describes the vertical transport of water vapor and the heat required for evaporation at the ground surface. This heat will be deposited later in the atmosphere when condensation occurs, e.g. in clouds. The relevant equations are:

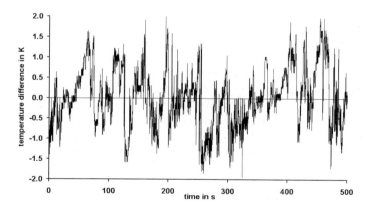

Fig. 1.13. Time series of the air temperature above a spruce forest (University of Bayreuth, Waldstein/Weidenbrunnen site), August 19, 1999, 11:51–12:00 UTC, 500 s measuring period (Wichura et al. 2001)

$$Q_H = -\rho c_p K_H \frac{\partial T}{\partial z},$$ (1.20)

$$Q_E = -\rho \lambda K_E \frac{\partial q}{\partial z}$$ (1.21)

with the air density ρ, the specific heat for constant pressure c_p, and evaporation heat of water λ. The turbulent diffusion coefficients K_H and K_E are normally complicated functions of the wind speed, the stratification, and the properties of the underlying surface. Their evaluation is a special issue of micrometeorology. In Chap. 2.3, several possible calculations of the diffusion coefficients are discussed. Also common is the *Austausch coefficient* (Schmidt 1925), which is the product of the diffusion coefficient and the air density:

$$A = \rho K$$ (1.22)

An example of the daily cycle of the turbulent fluxes including the net radiation and the ground heat flux is shown in Fig. 1.14. It is obvious that during the night all fluxes have the opposite sign to that during the day, but the absolute values are much less. After sunrise, the signs change and the turbulent fluxes increase rapidly. The time shift between the irradiation and the beginning of turbulence is only a few minutes (Foken *et al.* 2001). The time shift of the ground heat flux depends on the depth of the soil layer. The maximum of the turbulent fluxes on clear days occurs shortly after midday.

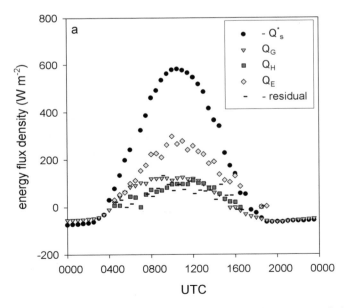

Fig. 1.14. Typical daily cycle of the components of the energy balance measured above a corn field on June 07, 2003 during the experiment LITFASS–2003 (Mauder *et al.* 2006)

In the case of optimal conditions for evaporation, *i.e.* high soil moisture and strong winds, the evaporation process will be greater than the sensible heat flux, if not enough radiation is available. In such cases, the sensible heat flux changes its sign 1–3 hours before sunset, and sometimes shortly after noon. This case is called the "oasis effect", and is also found in the temperate latitudes. The latent heat flux has also large values after sunrise, and changes sign after midnight (dewfall). From Fig. 1.14, it is obvious that using measured values the energy balance according to Eq. (1.1) is not closed. The missing value is called the residual. This very complex problem is discussed in Chap. 3.7.

While over land in the temperate latitudes the sensible and latent heat fluxes are of the same order, over the ocean the evaporation is much greater. In some climate regions and under extreme weather conditions, significant deviations are possible.

1.5 Water Balance Equation

The energy balance Eq. (1.1) is connected to the evaporation through the water balance equation:

$$0 = N - Q_E - A \pm \Delta S_W , \qquad (1.23)$$

Supplement 1.5.   Water cycle of Germany

Mean annual data of the water cycle of Germany (Source: German Meteorological Service, Hydrometeorology) in mm (1 mm = 1 L m^{-2})

precipitation		evaporation	
779 mm		463 mm	
into evaporation	463 mm	from transpiration	328 mm
into ground water	194 mm	from interception	72 mm
into runoff	122 mm	from soil evaporation	42 mm
		from surface water evaporation	11 mm
		from service water evaporation	11 mm

Supplement 1.6.   Water cycle of the earth

Mean water balance of the earth in 10^3 km^3 a^{-1} (Houghton 1997)

surface	precipitation	evaporation	runoff
land surface	111	71	40
ocean surface	385	425	40*

* water vapor transport in the atmosphere from the ocean to the land, for instance as cloud water

where N is the precipitation, A the runoff, and ΔS_W the sum of the water storage in the soil and ground water. Evaporation is often divided into the physically-caused part, the *evaporation*, which is dependent on the availability of water, the energy input, and the intensity of the turbulent exchange process; and the *transpiration* which is caused by plant-physiology, the water vapor saturation deficit, and the photosynthetic active radiation. The sum of both forms is called *evapotranspiration*. Evaporation occurs on the ground, on water surfaces, and on wetted plant surfaces. The latter is the evaporation of precipitation water held back at plant surfaces (interception). Micrometeorology plays an important role for the determination of evapotranspiration and the investigation of the water cycle (Supplement 1.5, Supplement 1.6). Data show that the precipitation-evaporation cycle over the land is not as strongly coupled as they are over the ocean.

The water balance equation is widely used in hydrological investigations. There, the evaporation connects meteorology with hydrology. This field of investigations is often called hydrometeorology.

2 Basic Equations of Atmospheric Turbulence

Before starting the derivation of the equations for the turbulent fluxes of momentum, heat and trace gases (Chap. 2.3), we present a short introduction into the basic equation. To illustrate the importance of micrometeorological equations and parameterizations for modeling on all scales, different closure techniques of the turbulent differential equations are described (Chap. 2.1.3). The more practical user of this book can proceed directly to Chap. 2.3.

2.1 Equation of Motion

2.1.1 Navier-Stokes Equation of Mean Motion

The Navier-Stokes equations describe the balance of all the forces in the earth's atmosphere without consideration of the centrifugal force (Arya 1999; Etling 2002; Salby 1995; Stull 1988):

$$\frac{\partial u}{\partial t} = -u\frac{\partial u}{\partial x} - v\frac{\partial u}{\partial y} - w\frac{\partial u}{\partial z} - \frac{1}{\rho}\frac{\partial p}{\partial x} + f v + \nu \nabla^2 u$$

$$\frac{\partial v}{\partial t} = -u\frac{\partial v}{\partial x} - v\frac{\partial v}{\partial y} - w\frac{\partial v}{\partial z} - \frac{1}{\rho}\frac{\partial p}{\partial y} - f u + \nu \nabla^2 v \qquad (2.1)$$

$$\frac{\partial w}{\partial t} = -u\frac{\partial w}{\partial x} - v\frac{\partial w}{\partial y} - w\frac{\partial w}{\partial z} + g + \nu \nabla^2 w$$

where u is the horizontal wind in the x-direction (east); v is the horizontal wind in the y-direction (north), and w is the vertical wind; p is the atmospheric pressure; f is the Coriolis parameter; g is the acceleration of gravity; ρ is the air density; ν is the kinematic viscosity, and ∇^2 is the Laplace operator. From left-to-right, the terms of the equation are the tendency, the advection, the pressure gradient force, the Coriolis force, and the (molecular) stress. In a turbulent atmosphere, a turbulent stress term, the Reynolds stress, must be applied. All the terms in the horizontal motion equations are of the order of 10^{-4}–10^{-3} m s^{-2}. Under certain condition, some terms are very small and can be neglected. For example, for steady-state flow, the tendency can be neglected; above horizontally homogeneous surfaces, the advection can be neglected; in the center of high and low pressure areas or for small scale processes the pressure gradient force can be neglected; at the equator or for small scale processes the Coriolis force can be neglected, and above the atmospheric boundary layer the stress terms can be neglected.

The three equations of the wind components can be combined applying Einstein's summation notation:

$$\frac{\partial u_i}{\partial t} = -u_j \frac{\partial u_i}{\partial x_j} - \delta_{i3}g + f\varepsilon_{ij3}u_j - \frac{1}{\rho}\frac{\partial p}{\partial x_i} + \frac{1}{\rho}\left(\frac{\partial \tau_{ij}}{\partial x_j}\right) \tag{2.2}$$

The shear stress tensor with dynamic viscosity μ is given in the form (Stull 1988):

$$\tau_{ij} = \mu\left(\frac{\partial u_i}{\partial x_j} + \frac{\partial u_j}{\partial x_i}\right) - \frac{2}{3}\mu\frac{\partial u_k}{\partial x_k}\delta_{ij} \tag{2.3}$$

The generalizations and applications of the Einstein summation operators are summarized in Table 2.1.

Table 2.1. Definitions of Einstein's summation notation

running index of	$i = 1,2,3$	$j = 1,2,3$	$k = 1,2,3$
the velocity components	$u_1 = u$	$u_2 = v$	$u_3 = w$
length components	$x_1 = x$	$x_2 = y$	$x_3 = z$
variables	no free index: scalar	one free index: vector	two free indexes: tensor
Kronecker delta-operator δ_{ij}	$= +1$, for $i = j$	$= 0$, for $i \neq j$	
alternating unit tensor ε_{ijk}	$= +1$, for ijk =123, 231 or 312	$= -1$, for ijk = 321, 213 or 132	$= 0$, for ijk = all other combinations

2.1.2 Turbulent Equation of Motion

The modification of the Navier-Stokes equations to include turbulent motions requires the decomposition of all the variables into a mean part, $\overline{x}$, and a random fluctuating part, x'. This is called the Reynolds's decomposition (Fig. 2.1), and is represented by:

$$x = \overline{x} + x'. \tag{2.4}$$

The application of Reynolds's decomposition requires some averaging rules for the turbulent value x' (a represents a constant), which are termed Reynolds's postulates:

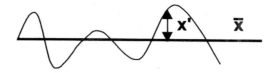

Fig. 2.1. Schematic presentation of Reynolds's decomposition of the value x

$$
\begin{array}{ll}
I & \overline{x'}=0 \\
II & \overline{x\,y}=\overline{x}\,\overline{y}+\overline{x'\,y'} \\
III & \overline{\overline{x}\,y}=\overline{x}\,\overline{y} \\
IV & \overline{a\,x}=a\,\overline{x} \\
V & \overline{x+y}=\overline{x}+\overline{y}
\end{array}
\qquad (2.5)
$$

It is assumed that the postulates are universal, but for special spectral regions or for intermitted turbulence this is not valid (Bernhardt 1980). The second postulate is the basis for the determination of turbulent fluxes according to the direct eddy-covariance method (see Chap. 4.1).

The turbulent equations of motion follow after application of Reynolds's decomposition and postulates into Eq. (2.2). It is also assumed (Businger 1982; Stull 1988), that:

$$
\begin{array}{l}
|\,p'/\overline{p}\,|<<|\,\rho'/\overline{\rho}\,| \\
|\,p'/\overline{p}\,|<<|\,T'/\overline{T}\,| \\
|\,\rho'/\overline{\rho}\,|<<1 \\
|\,T'/\overline{T}\,|<<1
\end{array}
\qquad (2.6)
$$

These assumptions are not trivial and need further inspections for individual cases.

A very important simplification results from the Boussinesq-approximation (Boussinesq 1877), which neglects density fluctuations but not in the buoyancy (gravitation) term. This is because the acceleration of gravity is relatively large in comparison with the other accelerations in the equation. Therefore, the shallow convective conditions (Stull 1988) are permitted. This form of averaging is not without consequences for the determination of turbulent fluxes (see Chap. 4.1.2). Applying all these simplifications it follows that:

$$
\frac{\partial \overline{u_i}}{\partial t}+\frac{\partial}{\partial x_j}\left(\overline{u_j}\,\overline{u_i}+\overline{u_j'\,u_i'}\right)=-\frac{1}{\rho}\frac{\partial \overline{p}}{\partial x_i}+\nu\frac{\partial^2 \overline{u_i}}{\partial x_i^{\,2}}+g\delta_{i3}+\varepsilon_{ijk}\,f\,\overline{u_k}
\qquad (2.7)
$$

Completely analogous equations for the heat transfer and the transfer of trace gases such as water vapor can be derived

$$\frac{\partial \overline{T}}{\partial t} + \frac{\partial}{\partial x_i}\left(\overline{u_i}\ \overline{T} + \overline{u_i'T'}\right) = a_T \frac{\partial^2 \overline{T}}{\partial x_i^2} + R,$$ (2.8)

$$\frac{\partial \overline{c}}{\partial t} + \frac{\partial}{\partial x_i}\left(\overline{u_i}\ \overline{c} + \overline{u_i'c'}\right) = D \frac{\partial^2 \overline{c}}{\partial x_i^2} + S,$$ (2.9)

where R and S are source and sink terms, and a_T and D are the molecular heat conduction and diffusion coefficients, respectively.

An important simplification is possible in the atmospheric boundary layer where only the equations for j=3, *i.e.* $u_3 = w$, are important, and steady state conditions $(\partial/\partial t = 0)$ and horizontal homogeneity $(\partial/\partial x_i = 0, \partial/\partial x_j = 0)$ are assumed. This assumption is far reaching because all the following applications are valid only under these conditions. For instance, for all micrometeorological measurements steady state conditions are implied (see Chap. 4.1.3), and a mostly homogeneous surface is necessary. Under these assumptions and including the components u_g und v_g of the geostrophic wind velocity and the angular velocity of the earth's rotation, Ω, the three equations of motion become:

$$\frac{\partial \overline{u'w'}}{\partial z} = f\left(\overline{v} - \overline{v_g}\right) + v \frac{\partial^2 \overline{u}}{\partial z^2}, \quad \overline{v_g} = \frac{1}{\rho f}\frac{\partial \overline{p}}{\partial x}$$ (2.10)

$$\frac{\partial \overline{v'w'}}{\partial z} = -f\left(\overline{u} - \overline{u_g}\right) + v \frac{\partial^2 \overline{v}}{\partial z^2}, \quad \overline{u_g} = -\frac{1}{\rho f}\frac{\partial \overline{p}}{\partial y}$$ (2.11)

$$\frac{\partial \overline{w'^2}}{\partial z} = \frac{1}{\rho}\frac{\partial \overline{p}}{\partial z} - g - 2\left[\Omega_u\ \overline{v} - \Omega_v\ \overline{u}\right], \quad f = 2\Omega \sin\varphi$$ (2.12)

Equations (2.10) and (2.11) are the basis of the so-called ageostrophic method for the determination of the components of the shear stress tensor using differences between the wind velocity in the atmospheric boundary layer and the geostrophic wind (Bernhardt 1970; Lettau 1957). The practical application of the ageostrophic method is limited because of baroclinicity, non steady-state conditions, and inhomogeneities (Schmitz-Peiffer *et al.* 1987). For example, they can be applied only for the determination of the shear stress at the ground surface using a large number of aerological observations (Bernhardt 1975). In addition, the continuity equation in the incompressible form is assumed:

$$\frac{\partial \overline{u_i}}{\partial x_i} = 0, \quad \frac{\partial \overline{w}}{\partial z} = 0, \quad \overline{w} = 0$$ (2.13)

The gas law with the specific gas constant for dry air R_L and the virtual temperature T_v completes the system of equations:

$$p = \rho R_L T_v$$ (2.14)

In an analogous way, the equations for heat and trace gas transfer are:

$$\frac{\partial \overline{w'T'}}{\partial z} = a_T \frac{\partial^2 \overline{T}}{\partial z^2}, \quad for \;\; R = 0 \qquad (2.15)$$

$$\frac{\partial \overline{w'c'}}{\partial z} = D \frac{\partial^2 \overline{c}}{\partial z^2}, \quad for \;\; S = 0 \qquad (2.16)$$

The influence of the individual terms in the different layers of the atmospheric boundary layer can be estimated using similarity numbers. These numbers are dimensionless values describing the relations between characteristic scales of the forces. Two physical systems are similar if the similarity numbers of both systems are on the same order. This is imported if atmospheric processes are investigated in a wind tunnel.

The ratio of the inertia to the pressure gradient force is called the Euler number

$$Eu = \frac{\rho V^2}{\Delta P}, \qquad (2.17)$$

where V is the characteristic velocity, and ΔP is the characteristic pressure gradient force.

The ratio of the inertia force to the Coriolis force is the Rossby number

$$Ro = \frac{V}{f\,L_h}, \qquad (2.18)$$

where L_h is the characteristic large-area horizontal length scale.

The ratio of the inertia force to the molecular stress is the Reynolds number

$$Re = \frac{L_z V}{\nu}, \qquad (2.19)$$

where L_z is the characteristic small-scale vertical length scale.

The ratio of shear production of turbulence energy to the buoyancy production or destruction of turbulence energy is the Richardson number

$$Ri = -\frac{g}{\overline{T}} \frac{\Delta_z T}{\left(\Delta_z u\right)^2}, \qquad (2.20)$$

where $\Delta_z T$ is the characteristic temperature gradient, and $\Delta_z u$ is the characteristic vertical wind gradient (see Chap. 2.3.2). For heights above 10 m the temperature must be replaced by the potential (see Chap. 2.3.1).

The relevant processes in the atmospheric boundary layer can be identified using dimensional analysis and the similarity numbers. Using the logarithms of the similarity numbers the relevant processes can be identified for logarithms smaller than zero (Bernhardt 1972) as listed in Table 2.2. From the structure of the atmospheric boundary layer presented in Fig. 1.4, this is a logical organization.

Table 2.2. Order of the similarity numbers in the layers of the atmospheric boundary layer (**bold**: Processes characterized by the similarity numbers are relevant)

layer	height	lg Ro	lg Eu	lg Re	lg $|Ri|$
upper layer	~1000 m	**< 0**	**< 0**	> 0 $Re > 10^8$	**> − 2**
surface layer	~10...50 m	~ 0	**< 0**	> 0 $Re \sim 10^7...10^8$	**> − 2**
dynamical sub-layer	~ 1 m	**> 0**	~ 0	> 0 $Re < 10^7$	~ − 2
viscous sub-layer	~ 0.01 m	**> 0**	**> 0**	~ **0**	< − 2
molecular or laminar boundary layer	~ 0.001 m	**> 0**	**> 0**	**< 0**	< − 2

It can be shown that the pressure gradient force is important only in the upper boundary layer. Molecular viscosity is only relevant in the viscous and molecular sub-layer. The effect of the Coriolis force can be neglected in the flux gradient relationship (see Chap. 2.3) in the surface layer, but not in general (see Chap. 2.4). On the other hand, the turbulent stress is relevant in the whole boundary layer. In the dynamical and viscous sub-layers, the stratification does not play a role. Under these circumstances, it is possible that the vertical gradients are nearly zero:

$$\frac{\partial \overline{u'w'}}{\partial z} \approx 0 , \quad \frac{\partial \overline{v'w'}}{\partial z} \approx 0 , \quad \frac{\partial \overline{T'w'}}{\partial z} \approx 0 , \quad \frac{\partial \overline{c'w'}}{\partial z} \approx 0 \tag{2.21}$$

These equations mean that the covariances are constant with height in the surface layer. An error in this assumption of approximately 10% is typical.

The covariance of the vertical wind velocity, w, and a horizontal wind component or a scalar x can be determined by:

$$\overline{w'x'} = \frac{1}{N-1} \sum_{k=0}^{N-1} \left[\left(w_k - \overline{w_k} \right)\left(x_k - \overline{x_k} \right) \right] = \tag{2.22}$$

$$= \frac{1}{N-1} \left[\sum_{k=0}^{N-1} w_k \cdot x_k - \frac{1}{N} \left(\sum_{k=0}^{N-1} w_k \sum_{k=0}^{N-1} x_k \right) \right]$$

According to the second Reynolds's postulate (2.5, II), in the case of a negligible vertical wind the total flux is equal to the covariance. The implementation of this equation into the measuring technique is by the eddy-covariance method, a direct method to determine turbulent fluxes (see Chap. 4.1). The dimensions of the turbulent fluxes of momentum, sensible, and latent heat are in kinematic units m^2s^{-2}, Kms^{-1}, and $g\,g^{-1}m\,s^{-1}$, respectively. For water vapor flow, the units are hPa m s^{-1}:

$$u_*^2 = -\overline{u'w'}, \quad \frac{Q_H}{\rho \cdot c_p} = \overline{T'w'}, \quad \frac{Q_E}{\rho \cdot \lambda} = \overline{q'w'}, \quad \frac{Q_c}{\rho} = \overline{c'w'} \tag{2.23}$$

The equation for the friction velocity is valid only if u is the direction of the mean wind velocity. The covariance of u' and w' is negative except below the crown in a forest (Amiro 1990). For a Cartesian coordinate system it follows that:

$$u_* = \left[\left(\overline{u'w'}\right)^2 + \left(\overline{v'w'}\right)^2 \right]^{1/4} \tag{2.24}$$

The friction velocity is a generalized velocity, $i.e.$, it is the shear stress divided by the density

$$u_* = \left(\frac{\tau}{\rho} \right)^{1/2}. \tag{2.25}$$

2.1.3 Closure Techniques

The transition from the equation of motion for the mean stream to the turbulent flow gives a system of differential equations with more unknown parameters than equations. To solve the system of equations, assumptions have to be made to calculate the unknown parameters. These assumptions, for example the covariance terms, are called closure techniques.

The order of the closure refers to the highest order of the parameters that must be calculated with the prognostic equations. Therefore, the moments of the next higher order must be determined (Table 2.3). Simple closure techniques are bulk approaches (see Chap. 4.2.1). Closer techniques of higher order require extensive calculations (Table 2.4). The most important approaches are presented here.

Table 2.3. Characterization of closure techniques (Stull 1988)

order of closure	prognostic equation for	to be approximate in the equation	number of equations	number of unknown parameters
1st order	$\overline{u_i}$	$\overline{u_i u_j}$	3	6
2nd order	$\overline{u_i u_j}$	$\overline{u_i u_j u_k}$	6	10
3rd order	$\overline{u_i u_j u_k}$	$\overline{u_i u_j u_k u_l}$	10	15

Table 2.4. Realization of closure techniques

order of closure	realization	
0. order	no prognostic equation	
	(bulk and similarity approaches)	
½ order	forecast with simple bulk approaches	
1. order	K-approach	transilient closure
	(local)	(non-local)
1½ order	TKE equation with variance terms	
2. order	prognostic equation for fluxes	
3. order	prognostic equation for triple correlations	

Local or First–Order Closure

First-order closure is analogous to molecular diffusion approaches, *i.e.* an assumed proportionality between the vertical flux and the vertical gradient of the relevant state parameter ξ. In the turbulent case, the proportionality factor is the eddy diffusion coefficient, K, and the approach is called K-theory. The gradient will be determined at the same place where the flux is to be calculated; therefore it is a local closure:

$$\overline{u_i'\xi'} = -K\frac{\partial\overline{\xi}}{\partial z} \tag{2.26}$$

The turbulent diffusion coefficients are formed for momentum, sensible heat, water vapor (latent heat), *etc.* The turbulent diffusion coefficient for momentum is proportional to that for sensible heat

$$K_m = \text{Pr}_t \cdot K_H, \tag{2.27}$$

where $\text{Pr}_t \sim 0.8$ is the turbulent Prandtl number for air. For the water vapor flux and the transport of other trace gases in general, the diffusion coefficient for sensible heat is used, although this is not required (see Chap. 2.3.1). For the single fluxes in kinematic units, the following relations are used:

$$\overline{u'w'} = K_m\frac{\partial\overline{u}}{\partial z} \tag{2.28}$$

$$\overline{w'T'} = -K_H\frac{\partial\overline{T}}{\partial z} \tag{2.29}$$

$$\overline{w'q'} = -K_E\frac{\partial\overline{q}}{\partial z} \tag{2.30}$$

For the determination of the turbulent diffusion coefficients, the mixing length parameterization is used, which is based on the work of Prandtl (1925). This approach describes the turbulent diffusion coefficient in terms of geometric and flux parameters. This is illustrated in the following example for the moisture flux. If an air parcel moves to a slightly different height, it will have a humidity and velocity slightly different than its new environment. This change of the environmental velocity and humidity can be described with gradients (Fig. 2.2):

$$u' = -\left(\frac{\partial \overline{u}}{\partial z}\right) z' \qquad q' = -\left(\frac{\partial \overline{q}}{\partial z}\right) z' \qquad (2.31)$$

The vertical velocity, which is necessary for the movement of the parcel, is assumed proportional to the horizontal velocity:

$$w' = \pm c\, u' = c\left|\frac{\partial \overline{u}}{\partial z}\right| z' \qquad (2.32)$$

The water vapor flux and the turbulent diffusion coefficient for evaporation are then given by

$$\overline{w'q'} = -c\,\overline{(z')^2}\left|\frac{\partial \overline{u}}{\partial z}\right| \cdot \left(\frac{\partial \overline{q}}{\partial z}\right) = -l^2 \left|\frac{\partial \overline{u}}{\partial z}\right| \cdot \left(\frac{\partial \overline{q}}{\partial z}\right), \qquad (2.33)$$

$$K_E = l^2 \left|\frac{\partial \overline{u}}{\partial z}\right|, \qquad (2.34)$$

where l is called the mixing length. The mixing length is usually assumed to be

$$l = \kappa\, z, \qquad (2.35)$$

where κ is the von-Kármán constant and z is the height above the ground surface. The value of κ is currently taken to be 0.4 (see Chap. 2.3.2).

The turbulent diffusion coefficient of momentum is:

$$K_m = \kappa^2 z^2 \left[\left(\frac{\partial \overline{u}}{\partial z}\right)^2 + \left(\frac{\partial \overline{v}}{\partial z}\right)^2\right]^{\frac{1}{2}} \qquad (2.36)$$

With Eqs. (2.23) and (2.28), the widely used exchange approach (K-approach) for neutral stratification in the surface layer is:

$$K_m = \kappa\, z\, u_* \qquad (2.37)$$

Applications of local closure for the stratified surface layer are discussed in Chap. 2.3.2.

Outside of the surface layer, closure parameterizations for greater heights in the atmospheric boundary layer are used. Overviews are given by, for example, Garratt (1992), Jacobson (2005)), and Stull (1988). The K-theory approach can be always applied when the exchange process takes place between direct neighboring atmospheric layers or turbulence elements. No application is possible for convective conditions. Also, K-theory should not be used within high vegetation (see Chap. 3.5) or for very stable stratification (see Chap. 3.6).

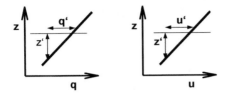

Fig. 2.2. Schematic view of the movement of an air parcel z' to explain the mixing length approach

Non-Local First–Order Closure

If the larger eddies contribute to the exchange process, and if the over all flux is greater than the flux due to the smaller eddies alone, then this must be considered in the closing approach and can no longer be locally closed. This is the case for the transport process in tall vegetation. Possible solutions are given by either the transilient theory or the spectral diffusion theory.

The transilient theory (Stull 1984) approximates the turbulent exchange process between adjoining atmospheric layers or boxes, and also admits the exchange between non-adjoining boxes The change of a scalar ξ with time in a box i is given by a fixed matrix of exchange coefficients, c_{ij} between the boxes i and j and the size of the scalar in the box j:

$$\overline{\xi}_i(t + \Delta t) = \sum_{j=1}^{N} c_{ij}(t, \Delta t)\overline{\xi}_j(t) \tag{2.38}$$

The matrix of mixing coefficients is called a transilient matrix. The flux in the kth layer is given by:

$$\overline{w'\xi'}(k) = \left(\frac{\Delta z}{\Delta t}\right) \sum_{i=1}^{k} \sum_{j=1}^{N} c_{ij}\left(\overline{\xi}_i - \overline{\xi}_j\right) \tag{2.39}$$

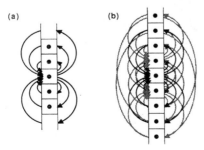

Fig. 2.3. Schematic view of the mixing of eddies from the central layer (**a**) and by superposition of similar mixing of the three layers in the centre (**b**) (Stull 1984)

Figure 2.3 illustrates the possibilities for mixing between boxes. In principle all possibilities of mixing can be realized by the definition of the transilient matrix, which must be parameterized using external parameters such as the wind field or the radiation (Stull 1988). The difficulties of getting these parameters are the reasons why this form of the closure is not widely used.

Higher Order Closure

Closures of order higher than first-order are now usual. But most of the parameterizations are experimental not validated. Commonly used, is a closure of 1.5th order. This is a closure using variances, which can be partly determined with the equation of the turbulent kinetic energy (*TKE*, see Chap. 2.2).

2.2 Equation of the Turbulent Kinetic Energy

The equation of the turbulent kinetic energy, (*TKE*) in kinematic form is obtained by multiplication of the Navier-Stokes equation for turbulent flow Eq. (2.7) with u_i'. With the kinetic energy defined as (Etling 2002; Stull 1988)

$$\bar{e} = 0.5\left(\overline{u'^2} + \overline{v'^2} + \overline{w'^2}\right) = 0.5\,\overline{u_i'^2} \tag{2.40}$$

it follows:

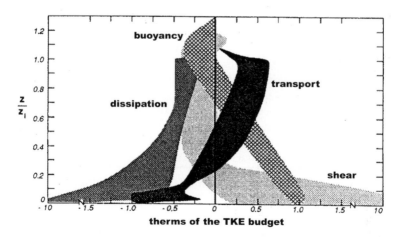

therms of the TKE budget

Fig. 2.4. Order of the terms of the *TKE* equation in the atmospheric boundary layer at daytime (Stull 1988) normalized with $w_*^3 z_i^{-1}$ (about $6\cdot10^{-3}$ m^{2}s^{-3})

Table 2.5. Meaning of the terms of the *TKE* equation

term	process
I	local *TKE* storage or tendency
II	*TKE* advection
III	buoyancy production or consumption
IV	product from momentum flux (<0) and wind shear (>0)
	mechanical (or shear) production or loss term of turbulent energy
V	turbulent *TKE* transport
VI	pressure correlation term
VII	energy dissipation

$$\frac{\partial \overline{e}}{\partial t} = -\overline{u}_j \frac{\partial \overline{e}}{\partial x_j} + \delta_{i3} \frac{g}{\theta_v}\left(\overline{u_i'\theta_v'}\right) - \overline{u_i'u_j'}\,\frac{\partial \overline{u_i}}{\partial x_j} - \frac{\partial \left(\overline{u_j'e}\right)}{\partial x_j} - \frac{1}{\rho}\frac{\partial \left(\overline{u_i'p'}\right)}{\partial x_i} - \varepsilon \qquad (2.41)$$

$$\quad I \qquad\quad II \qquad\qquad III \qquad\qquad IV \qquad\quad V \qquad\quad VI \qquad VII$$

Descriptions of the terms in Eq. (2.41), which are in the order of $10^{-4}\ \mathrm{m^2 s^{-3}}$, are given in Table 2.5. The changes of the magnitudes of the terms in Eq. (2.41) with height in the boundary layer are shown in Fig. 2.4. In a boundary layer that is strongly influenced by convective processes, the terms are usually normalized by the characteristic convective or Deardorff velocity:

$$w_* = \left[\frac{g \cdot z_i}{\theta_v}\left(\overline{w'\theta_v'}\right)\right]^{1/3} \qquad (2.42)$$

> **Convection** is the vertical transport or mixing of properties of the air (horizontal transport: advection). Forced convection results from mechanical forces (wind field) and inhomogeneities of the ground surface. The scaling is made with u_* and T_* and occurs for $1 > z/L > -1$ (L: Obukhov-length, T_*: dynamical temperature, see Chap. 2.3.2). In contrast, free convection is caused by density differences and occurs for $z/L < -1$, and the scaling parameter is w_*. The fluxes in the case of free convection are often directed contrary to the gradient (counter gradient).

Comparing the magnitudes of the terms of the *TKE* equation near the surface, terms I, II, V, and VI can be neglected relative to terms III, IV, and VII. The resulting equation is:

$$0 = \delta_{i3} \frac{g}{\theta_v}\left(\overline{u_i'\theta_v'}\right) - \overline{u_i'u_j'}\,\frac{\partial \overline{u_i}}{\partial x_j} - \varepsilon \qquad (2.43)$$

This equation can be used in the surface layer to determine the energy dissipation ε, *i.e.*, the decay of turbulent eddies into heat:

$$\varepsilon = \frac{g}{\theta_v}\left(\overline{w'\theta_v'}\right) - \overline{w'u'}\frac{\partial \overline{u}}{\partial z} \tag{2.44}$$

2.3 Flux-Gradient Similarity

2.3.1 Profile Equations for Neutral Stratification

In Chap. 2.1.3, it was shown that the flux could be determined by the vertical gradient of the state variable and a diffusion coefficient. These relations are called flux-gradient similarities. Thus, the turbulent diffusion coefficient for momentum can be parameterized in a simple way using Eq. (2.37). For the shear stress, it follows:

$$\tau = \rho \, K_m \frac{\partial u}{\partial z} \tag{2.45}$$

Often not the shear stress but the generalized velocity, the so-called friction velocity Eq. (2.25), is used.

The turbulent fluxes of momentum, Eq. (2.28), sensible heat, Eq. (2.29), and latent heat, Eq. (2.30), can be calculated using the turbulent diffusion coefficient for momentum in the case of neutral stratification, Eq. (2.37), as the profile equations:

$$u_* = \sqrt{-\overline{u'w'}} = \kappa z \frac{\partial u}{\partial z} = \kappa \frac{\partial u}{\partial \ln z} \tag{2.46}$$

$$\overline{w'T'} = -\alpha_0 \, \kappa \, u_* \frac{\partial T}{\partial \ln z} \tag{2.47}$$

$$\overline{w'q'} = -\alpha_{0E} \, \kappa \, u_* \frac{\partial q}{\partial \ln z} \tag{2.48}$$

In Eq. (2.46) the friction velocity was defined in a simplified way in comparison to Eq. (2.24) by using the mean horizontal wind u. This first-order approximation is possible in the case of small wind fluctuations as shown by (Foken 1990).

Because the diffusion coefficients for momentum, sensible and latent heat are not identical, the coefficients α_0 and α_{0E} are introduced, which are the ratios of the diffusion coefficients of heat and moisture to the coefficient for momentum respectively. The reciprocal value of α_0 is analogous to the Prandtl number of molecular exchange conditions

$$Pr = \frac{v}{v_T}.$$ (2.49)

and is called the turbulent Prandtl number, Eq. (2.27):

$$\alpha_0 = \frac{1}{Pr_t} = \frac{K_H}{K_m} \approx 1.25$$ (2.50)

Similarly, the reciprocal value α_{0E} is analogous to the Schmidt number

$$Sc = \frac{v}{D},$$ (2.51)

of the molecular diffusion coefficient for water vapor, D, and is called the turbulent Schmidt number:

$$\alpha_{0E} = \frac{1}{Sc_t} = \frac{K_E}{K_m} \approx 1.25$$ (2.52)

The coefficients α_0 and. α_{0E} can be determined only by the comparisons of profile measurements (see Chap. 4.2) and flux measurements with the eddy-covariance method (see Chap. 4.1) for neutral conditions. Because of the inaccuracies in these methods the coefficients contain remarkable defects. Table 2.6 gives an overview of the currently available data.

Transferring the given equations in kinematic units into energetic units requires multiplication by the air density, which can be determined according to the ideal gas law and either the specific heat for constant pressure c_p (for sensible heat flux) or the latent heat of evaporation λ (for latent heat flux). These values are temperature and pressure dependent (see Appendix A3):

$$\rho = \frac{p[hPa] \cdot 100}{R_L \cdot T_v} \ \left[kg \ m^{-3} \right]$$ (2.53)

$$c_p = 1004.834 \ \left[J \ K^{-1} \ kg^{-1} \right]$$ (2.54)

$$\lambda = 2500827 - 2360(T - 273.15) \ \left[J \ kg^{-1} \right]$$ (2.55)

For the case of the latent heat flux in kinematic units, the water vapor pressure, with an additional factor $0.62198/p$ (p in hPa), the specific humidity must be calculated in kg kg^{-1}. To achieve an accuracy lower than 1% of the fluxes, the temperature must be determined with an accuracy of 1 K and the pressure should be determined as a mean value with the barometric equation (Supplement 2.1) for the height above sea level. The following transformation relations are given:

$$Q_H [W \ m^{-2}] = c_p \ \rho \ \overline{w'T'} [K \ m \ s^{-1}]$$

$$= 1004.832 \ \frac{p[hPa] \cdot 100}{287.0586 \cdot T} \ \overline{w'T'} [K \ m \ s^{-1}]$$ (2.56)

Table 2.6. Coefficients α_0 according to different authors

Author	α_0
Businger *et al.* (1971)	1.35
– correction according to Wieringa (1980)	1.00
– correction according to Högström (1988)	1.05
Kader and Yaglom (1972)	1.15 – 1.39
Foken (1990)	1.25
Högström (1996)	1.09 ± 0.04

Supplement 2.1.   Barometric equation

The pressure at some height Z can be calculated from the pressure at sea level $p(Z{=}0)$ and the mean virtual temperature, T_v between sea level and Z,

$$p(Z) = p(Z = 0) \, e^{\frac{g_0}{R_L \, T_v} Z} \, ,$$

(2.57)

where T_v is given by Eq. (2.69).

In Eq. (2.57), Z is the geopotential height (Stull 2000),

$$Z = \frac{\Phi}{g_0} ,$$

(2.58)

where $g_0 = 9.80$ m s^{-2} and Φ is the geopotential. In the lower troposphere, the geopotential height differs only slightly from the geometric height. Also, depending on the required accuracy the virtual temperature can be replaced by the actual temperature.

$$Q_E \, [W \, m^{-2}] = \rho \, \lambda \, \overline{w'q'} \, [kg \, kg^{-1} \, m \, s^{-1}]$$
$$= \frac{p[hPa] \cdot 100}{287.0586 \; T} \frac{0.62198}{p[hPa]} [2500827 - 2360(T - 273.15)]$$
$$\cdot \overline{w'e'} \, [hPa \, m \, s^{-1}]$$

(2.59)

The profile equations Eqs. (2.46) to (2.48) give the opportunity to determine the fluxes in a simple way using either semi-log paper or similar computer outputs as shown in Fig. 2.5. Such graphs are helpful for a quick check of the measuring results or the functioning of the sensors. The humidity should not be confused with relative humidity, because then for calculations the temperature is necessary. An overview of applicable humidity units is given in Supplement 2.2 for calculations. Thus, using relative humidity can generate a cross correlation between the latent and sensible heat flux.

In Eq. (2.47), the air temperature is used; however because the temperature changes with height due to decreasing pressure, the temperature must be replaced by the potential temperature

$$\theta = T\left(\frac{1000}{p}\right)^{R_L/c_p},$$

(2.60)

which can be calculated with this special form of the Poisson equation. The use of the air temperature up to heights of 10 m is possible with an error of about 0.1 K.

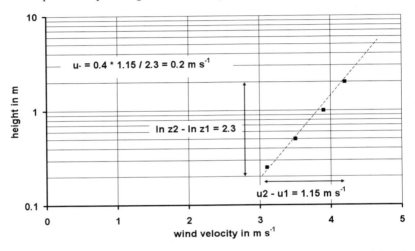

Fig. 2.5. Determination of the friction velocity from the wind profile with a semi-log plot

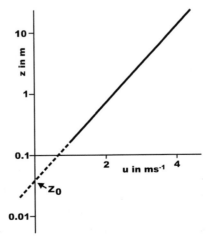

Fig. 2.6. Determination of the roughness height z_0 by extrapolation of the log-linear wind profile to the point $u(z_0) = 0$

Supplement 2.2.   Humidity units

humidity unit	equation
water vapor pressure: partial pressure of the water vapor in hPa	e
relative humidity: ratio of the water vapor pressure and the water vapor pressure at saturation in %	$R = \dfrac{e}{E}\,100\%$
dew point τ: temperature, at which the water vapor pressure for saturation can be reached in °C	$E(\tau)$
water vapor pressure for saturation in with Tetens' equation over water (Stull 2000)	$E = 6.11\,e^{\frac{17.6294\cdot(T-273.16\,K)}{T-35.86\,K}}$
water vapor pressure for saturation with Magnus's equation (–45–60°C over water) according to Sonntag (1990) in hPa	$E = 6.112\,e^{\frac{17.62\cdot t}{243.12+t}}$
water vapor pressure for saturation with Magnus's equation (–65–0.01°C over ice) according to Sonntag (1990) in hPa	$E = 6.112\,e^{\frac{22.46\cdot t}{272.62+t}}$
absolute humidity: mass water vapor per volume moist air in kg m^{-3}	$a = \dfrac{0.21667\,e}{T}$
specific humidity: mass water vapor per mass moist air in kg kg^{-1}, can be replaced with sufficient accuracy by the mixing ratio or visa versa	$q = 0.622\,\dfrac{e}{p-0.378\,e}$
mixing ratio: mass water vapor per mass dry air in kg kg^{-1}	$m = 0.622\,\dfrac{e}{p}$

The integration of the profile equation for the momentum flux Eq. (2.46) from height z_0 up to height z is

$$u(z) - u(z_0) = u(z) = \frac{u_*}{\kappa}\ln\frac{z}{z_0},$$

(2.61)

where z_0 is the height of the extrapolated logarithmic wind profile where $u(z_0) = 0$ as illustrated in Fig. 2.6. Thus, z_0 is only an integration constant. Because this parameter is dependant on the state of the surface it is called roughness parameter or roughness length. It varies from 10^{-3}–10^{-6} m for water and ice, 10^{-2} m for grassland, and up to 0.2 m for small trees. More values are given in Table 3.1.

The integration of the equations for the sensible and the latent heat flux is formally identical to those of the momentum flux. The integration constants are the so-called roughness temperature and roughness humidity. At these heights, the temperature and the humidity are assumed to have approximately the same values as at the ground surface. That this cannot be true was already shown in Fig. 1.5, because near the surface large temperature gradients occur. Therefore, no typical parameterizations are available and approximately 10% of the surface roughness length z_0 assumed.

For atmospheric models these roughness lengths are parameterized (see Chap. 5.5) and used in the following equations

$$T(z) - T(z_{0T}) = \frac{T_*}{\alpha_0 \, \kappa} \ln \frac{z}{z_{0T}}, \qquad (2.62)$$

$$q(z) - q(z_{0q}) = \frac{q_*}{\alpha_{0E} \, \kappa} \ln \frac{z}{z_{0q}}, \qquad (2.63)$$

where

$$T_* = -\frac{\overline{w'T'}}{u_*} \qquad (2.64)$$

is the dynamic temperature or temperature scale and

$$q_* = -\frac{\overline{w'q'}}{u_*} \qquad (2.65)$$

is the humidity scale.

2.3.2 Monin-Obukhov's Similarity Theory

The equations given in Chap. 2.3.1 are strictly speaking only valid in the dynamic sub-layer in which the influence of thermal stratification can be neglected. Monin and Obukhov (1954) used dimensional analysis according to Buckingham's Π-theorem (Supplement 2.3), to extend these equations to the non-neutral (diabatic) case (Foken 2006a). In this analysis, the dependent parameters in the surface layer are the height z in m, the friction velocity u_* in ms^{-1}, the kinematic heat flux

$$\overline{w'T'} = \frac{Q_H}{\rho \, c_p} \qquad (2.66)$$

in Kms^{-1}, and the buoyancy parameter g/T in ms^{-2}K^{-1}. The independent dimensions are the length in meters, the time in seconds, and the temperature in K degrees. The dimensionless parameter that characterizes the processes in the surface layer is:

$$\varsigma = z/L, \qquad (2.67)$$

where

Supplement 2.3.   Buckingham's Π-theorem

Buckingham's Π-theorem (Kantha and Clayson 2000) states that for n+1 dependent parameters and k independent dimensions, there exist exactly n+1-k dimensionless parameters which characterize the process. This can be demonstrated for the free throw of an object (Kitajgorodskij 1976), where z, u_0, g and x are the n+1 dependent parameters corresponding to the dropping height, velocity of the throw, gravity acceleration, and distance of the impact point. The independent dimensions are the length in m and the time in s. The benefit of the dimension analysis is shown stepwise:

- The determination of x=f (z, u_0, g) from many single experiments is very expensive.
- The assumption g = const gives an array of curves for different dropping heights.
- Using Buckingham's Π-theorem one can determine both dimensionless values $x^+ = x/z$ and $z^+ = g z u_0^2$, which gives a direct functional relation between x^+ and z^+.
- If you increase the number of dimensions by separation of the length scale for the horizontal and vertical direction, the process can be described with only one dimensionless parameter $x^* = c u_0 (z g^{-1})^{1/2}$. With some experiments it is even possible to determine for the constant c the standard deviation.

The difficulty in the application of Buckingham's Π-theorem is the selection of the parameters, the dimensions, and the determination of the suitable dimensionless parameters. Because of the many influencing parameters in meteorology, this theorem is very important.

$$L = -\frac{u_*^3}{\kappa \dfrac{g}{T} \dfrac{Q_H}{\rho \cdot c_p}} . \tag{2.68}$$

The characteristic length scale L is called the Obukhov length (Obukhov 1946). Initially, the notation Monin-Obukhov-length was used, but this is, in the historical sense, not exact (Businger and Yaglom 1971). The Obukhov length gives a relation between dynamic, thermal, and buoyancy processes, and is proportional to the height of the dynamic sub-layer (Obukhov 1946), but is not identical with it (Monin and Yaglom 1973; 1975).

Currently (Stull 1988), the Obukhov length is derived from the TKE-equation Eq. (2.41). A physical interpretation of L was made by Bernhardt (1995), *i.e.*, the absolute value of the Obukhov length is equal to the height of an air column in which the production ($L < 0$) or the loss ($L > 0$) of TKE by buoyancy forces is equal to the dynamic production of TKE per volume unit at any height z multiplied by z.

It is more accurate to define the Obukhov length using the potential temperature. Furthermore, for buoyancy considerations the content of water vapor is important, which changes the air density. This can be considered using the virtual temperature:

$$T_v = T(1 + 0.61q)$$ (2.69)

The Obukhov length now can define as:

$$L = -\frac{u_*^3}{\kappa \dfrac{g}{\theta_v} \overline{w'\theta_v'}}$$ (2.70)

This is precise, but is not often used. The universal functions, discussed below, were determined in the lower surface layer and often in dry regions. Observations made in other regions showed within the accuracy of the measurements no significant differences.

The application of Monin-Obukhov similarity theory on the profile equations Eqs. (2.46) to (2.48) is done using the so-called universal functions $\varphi_m(\varsigma)$, $\varphi_H(\varsigma)$ and $\varphi_E(\varsigma)$ for the momentum, sensible and latent heat exchange respectively. Therefore, a new functional dependence on the dimensionless parameter ζ, is given along with Eqs. (2.50)–(2.52):

$$u_* = \sqrt{-\overline{u'w'}} = \frac{\kappa z}{\varphi_m(\varsigma)} \frac{\partial u}{\partial z} = \frac{\kappa}{\varphi_m(\varsigma)} \frac{\partial u}{\partial \ln z}$$ (2.71)

$$\overline{w'T'} = -\frac{\alpha_0 \, \kappa \, u_*}{\varphi_H(\varsigma)} \frac{\partial T}{\partial \ln z}$$ (2.72)

$$\overline{w'q'} = -\frac{\alpha_{0E} \, \kappa \, u_*}{\varphi_E(\varsigma)} \frac{\partial q}{\partial \ln z}$$ (2.73)

The universal functions can be approximated by a Taylor series (Monin and Obukhov 1954) with the argument ς:

$$\varphi(\varsigma) = 1 + \beta_1 \, \varsigma + \beta_2 \, \varsigma^2 + \dots$$ (2.74)

Based on Obukhov's (1946) investigations, the so-called O'KEYPS-function (Kaimal, Elliot, Yamamoto, Panofsky, Sellers) followed from the studies made in the 1950s and 1960s (Businger 1988; Panofsky 1963):

$$[\varphi_m(\varsigma)]^4 - \gamma\varsigma [\varphi_m(\varsigma)]^3 = 1$$ (2.75)

Solutions of this equation are of the form

$$\varphi_m(\varsigma) = (1 + \gamma\varsigma)^{-\frac{1}{4}},$$ (2.76)

with the coefficient γ determined experimentally. This form is called Dyer-Businger relation, and is the most often used form of the universal functions. Therefore, the relation of the universal functions of momentum and sensible heat is given by:

$$\varphi_H \approx \varphi_m^2 \quad for \quad \varsigma < 0$$

$$\varphi_H \approx \varphi_m \quad for \quad \varsigma \geq 0$$

(2.77)

Consequently, the influence of stratification in the surface layer on the universal functions can be described with the stability parameter ς. The universal functions are generally defined in the range of $-1 < \varsigma < 1$. In the stable case ($\varsigma > 1$) a height-independent scaling occurs, the so-called z-less scaling. In this case, the eddy sizes do not depend on the height above the ground surface but on the Obukhov-length (Table 2.7, Fig. 2.7).

Presently the universal functions derived from the Kansas-experiment of 1968 by Businger et al. (1971) and later modified by Högström (1988) are widely used. Högström (1988) considered important criticisms of the Kansas-experiment, for example of the wind measurements (Wieringa 1980), and has re-calculated the von-Kármán constant from $\kappa = 0.35$ to the presently used value of $\kappa = 0.40$ (Table 2.8).

The universal functions given in Table 2.8 are recommended for use. In the last 40 years, many universal functions where determined. The most important of these are given in Appendix A4. A difficulty in applying the universal functions is the normalization. The basics were discussed by Yaglom (1977). For applying the universal functions, one must consider if the parameter α_0 is part of the universal function, is used in the profile equation, or even included in the Obukhov length (Skeib 1980). Furthermore, all universal functions must be recalculated for the updated value of the von-Kármán constant of 0.4 (Högström 1988). It must be noted that the universal functions by Zilitinkevich and Tschalikov (1968) and Dyer (1974) are widely used in the Russian and English literature respectively. Also interesting, is the universal function by Skeib (1980) which is based on the separation of the atmospheric boundary layer into a dynamical sub-layer (without influence of stratification) and a surface layer. In analogy with hydrodynamics, critical values for the separation of both layers are used. A disadvantage is the non-continuous function, but the integration gives a physically rational function.

There are only a few papers with results using the universal function for the water vapor exchange. Therefore, the universal function for the sensible heat flux is used widely for the latent heat flux and the fluxes of trace gases.

Table 2.7. Determination of the stratification in the surface layer dependent on the dimensionless parameter ς and the universal function $\varphi(\varsigma)$

stratification	remark	ς	$\varphi(\varsigma)$
unstable	free convection, independent from u_*	$-1 > \varsigma$	no definition
	dependent from u_*, T_*	$-1 < \varsigma < 0$	$\varphi(\varsigma) < 1$
neutral	dependent from u_*	$\varsigma \sim 0$	$\varphi(\varsigma) = 1$
stable	dependent from u_*, T_*	$0 < \varsigma < 0.5...2$	$1 < \varphi(\varsigma) < 3 ... 5$
	independent from z	$0.5...1 < \varsigma$	$\varphi(\varsigma) \sim const \sim 3 ... 5$

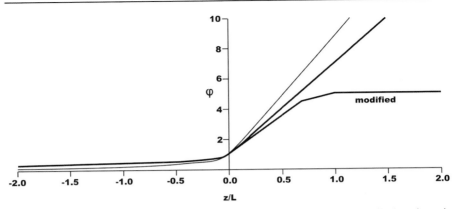

Fig. 2.7. Typical graph of the universal functions for momentum flux (bold line) and sensible and latent heat flux (fine line); 'modified' indicates the function is characterized under the conditions of z-less-scaling

Table 2.8. Universal function according to Businger *et al.* (1971), recalculated by Högström (1988)

stratification	$\varphi_m(\varsigma)$	$\varphi_H(\varsigma) \sim \varphi_E(\varsigma), \alpha_0 = \alpha_{0E} = 1$
unstable	$(1 - 19.3 \, \varsigma)^{-1/4}$	$0.95 \, (1 - 11.6 \, \varsigma)^{-1/2}$
stable	$1 + 6.0 \, \varsigma$	$0.95 + 7.8 \, \varsigma$

Applying the universal function in the stable case is difficult. It is already well known that the universal function in the above form underestimates the turbulent exchange process. The assumption of a nearly constant universal function, which also supports the z-less scaling is obvious. The lack of measurements has not allowed a general applied formulation (Andreas 2002). Handorf *et al.* (1999) based on measurements in Antarctica that for $\zeta > 0.6$ a constant universal function $\varphi_m \sim 4$.

The uncertainty of the universal functions is similar to those of the parameter α_0 (Table 2.6), and is also dependent on the accuracy of the measuring methods. Furthermore, the determination of α_0 and κ for neutral conditions is relevant for the universal function at $\zeta=0$. The von-Kármán constant is presently accepted as $\kappa = 0.40 \pm 0.01$ (Högström 1996). But a slight dependency on the Rossby and Reynolds numbers was discovered (Oncley *et al.* 1996), which is probably a self-correlation effect (Andreas *et al.* 2004). An overview of values of the von-Kármán constant appearing in the literature is shown in Table 2.9. Regarding the universal function, the following accuracies are given by Högström (1996), where normally the virtual temperature is not applied:

Table 2.9. The von-Kármán constant according to different authors (Foken 1990, 2006a), where the value by Högström (1996) is recommended

author	κ
Monin and Obukhov (1954)	0.43
Businger *et al.* (1971)	0.35
Pruitt *et al.* (1973)	0.42
Högström (1974)	0.35
Yaglom (1977)	0.40
Kondo and Sato (1982)	0.39
Högström (1985; 1996)	0.40 ± 0.01
Andreas *et al.* (2004)	0.387 ± 0.004

Table 2.10. Overview about different stability parameters

stratification	temperature	Ri	L	$\varsigma = z/L$
unstable	$T(0) > T(z)$	< 0	< 0	< 0
neutral	$T(0) \sim T(z)$	~ 0	$\pm \infty$	~ 0
stable	$T(0) < T(z)$	$0 < Ri < 0.2 = Ri_c$	> 0	$0 < \varsigma < \sim 1$

$$|z/L| \leq 0.5: \quad |\delta\varphi_H| \leq 10\%$$
$$|z/L| \leq 0.5: \quad |\delta\varphi_m| \leq 20\% \qquad (2.78)$$
$$z/L > 0.5: \quad \varphi_m, \varphi_H = const \,?$$

The discussion of the accuracy of the parameters and functions is not yet finished. For example, a dependence on the mixed layer height cannot be excluded (Johansson *et al.* 2001). This means that processes in the surface layer may be influenced by the whole boundary layer

In addition to the parameter ς, the Richardson number Eq. (2.20) is used to determine the stratification. The definitions of the gradient, bulk, and flux Richardson numbers are:

Gradient Richardson number:

$$Ri = -\frac{g}{\overline{\overline{T}}} \cdot \frac{\partial T / \partial z}{\left(\partial u / \partial z\right)^2} \qquad (2.79)$$

Bulk Richardson number:

$$Ri_B = -\frac{g}{\overline{\overline{T}}} \frac{\Delta T \, \Delta z}{\left(\Delta u\right)^2} \qquad (2.80)$$

Flux Richardson number:

$$Rf = \frac{g}{T} \frac{\overline{w'T'}}{\overline{w'u'}\left(\partial u \middle/ \partial z\right)} \tag{2.81}$$

Analogous to the Obukhov length, the temperature can be replaced by the virtual potential temperature.

The critical Richardson numbers are $Ri_c = 0.2$ and $Rf_c = 1.0$, for which in the case of stable stratification the turbulent flow changes suddenly to a quasi laminar, non-turbulent, flow. The conversion of ς into Ri is stratification-dependent according to (Arya 2001):

$$\varsigma = Ri \quad for \quad Ri < 0$$

$$\varsigma = \frac{Ri}{1 - 5Ri} \quad for \quad 0 \le Ri \le 0.2 = Ri_c \tag{2.82}$$

An overview about the ranges of different stability parameters is shown Table 2.10.

The integration of the profile equations (2.71) to (2.73) using the universal functions in the form of Eq. (2.76) for the unstable case is not trivial and was first presented by Paulson (1970). For the wind profile, the integration from z_0 to z using the definition of the roughness length $u(z_0) = 0$ is

$$u(z) - u(z_0) = u(z) = \frac{u_*}{\kappa}\left[\ln\frac{z}{z_0} - \int\phi_m(\varsigma)\,d\varsigma\right], \tag{2.83}$$

$$u(z) = \frac{u_*}{\kappa}\left[\ln\frac{z}{z_0} - \psi_m(\varsigma)\right],$$

where the integral of the universal function, $\psi_m(\varsigma)$, is:

$$\psi_m(\varsigma) = \int_{z_0/L}^{z/L}\left[1 - \phi_m(\varsigma)\right]\frac{d\varsigma}{\varsigma} \tag{2.84}$$

The integration of the universal function for the momentum and sensible heat flux for unstable conditions according to Businger et al. (1971) in the form of Högström (1988) is

$$\psi_m(\varsigma) = \ln\left[\left(\frac{1+x^2}{2}\right)\left(\frac{1+x}{2}\right)^2\right] - 2\tan^{-1}x + \frac{\pi}{2} \quad for \quad \frac{z}{L} < 0, \tag{2.85}$$

$$\psi_h(\varsigma) = 2\ln\left(\frac{1+y}{2}\right) \quad for \quad \frac{z}{L} < 0, \tag{2.86}$$

with

$$x = (1 - 19.3\varsigma)^{1/4} \quad y = 0.95(1 - 11.6\varsigma)^{1/2}. \tag{2.87}$$

It is obvious that the cyclic term in Eq. (2.85) is not physically realistic. But this term is relatively small and has no remarkable influence on the result. The universal function according to the Dyer-Businger relation Eq. (2.76) shows a good asymptotic behavior for neutral stratification. Physically more accurate, is the integration for different layers according to Skeib (1980). However, in the stable case there are very simple solutions for the integrals of the universal functions:

$$\psi_m(\varsigma) = -6\frac{z}{L} \quad for \quad \frac{z}{L} \geq 0 \tag{2.88}$$

$$\psi_h(\varsigma) = -7.8\frac{z}{L} \quad for \quad \frac{z}{L} \geq 0 \tag{2.89}$$

2.3.3 Bowen-Ratio Similarity

The Bowen ratio (Bowen 1926) is defined as the ratio of the sensible to the latent heat flux:

$$Bo = \frac{Q_H}{Q_E} \tag{2.90}$$

Using Eqs. (2.72) and (2.73) in Eq. (2.90), and taking the conversion between kinematic and energetic units into consideration according to Eqs. (2.56) and (2.59) a very simple relation develops. With the assumption $\varphi_H(\varsigma) \sim \varphi_E(\varsigma)$, for neutral stratification or restriction to the dynamic sub-layer, $\alpha_0 \sim \alpha_{0E}$, and replacing the partial derivatives by finite-differences gives

$$Bo = \frac{c_p}{\lambda}\frac{\Delta T}{\Delta q} = \frac{c_p}{\lambda}\frac{p}{0.622}\frac{\Delta T}{\Delta e} = \gamma\frac{\Delta T}{\Delta e}, \tag{2.91}$$

where the psychrometric constant is $\gamma = 0.667$ for $p = 1000$ hPa and $t = 20°C$.

A special case of the flux-gradient similarity Eq. (2.91), is the so-called Bowen-ratio similarity. The ratio of the gradients of temperature and humidity between two heights behaves like the Bowen ratio. This simplification is the basis of the Bowen-ratio method (see Chap. 4.2.2). It must be noted that the simplifications used are also limitations of the method.

The generalization of this similarity is

$$\frac{F_x}{F_y} \approx \frac{\Delta x}{\Delta y}, \tag{2.92}$$

i.e., the ratio of two fluxes is proportional to the ratio of the differences of the relevant state parameters between two heights.

This equation opens the possibility of determining the flux of an inert gas if the energy flux is known, if the differences of the trace gas concentrations can be determined with a high accuracy, and if the above made simplifications can be accepted. This method was proposed as a modified Bowen-ratio method by Businger (1986), compare with Chap. 4.2.3.

2.4 Flux-Variance Similarity

Analogous to the derivation of the *TKE* equation, Eq. (2.41), the balance equations for the momentum and sensible heat flux can be derived (Foken *et al.* 1991; Wyngaard *et al.* 1971a):

$$\frac{\partial \overline{w'u'}}{\partial t} = -\overline{w'^2}\frac{\partial \overline{u}}{\partial z} - \frac{g}{\overline{T}}\left(\overline{u'T'}\right) - \frac{\partial}{\partial z}\overline{u'w'^2} - \frac{1}{\rho}\frac{\partial\left(\overline{w'p'}\right)}{\partial z} - \varepsilon \qquad (2.93)$$

$$\underset{II}{} \qquad\qquad\qquad\qquad\qquad\qquad \underset{VII}{}$$

$$\frac{\partial \overline{w'T'}}{\partial t} = -\overline{w'^2}\frac{\partial \overline{T}}{\partial z} - \frac{g}{\overline{T}}\left(\overline{T'^2}\right) - \frac{\partial}{\partial z}\overline{T'w'^2} - \frac{1}{\rho}\frac{\partial\left(\overline{T'p'}\right)}{\partial z} - N_T \qquad (2.94)$$

$$\underset{III}{} \qquad\qquad\qquad\qquad\qquad\qquad \underset{VII}{}$$

These equations include the standard deviations of the vertical wind component and the temperature:

$$\sigma_w = \sqrt{\overline{w'^2}} \quad und \quad \sigma_T = \sqrt{\overline{T'^2}} \qquad (2.95)$$

For steady state conditions and after estimation of the size of the terms II and VII in Eq. (2.93) and terms III and VII in Eq. (2.94) it follows for the surface layer (Foken *et al.* 1991; Wyngaard *et al.* 1971a):

$$\sigma_w \big/ u_* \cong const. \quad und \quad \sigma_T \big/ T_* \cong const \qquad (2.96)$$

These normalized standard deviations are also called integral turbulence characteristics (Tillman 1972), because they characterize the state of turbulence integral over all frequencies. For the integral turbulence characteristics of the three wind components, the following values are given (Lumley and Panofsky 1964; Panofsky and Dutton 1984):

$$\sigma_w \big/ u_* \cong 1.25$$

$$\sigma_u \big/ u_* \cong 2.45$$

$$\sigma_v \big/ u_* \cong 1.9$$

$$(2.97)$$

The constancy is only valid for neutral stratification. From similarity relations for diabatic conditions follows (Foken *et al.* 1991):

$$\sigma_w / u_* = a \left[\varphi_m (z/L) \right]^{-0,5} \tag{2.98}$$

$$\sigma_T / T_* = b \left[\frac{z}{L} \varphi_h (z/L) \right]^{-0,5}. \tag{2.99}$$

Many dependencies are reported in the literature for the integral turbulence characteristics under diabatic conditions (see Appendix A4). An experimentally verified relation is, for example, given in Table 2.11. Therefore, for the integral turbulence characteristics a general form is used for the wind components

$$\sigma_{u,v,w} \Big/ u_* = c_1 \left(z \Big/ L \right)^{c_2}, \tag{2.100}$$

and for temperature and other scalars

$$\sigma_T \Big/ T_* = c_1 \left(z \Big/ L \right)^{c_2}. \tag{2.101}$$

At least for the vertical wind, there are no significant differences between the available parameterizations. The most common parameterization by Panofsky et al. (1977) can be applied in the neutral and unstable case:

$$\sigma_w \Big/ u_* = 1.3 \left(1 - 2 \frac{z}{L} \right)^{\frac{1}{3}} \tag{2.102}$$

The integral turbulence characteristics for temperature and other scalars in the case of neutral stratification are not exact because $T_* \to 0$. In the case of unstable stratification for wind components and scalars, dependencies on stratification were found. Thus, some authors (Johansson et al. 2001; Panofsky et al. 1977; Peltier et al. 1996) reported a dependency on the mixed layer height in the case of strong unstable stratification. But the difference is relevant only for free convection (Thomas and Foken 2002). For stable stratification, only a few verified measurements are available. Therefore, the use of the given parameterizations for the unstable case with the argument $|(z-d)/L|$ is recommended as a first approximation. For temperature, a slightly modified approach in Table 2.11 is given (Thomas and Foken 2002).

Based on Rossby similarity (Garratt 1992) some authors (Högström 1990; Smedman 1991; Tennekes 1982; Yaglom 1979) assumed, at least in the neutral case, a visible dependency on the Coriolis parameter. The verification was developed by Högström et al. (2002). A parameterization according to this finding for the neutral and slightly unstable and stable range is given in Table 2.12. For scalars, such a parameterization cannot be found due to the high dependency on the stratification in this range.

Table 2.11. Integral turbulence characteristics for diabatic stratification (Foken *et al.* 1991; Foken *et al.* 1997a; Thomas and Foken 2002)

parameter	z/L	c_1	c_2
$\sigma_w/u*$	$0 > z/L > -0.032$	1.3	0
	$-0.032 > z/L$	2.0	1/8
$\sigma_u/u*$	$0 > z/L > -0.032$	2.7	0
	$-0.032 > z/L$	4.15	1/8
$\sigma_T/T*$	$0.02 < z/L < 1$	1.4	$-1/4$
	$0.02 > z/L > -0.062$	0.5	$-1/2$
	$-0.062 > z/L > -1$	1.0	$-1/4$
	$-1 > z/L$	1.0	$-1/3$

Table 2.12. Parameterization of the integral turbulence characteristics for neutral and slightly unstable and stable stratification (Thomas and Foken 2002)

Parameter	$-0.2 < z/L < 0.4$
$\sigma_w \big/ u*$	$0.21 \ln\left(\dfrac{z_+ \, f}{u_*} \right) + 3.1 \qquad z_+ = 1\,m$
$\sigma_u \big/ u*$	$0.44 \ln\left(\dfrac{z_+ \, f}{u_*} \right) + 6.3 \qquad z_+ = 1\,m$

For free convection ($z/L < -1$), a scaling with the Deardorff velocity Eq. (2.42) and the mixed layer height, z_i, is necessary (Garratt 1992). Such parameterizations must show the decrease of the integral characteristics with increasing height as well as the increase in the Entrainment layer. The following parameterizations were given by Sorbjan (1989):

$$\sigma_w \big/ w_* = 1.08 \left(\frac{z}{z_i} \right)^{1/3} \left(1 - \frac{z}{z_i} \right)^{1/3} \tag{2.103}$$

$$\sigma_\theta \big/ T_* = 2 \left(\frac{z}{z_i} \right)^{-2/3} \left(1 - \frac{z}{z_i} \right)^{4/3} + 0.94 \left(\frac{z}{z_i} \right)^{4/3} \left(1 - \frac{z}{z_i} \right)^{-2/3} \tag{2.104}$$

2.5 Turbulence Spectrum

Knowledge of the turbulence spectrum (see Chap. 1.4.3) is a basic assumption for the optimal adaptation of measuring sensors and measuring conditions in the atmosphere. Many measuring methods are possible only in certain ranges of the turbulence spectra, and many modeling approaches are based on the distribution of spectral energy density. Turbulence spectra distinguish between different state parameters and fluxes, and are dependent on the micrometeorological conditions. In the frequency range of interest to micrometeorology, *i.e.* periods shorter than about 30 min, three ranges can be identified. The range of production of the turbulent energy by the mean flow is characterized by the integral turbulent length scale Λ, which is in the order of 10 to 500 m (Kaimal and Finnigan 1994). The typical frequency range is $f \sim 10^{-4}$ Hz (note that here f is not the Coriolis parameter). This range is followed by the inertial sub-range in which turbulence is assumed to be isotropic, *i.e.* the turbulent movements have no preferential direction. In this range, a well defined energy decrease with increasing frequency occurs according to Kolmogorov's laws (Kolmogorov 1941a; 1941b). The decrease of energy is proportional to $f^{-5/3}$. After multiplying the energy by the frequency the Kolmogorov laws predict an $f^{-2/3}$ decrease for state parameters and an $f^{-4/3}$ decrease for the fluxes. In the third range, with frequencies of $10 - 30$ Hz the kinetic energy of the small eddies are transformed into thermal energy (energy dissipation). The typical dissipation length scale is the Kolmogorov's microscale

$$\eta = \left(\frac{v^3}{\varepsilon} \right)^{\frac{1}{4}},$$ (2.105)

which is about of about 10^{-3} m.

The three parts of the turbulence spectrum in the micrometeorological range are plotted against wave number in Fig. 2.8. Maximum energy occurs at the integral turbulence length $\sim 1/k$. Note that $\Lambda = \pi/k$, is the Eulerian integral turbulence length scale. This length scale can be given for all wind components and scalars. According to Taylor's frozen turbulence hypothesis (Taylor 1938):

$$k = 2\pi \, f \big/ \overline{u}$$ (2.106)

Using Eq. (2.106), the integral turbulence length scale can be transferred into an integral turbulence time scale using the autocorrelation function for the horizontal velocity perturbation, ρ_u,

$$\Lambda_u = \overline{u} \cdot T_u = \overline{u} \int_0^\infty \rho_u(\xi) \, d\xi = \overline{u} \int_0^\infty \frac{\overline{u'(t) \, u'(t+\xi)}}{\sigma_u^2} \, d\xi .$$ (2.107)

Because the autocorrelation function is usually an exponential function, the integral time scale, τ, is given $\rho(\tau) = 1/e \sim 0.37$. This is schematically illustrated in Fig. 2.9.

The energy spectrum under the assumption of local isotropy along with the Kolmogorov constant $\beta_u \sim 0.5$–0.6 (Kolmogorov 1941a) can be given in the inertial sub-range by the so-called –5/3-law, here presented for the horizontal wind component:

$$E_u(k) = \beta_u \, \varepsilon^{2/3} \, k^{-5/3}$$ (2.108)

In an analogous way, the energy spectrum of the temperature in the inertial sub-range is

$$E_T(k) = \beta_T \, N_T \, \varepsilon^{-1/3} \, k^{-5/3},$$

(2.109)

where $\beta_T \sim 0.75$–0.85 (Corrsin 1951), and N_T is the dissipation rate of the variance of the temperature Eq. (2.94). For humidity, the inertial sub-range spectrum is

$$E_q(k) = \beta_q \, N_q \, \varepsilon^{-1/3} \, k^{-5/3}$$

(2.110)

with $\beta_q \sim 0.8$–1.0. The mathematical background for spectral diagrams is briefly given in Supplement 2.4.

Multiplying the *TKE* equation, Eq. (2.41), by the factor $\kappa z / u_*^3$ and assuming steady state conditions ($\partial e / \partial t = 0$), gives the dimensionless *TKE* equation for the surface layer (Kaimal and Finnigan 1994; Wyngaard and Coté 1971):

$$\varphi_m - \frac{z}{L} - \varphi_t - \varphi_\varepsilon + I = 0$$

(2.111)

The imbalance term I is based on the pressure term. I disappears in the unstable case and is of order z/L in the stable case. The transport term, φ_t, is of order $-z/L$ in the unstable case, and disappears in the stable case.

The dimensionless energy dissipation term is

$$\varphi_\varepsilon = \frac{kz\varepsilon}{u_*^3},$$

(2.112)

which can be described by a universal function (Kaimal and Finnigan 1994):

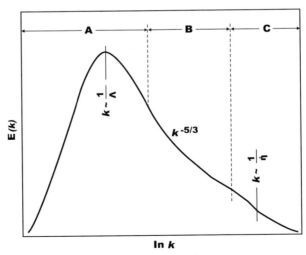

Fig. 2.8. Schematic plot of the turbulence spectra and the ranges of energy production (A), the inertial sub-range (B) and the dissipation range (C) dependent on the wave number k (Kaimal and Finnigan 1994)

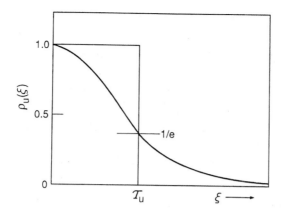

Fig. 2.9. Autocorrelation function and its relation to the integral time scale. The value 1/e is a good approximation for which the area of the shown rectangle is equal to the area below the exponential plot (Kaimal and Finnigan 1994)

Supplement 2.4.   Fourier series and frequency spectrum

A function f, e.g., a time series of a meteorological parameter with turbulent fluctuations (Fig. 1.13), can be represented by a system of orthogonal functions:

$$f(x) = \frac{a_0}{2} + \sum_{k=1}^{\infty} \left(a_k \cos kx + b_k \sin kx \right) \tag{2.113}$$

$$a_k = \frac{1}{\pi} \int_0^{2\pi} f(x) \cos kx \, dx \qquad b_k = \frac{1}{\pi} \int_0^{2\pi} f(x) \sin kx \, dx \tag{2.114}$$

In an analogous way, its representation by an exponential function is possible:

$$f(x) = \sum_{k=-\infty}^{\infty} \left(c_k \, e^{ikx} \right) \qquad c_k = \frac{1}{2\pi} \int_{-\pi}^{\pi} f(x) e^{-ikx} \, dx \tag{2.115}$$

The Fourier transformation is an integral transformation, which converts a function of time into a function of frequency:

$$F(\omega) = \frac{1}{\sqrt{2\pi}} \int_{-\infty}^{\infty} f(t) e^{i\omega t} \, dt \tag{2.116}$$

The square root of the time integral over the frequency spectrum (energy spectrum, power spectrum) corresponds to the standard derivation. The frequency spectrum of two time series is called the cross spectrum. Its real part is the co-spectrum, and the time integral of the co-spectrum corresponds to the covariance.

$$\varphi_\varepsilon^{2/3} = \begin{cases} 1+0.5\left|z/_L\right|^{2/3} & for\, z/_L \le 0 \\ \left(1+5\, z/_L\right)^{2/3} & for\, z/_L \ge 0 \end{cases} \tag{2.117}$$

Further universal functions for the energy dissipation are given in Appendix A4. A practical application of these equations is the determination of the friction velocity using a scintillometer (see Chap. 6.2.5). Also, the energy spectrum of the horizontal wind can be derived in the inertial sub-range (Kaimal and Finnigan 1994):

$$\frac{f \cdot S_u(f)}{u_*^2} = \frac{0.55}{(2\pi)^{2/3}} \left(\frac{\varepsilon^{2/3} z^{2/3}}{u_*^2}\right) \left(\frac{fz}{\overline{u}}\right)^{-2/3} \tag{2.118}$$

$$= \frac{0.55}{(2\pi\kappa)^{2/3}} \varphi_\varepsilon^{2/3} \left(\frac{fz}{\overline{u}}\right)^{-2/3}$$

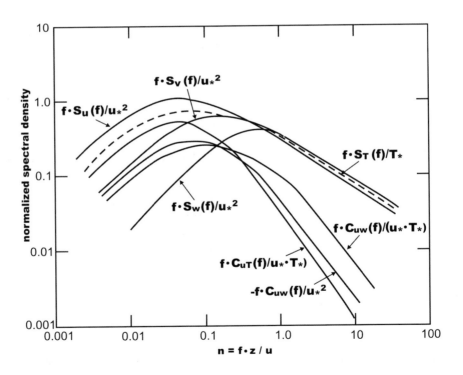

Fig. 2.10. Models of the energy density spectrum of state variables and fluxes (Kaimal *et al.* 1972)

The turbulence spectra of various parameters differ significantly in peak frequency (maximum of the energy density) and their dependence on the stratification. The peak frequency of the vertical wind corresponds to the highest frequencies (0.1–1 Hz), and those of the horizontal wind are one order of magnitude lower (Fig. 2.10). The vertical wind shows this typical form of spectrum for stable (peak frequency shifted to higher frequencies, neutral, and unstable stratification. For the other wind components and scalars, the peak frequencies are shifted to lower frequencies, and because of large scatter in the data, they are often not clearly seen.

Energy density spectra are often used for the correction of measurement data (see Chap. 4.1.2). Because of the difficulty in calculating spectra, models of spectra are often applied. Such models were parameterized by Kaimal *et al.* (1972) on the basis of the data of the Kansas experiment. With the usual normalization of the frequency according to Eq. (1.19)

$$n = f\,\frac{z}{u},$$
(2.119)

the wind components for neutral and slightly unstable stratification (Kaimal and Finnigan 1994) are:

$$\frac{f\,S_u(f)}{u_*^2} = \frac{102\,n}{(1+33\,n)^{5/3}}$$
(2.120)

$$\frac{f\,S_v(f)}{u_*^2} = \frac{17n}{(1+9.5n)^{5/3}}$$
(2.121)

$$\frac{f\,S_w(f)}{u_*^2} = \frac{2,1n}{(1+5.3n)^{5/3}}$$
(2.122)

Højstrup (1981) proposed a stability-dependent parameterization for the vertical wind for unstable stratification:

$$\frac{f\,S_w(f)}{u_*^2} = \frac{2n}{(1+5.3n)^{5/3}} + \frac{32n\,\varsigma}{(1+17n)^{5/3}}$$
(2.123)

For the temperature spectrum, which can also be used for other scalars, Kaimal *et al.* (1972) published the following parameterization:

$$\frac{f\,S_T(f)}{T_*^2} = \begin{cases} \dfrac{53.4n}{(1+24n)^{5/4}} & for \quad n < 0.5 \\[3mm] \dfrac{24.4n}{(1+12.5n)^{5/3}} & for \quad n \geq 0.5 \end{cases}$$
(2.124)

For the co-spectra, (Kaimal and Finnigan 1994) proposed for the unstable case $(-2 < z/L < 0)$

$$-\frac{f\,C_{uw}(f)}{u_*^2}=\frac{12n}{(1+9.6n)^{7/3}},\qquad(2.125)$$

$$-\frac{f\,C_{wT}(f)}{u_*\,T_*}=\begin{cases}\dfrac{11\,n}{(1+13.3\,n)^{7/4}}&\text{for}\quad n\le1.0\\[4mm]\dfrac{4\,n}{(1+3.8\,n)^{7/3}}&\text{for}\quad n>1.0\end{cases},\qquad(2.126)$$

and for the stable case ($0 < z/L < 2$):

$$-\frac{f\,C_{uw}(f)}{u_*^2}=0.05\,n^{-4/3}\left(1+7.9\frac{z}{L}\right)\qquad(2.127)$$

$$-\frac{f\,C_{wT}(f)}{u_*\,T_*}=0.14\,n^{-4/3}\left(1+6.4\frac{z}{L}\right)\qquad(2.128)$$

The log–log plot of ln ($f\,S_A(f)$) versus ln f show the graphs of the power laws in the form $f\,S_A(f)\approx f^{-2/3}$ and The plot ln $S_A(f)$ versus ln f in Fig. 2.10 shows $S_A(f)\approx f^{-5/3}$. In a plot of $S_A(f)$ versus f, the area below the graph in the range Δf is equal to the standard deviation $\sigma_A(\Delta f)$. The resolution at low frequencies is often bad. The area below the graph of $f\,S_A(f)$ versus ln f, is equal to the energy density. With the multiplication of the values of the ordinate by f and logarithmic abscissa, a good representation of the low frequencies is possible.

The energy density in the inertial sub-range according to Eq. (2.108) can also be determined using the structure function (Tatarski 1961, Supplement 2.5)

$$D_u(r)=c_u\,\varepsilon^{2/3}\,r^{2/3},\qquad(2.129)$$

with the structure constant $c_u\sim2$ and the structure function parameter

$$C_u^2=c_u\,\varepsilon^{2/3}.\qquad(2.130)$$

The structure function for the temperature is given by

$$D_T(r)=c_T\,N_T\,\varepsilon^{-1/3}\,r^{2/3},\qquad(2.131)$$

with the structure constant $c_T\sim3.2$ and the structure function parameter

$$C_T^2=c_T\,N_T\,\varepsilon^{-1/3}.\qquad(2.132)$$

The structure function for moisture is given by

$$D_q(r)=c_q\,N_q\,\varepsilon^{-1/3}\,r^{2/3},\qquad(2.133)$$

with the structure constant $c_q\sim4.02\,\beta_q$ and the structure function parameter

$$C_q^2=c_q\,N_q\,\varepsilon^{-1/3}.\qquad(2.134)$$

Supplement 2.5.   Structure Function

An alternative statistical parameter to the autocorrelation function, Eq. (2.107), is the structure function. It is a relation between the value of a variable at time t, $A(t)$, and its value at a later time, $A(t+L)$, note L is not the Obukhov length. If the time between the measurements is ΔT, then $L = j\,\Delta T$ and

$$D_{AA}(L) = \frac{1}{N} \sum_{k=0}^{N-j} \left(A_k - A_{k+j} \right)^2 \qquad (2.135)$$

If instead of time, the measurements are made in distance, then $L=j\Delta r$ where Δr is the spatial separation between measurements.

The structure function parameters are very relevant for meteorological measuring techniques (Beyrich et al. 2005; Kohsiek 1982), because they are directly connected to the refraction structure function parameter C_n^2 (Hill et al. 1980), which is proportional to the backscatter echo of ground-based remote sensing technique:

$$C_n^2 = A^2 C_T^2 + 2ABC_{Tq} + B^2 C_q^2 \qquad (2.136)$$

The coefficients A and B are dependent on temperature, humidity, pressure and the electromagnetic wave length. Data are given by Hill et al. (1980) and Andreas (1989). While a dependence on temperature and humidity exists for microwaves, the humidity dependence on visible and near infrared light is negligible:

$$C_n^2 = \left(79.2 \cdot 10^{-6} \frac{p}{T^2} \right)^2 C_T^2 \qquad (2.137)$$

The stability dependence of the structure function parameter is given by a dimensionless function of the energy dissipation, which has the character of a universal function (Wyngaard et al. 1971b). For the surface layer, they are given in the form (Kaimal and Finnigan 1994):

$$\frac{C_u^2 \, z^{2/3}}{u_*^2} = \begin{cases} 4\left(1+0.5\dfrac{-z}{L}\right)^{2/3} & for \quad z/L \le 0 \\[4mm] 4\left(1+5\dfrac{z}{L}\right)^{2/3} & for \quad z/L > 0 \end{cases} \qquad (2.138)$$

$$\frac{C_T^2 \, z^{2/3}}{T_*^2} = \begin{cases} 5\left(1+6.4\dfrac{-z}{L}\right)^{-2/3} & for \quad z/L \le 0 \\[4mm] 5\left(1+3\dfrac{z}{L}\right) & for \quad z/L > 0 \end{cases} \qquad (2.139)$$

Other universal functions are listed in Appendix A4. These will be used later for the determination of the sensible heat flux with scintillometers (see Chap. 6.2.5).

The temperature structure function parameter C_T^2 can be determined according to Wyngaard $et\ al.$ (1971b) by the vertical temperature profile

$$C_T^2 = z^{-4/3} \left(\partial\theta \big/ \partial z \right)^2 f_T(Ri)$$

(2.140)

with the empirical stability function $f_T(Ri)$.

3 Specifics of the Near-Surface Turbulence

The equations given in Chap. 2 describe the atmospheric turbulence near the surface for ideal conditions such as horizontally homogeneous surfaces free of obstacles, steady-state conditions, and others. Such conditions are seldom seen in nature, and the Earth's surface is not completely flat with uniform properties. Furthermore, all processes are not steady-state because of changing cloudiness and the diurnal cycle of irradiation. In this Chapter, the properties of the plant covered heterogeneous surface and its influences on the exchange process are described. Because these effects are important in the generation of atmospheric turbulence, they must be taken into account for all measurements and modeling efforts.

3.1 Properties of the Underlying Surface

Underlying surfaces are distinguished according to their thermal properties, roughness, canopy height and other obstacles, and unevenness. For low vegetation, the surface can be approximated as a porous sponge-like layer, which can be described by simple mathematical equations; however, no special phenomena should be considered. Much more complicated to handle are obstacles and tall vegetation. Therefore a special paragraph on this subject is included in this section.

3.1.1 Roughness

As illustrated in Fig. 3.1, there are two different ways to determine roughness: the determination of the geometric surface structure, and the calculation of the roughness from the turbulence near the surface. There have been many investigations of both methods, but the second way has the highest practical relevance.

The possibility of the determination of the surface roughness as an integration constant of the profile equation Eq. (2.46) for neutral conditions Eq. (2.61) was shown in Chap. 2.3.1 with the roughness height z_0 defined by:

$$u(z) = \frac{u_*}{\kappa} \ln \frac{z}{z_0} \tag{3.1}$$

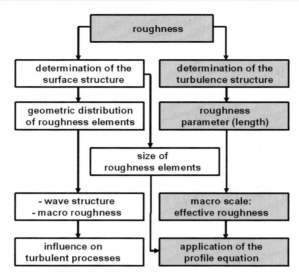

Fig. 3.1. Scheme of the possible ways to determine the roughness of the surface (Foken 1990)

Table 3.1. Roughness height (roughness length) of different surfaces (ESDU 1972; Oke 1987) and effective roughness height according to Fiedler and Panofsky (1972)

surface	roughness height in m
ice	0.00001
water	0.0001–0.001
wavy snow	0.002
short grass	0.005
long grass	0.02
grain	0.05
bushes	0.2
forest	1–2
suburban	0.5–2
	effective roughness height in m
flat lands	0.42
low hills	0.99
high mountains	1.42

The values of the roughness height (Table 3.1) given by various authors differ only slightly (Davenport *et al.* 2000; ESDU 1972; Wieringa 1992). The direct determination of the roughness height using the profile equation under neutral conditions is the most common method. The method is limited to roughness elements with small vertical extension (maximal forest and suburban with low houses).

Principally the method can also be applied on the profile equation for more complex landscapes, towns and others, which are higher than the surface layer. These heights are called effective roughness (Fiedler and Panofsky 1972). Effective roughness heights are used in connection with the *blending-height*-concept (see Chap. 3.2.4) and area averaging (see Chap. 5.7).

If the roughness values in Table 3.1 are not accurate enough, then estimates of the roughness height can be made using wind measurements at different levels (see Chap. 4.2.5). From a log–linear plot analogous to Fig. 2.6, the roughness height can be determined for neutral stratification. In the case of low roughness (ice or water) this method can be very inaccurate, because small errors in the measurements of the wind velocity can cause large changes in the estimated roughness height. This is the reason for the large scatter in the reported roughness heights over water.

Panofsky (1984), proposed an interesting way of determining the roughness height using the integral turbulence characteristics or turbulence intensities. From the combination of Eq. (2.61) and Eq. (2.97) for neutral stratification, one gets:

$$\frac{\sigma_w}{u} = \frac{1.25\,\kappa}{\ln\left(z/z_0\right)} \tag{3.2}$$

A more exact method is to use the morphological determination of the roughness from the surface structures. For solid surfaces, this method was used for a long time in hydrodynamics where the roughness elements are exactly measured. In hydrodynamical investigations, the so-called equivalent sand roughness k_s according to Nikuradse (1933) is applied, which can be determined in a way analogous to the profile method Eq. (3.1) with the integration constant B of about 2.5:

$$u(z) = \frac{u_*}{\kappa}\ln\frac{z}{k_s} + B \tag{3.3}$$

This method is the connection between the exact determination of the surface structures and the application of the profile equation. Errors in the determination of the surface structure and the influence of stratification are transferred into the integration constant B. This method has been used mostly in wind tunnel experiments (Raupach 1992); however, in urban meteorology it is a good method to estimate and classify roughness (Grimmond and Oke 1999, 2000; Lettau 1969).

In some applications, roughness parameters based on the statistical distribution of the roughness elements have been used. One example is the replacement of the roughness height in the roughness-Reynolds-number

$$\mathrm{Re}_s = \frac{z_0 \cdot u_*}{\nu} \tag{3.4}$$

with the standard deviation of the height of water waves (Kitajgorodskij and Volkov 1965), which was used successfully by Russian scientists. A further example is the so-called macro roughness which is based on the standard deviation of the topography or geometric structures within a certain horizontal scale (Zelený and Pretel 1986). For both cases, a direct connection with the profile equation is not possible, but *e.g.* integral turbulence characteristics have significant correlations with these parameters.

For real land surfaces, estimating roughness parameters is complicated because landscapes are often complex, and do not have uniform roughness which is comparable with the data given in Table 3.1. A further complication is that single roughness elements such as trees can have a significant influence on the roughness of the area. Furthermore, it is impossible to calculate a simple area-average roughness from the roughness of individual regions because Eq. (3.1) is non-linear. For example, an area comprised of 50% bush land and 50% open water is on average a bush land with a lower surface roughness height. The wind field calculated with this new roughness height will not be representative of the smooth water surface. In spite of these findings, the arithmetic averaging is still widely applied in modeling. This is because the heterogeneity of the landscape exists on spatial scales that are smaller than the grid size of most of the models (see Chap. 5.7).

The determination of surface roughness for wind power applications is of high practical significance. Therefore, for the European Wind Atlas an averaging scheme was developed (Petersen and Troen 1990) which takes into account the non-linear character of averaging and integrating low roughness surfaces into the area roughness. For simplicity, the authors classify only four types (0–3) of surface roughness (class 0: water; class 1: open landscape with flat meadows; class 2: croplands with bushes; class 3: forest and suburban). The whole area is divided into quarters; then depending on the combination of the surfaces of each quarter a mean surface roughness, z_0^R, is determined as in Table 3.2. This is a special form of parameter averaging, which is described in more detail in Chap. 5.7.1. More sophisticated approaches use flux averaging so that the roughness distributions can be determined, *e.g.*, by remote sensing (Hasager and Jensen 1999).

For the case of roughness calculations over vegetated land, the annual change of foliage and the growth of plants must also be taken into account. An example of these changes is given in Fig. 3.2.

For water surfaces, the roughness height z_0 is generally parameterized as a function of the wind velocity. The parameterization by Charnock (1955)

$$z_0 = \frac{u_*^2}{81.1\,g} \tag{3.5}$$

is often used in models. However, Eq. (3.5) underestimates z_0, for low wind velocities, because under these circumstances existing capillary waves are very rough. It is highly recommended to use the relation by Zilitinkevich (1969), which is a combination of Eq. (3.5) and the relation by Roll (1948) in the form

$$z_0 = c_1 \frac{v}{u_*} + \frac{u_*^2}{c_2\,g}, \tag{3.6}$$

where the coefficients, c_1 and c_2 are given in Table 3.3 and plotted in Fig. 3.3. Because of the remarkable scatter of experimental data for the determination of the roughness height (Kitajgorodskij and Volkov 1965), all combinations of both parameterizations as well the Roll-type for $u_* < 0.1$ ms^{-1} and the Charnock-type for $u_* > 0.1$ ms^{-1} are applicable.

Table 3.2. Averaging scheme for the mean roughness length z_0^R in the European Wind Atlas (Petersen and Troen 1990). In the Table, the different portions of the roughness classes 0–3 to the whole area are given

type	0	1	2	3	z_0^R
z_0 in m	0.0002	0.03	0.1	0.4	in m
	3	1			0.001
	3		1		0.002
	3			1	0.003
	2	2			0.004
	2	1	1		0.006
	2	1		1	0.010
	2		2		0.009
	2		1	1	0.015
	2				0.025
	1	3		1	0.011
	1	2	1	2	0.017
	1	2			0.027
	1	1	2		0.024
	1	1	1	1	0.038
	1	1		2	0.059
	1		3		0.033
	1		2	1	0.052
	1		1	2	0.079
	1			3	0.117
		3	1		0.042
		3		1	0.064
		2	2		0.056
		2	1	1	0.086
		2		2	0.127
		1	3		0.077
		1	2	1	0.113
		1	1	2	0.163
		1		3	0.232
			3	1	0.146
			2	2	0.209
			1	3	0.292

Table 3.3. Coefficients for Eq. (3.6)

author	c_1	c_2
Roll (1948)	0.48	∞
Charnock (1955)	0.0	81.1
Zilitinkevich (1969)	0.1	20.8
Brocks and Krügermeier (1970)	0.0	28.5
Foken (1990)	0.48	81.1
Beljaars (1995)	0.11	55.6
Zilitinkevich *et al.* (2002)	0.1	56 open ocean
		32 costal zone

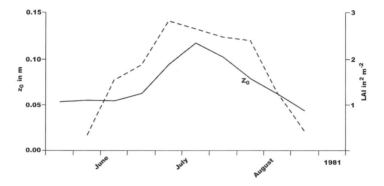

Fig. 3.2. Change of the roughness height and the leaf-area index (*LAI*) during the growth of a corn field according to Hurtalová *et al.* (1983), added by data given by Gurtalova *et al.* (1988)

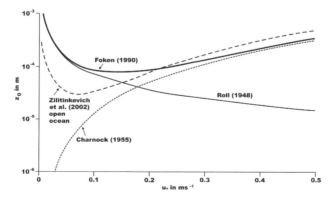

Fig. 3.3. Dependence of the roughness height over water on the friction velocity according to different authors

3.1.2 Zero-Plane Displacement

For dense vegetation (grass, grain) the zero-level for the wind field according to Eq. (3.1) is no longer the ground surface but is within the plant canopy, which can be simplified as a porous medium. For this level, with the height d (displacement height, zero-plane displacement) all equations given thus far are valid analogs to the bare soil (Paeschke 1937). The scale, which is based on this level, is called the aerodynamic scale with $z'(d) = 0$. In contrast, the geometric scale, which is measured from the ground surface, is $z = z' + d$ (Fig. 3.4). Because Eq. (3.1) is valid only for the aerodynamic scale, for low vegetation the geometric scale gives:

$$u(z) = \frac{u_*}{\kappa} \ln \frac{z - d}{z_0 - d} \qquad (3.7)$$

Consequently, all profile equations and equations for integral turbulence characteristics in Chap. 2 must be modified for low vegetation by replacing "z" with "$z+d$".

The log-linear extrapolated plot of the wind profile above a dense plant canopy with the geometric height as the ordinate cuts the ordinate at $u(z_0+d) = 0$. Thus, it is not possible to determine the roughness and displacement heights in a simple way. Instead, a system of a number of equations is required. However, it is possible to use a value of z_0 from Table 3.1 and then calculate d. A simple approximation is also given by

$$z_0 = 0.1\, z_B, \qquad (3.8)$$

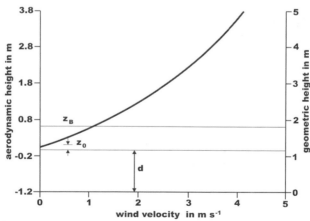

Fig. 3.4. Aerodynamic and geometric scale for dense vegetation ($d = 1.2$ m)

where z_B is the canopy height. Another method is to assume that the displacement height is about 2/3 of the canopy height. Therefore, with a known value of z_0+d the roughness height can be determined by:

$$z_0 = [z_0 + d] - \frac{2}{3} z_B \qquad (3.9)$$

The relation between zero-plane displacement and canopy height is independent of the plant species, but changes during the growth of the plants (Table 3.4). There are also approaches available which include the plant cover (Raupach 1994). There is a wind-velocity dependence caused by the changing of the canopy structure with wind speed (Marunitsch 1971). For low wind velocities over the forest, the ratio d/z_B increases up to values of about 0.8, while for high wind velocities values of about 0.5 are found. In the opposite way, the ratio z_0/z_B changes from about 0.01 to 0.2.

Of great difficulty, is the determination of the canopy height. This is because the top of the canopy often has a non-uniform height, and the highest parts of the canopy have a disproportionably large influence on the roughness (Foken 1990). It is therefore recommended to determine the canopy height using the largest canopy heights covering 10% of the area.

For a more exact determination of the zero-plane displacement, it is necessary to measure the friction velocity using the eddy-covariance method (see Chap. 4.1) and the wind profile above the canopy. With an iterative model, the zero-plane displacement is varied until the friction velocity of the profile method is equal to the friction velocity from the eddy-covariance.

Table 3.4. Dependence of the ratio of zero-plane displacement and canopy height on the canopy density for winter wheat field (Koitzsch *et al.* 1988)

canopy density stems with ears m^{-2}	canopy height in m	ratio d/z_B
390	0.80	0.60
660	0.97	0.73
570	0.40	0.76

3.1.3 Profiles in Plant Canopies

Meteorological measurements within low plant canopies are very rare because they cannot be made with the usual measuring techniques without disturbing the canopy. Temperature measurements are very costly because of the difficulties in measuring temperature without a radiation error (see Chap. 6.2.3). An introduction to the principle structures of temperature, wind speed, relative humidity, water vapor pressure, and the saturation deficit are given in Fig. 3.5.

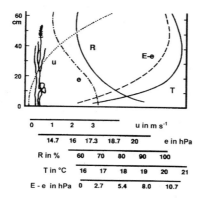

Fig. 3.5. Profiles of different meteorological elements in a dense grass canopy on a sunny day in June, 3–4 p.m., in Scotland (Waterhouse 1955)

By normalizing the profiles of turbulence parameters with their mean values at the canopy height, the profiles become very similar regardless of the character of the surface, *i.e.* obstacles in a wind tunnel, low vegetation, or forest (Fig. 3.6). For the determination of the wind profile, exponential expressions are used with coefficients dependent on the plant species and the leaf-area index:

$$\overline{u(z)} = \overline{u(z_B)}\, e^{\alpha\left(\frac{z}{z_B}-1\right)} \tag{3.10}$$

Table 3.5 contains the values of profile coefficient, α, for different plant canopies (Cionco 1978). An explicit expression for α was given by Goudriaan and cited by Campbell and Norman (1998),

$$\alpha \cong \left(\frac{0.2\, LAI\, z_B}{l_m} \right), \tag{3.11}$$

where LAI is the leaf-area index, l_m is the mean separation of leaves. For plant physiological investigations, the distribution of light is important in the calculation of photosynthesises. Relevant contributions to this topic were made by Ross (1981) and Jones (1992).

Table 3.5. Values of the profile parameter of the wind profile within the plant canopy in Eq. (3.10) according to Cionco (1978)

plant canopy	profile parameter α
wheat	2.45
rye	1.97
rice	1.62
sun flowers	1.32
larch tree plantation	1.00
fruit plantation	0.44

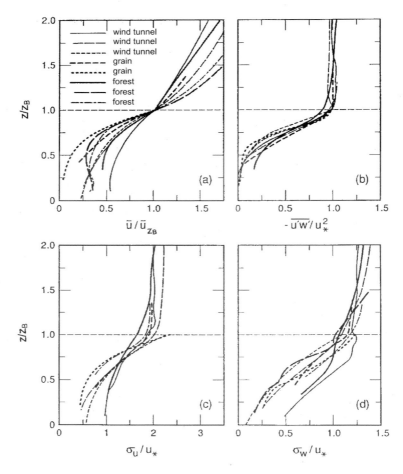

Fig. 3.6. Profiles of the mean wind velocity, the friction velocity and the standard deviations of the horizontal and vertical wind velocity normalized with their values at the canopy height according to different authors (Kaimal and Finnigan 1994)

3.2 Internal Boundary Layers

3.2.1 Definition

The statements so far are based on the assumption of a homogeneous surface. More typical for our landscapes are heterogeneous surfaces with a change in the surface characteristics within about 100 m. In Fig. 3.7, it is shown that a complicated flow system is developing above such an inhomogeneous surface. The reason is that over each surface a wind profile is generated dependent on the surface roughness or a temperature profile dependent on the surface temperature. Due to the horizontal wind field the different profiles are shifted downwind forming a layer of discontinuity, which is called an internal boundary layer. Therefore, neighboring areas can also affect the exchange of energy and matter above a different downwind area. The internal boundary layers are caused by different surface roughness and different source areas of surface temperature, moisture *etc.* Internal boundary layers can be found for neutral, stable, and unstable stratifications. In the case of free convection, the exchange process is more vertically dominated.

> Internal boundary layers are significantly developed disturbance layers in the near-surface layer, which are generated by horizontal advection over discontinuities of the surface properties (roughness, thermal properties, *etc.*).

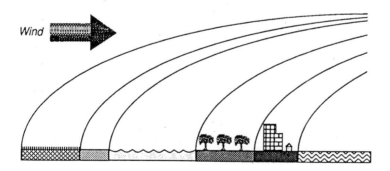

Fig. 3.7. Generation of internal boundary layers above an inhomogeneous surface (Stull 1988)

The internal boundary layer is sometimes composed of several layers each with a different thickness. The layer below the discontinuity layer is called new equilibrium layer (NEL). Their properties come from the new surface. Above the discontinuity layer (internal boundary layer, IBL) the layer is influenced by the surface upwind of the site (Fig. 3.8). Relevant to the development of mechanical internal boundary layers are the roughness heights upwind and downwind of the site of a sudden change of the surface roughness and the friction velocities related to the wind field on both sides of the change. The height of the internal boundary layer depends on the fetch (distance to the roughness change). For large fetches the differences between both sides of an internal boundary layer decreases. Overviews are given by Stull (1988), Garratt (1990, 1992), and Savelyev and Taylor (2001, 2005).

In Fig. 3.8 it is shown that above the new equilibrium layer the internal boundary layer has a large range of transition. To determine the height of the internal boundary layer several methods are available.

In the simplest case, the height of an internal boundary layer can be determined by extrapolation of the wind profiles from both sides of the internal boundary layer (Elliott 1958; Raabe 1983):

$$u_1(\delta) = u_2(\delta) \tag{3.12}$$

This method has the disadvantage that the point of intersection may be outside the internal boundary layer.

More successful is the assumption that if the undisturbed wind profiles can be measured below and above the internal boundary layer, than the point of intersection is taken to be the height of internal boundary layer:

$$\delta = \frac{\delta_1 + \delta_2}{2} \tag{3.13}$$

The height of the internal boundary layer can also be determined by the change of the friction velocity (Peterson 1969; Shir 1972; Taylor 1969):

$$z = \delta \quad for \quad \frac{u_*(x,z) - u_{*1}}{u_{*2} - u_{*1}} < c \sim 0.1 \tag{3.14}$$

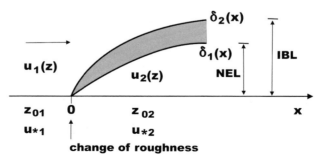

Fig. 3.8. Schematic structure of the internal boundary layer at a sudden change of the surface roughness according to Rao *et al.* (1974)

Equation (3.14) is often used by modelers. It should be mentioned that the internal boundary layer defined this way is higher than those derived from wind profiles (Shir 1972).

For practical reasons, it is useful to use the lower level of the layer of disturbances as the height of the internal boundary layer, $\delta = \delta_1$, because the new equilibrium layer can be assumed as undisturbed above the new surface (Rao *et al.* 1974), and is therefore relevant for making measurements.

Mechanical internal boundary layers occur for flow from rough to smooth or from smooth to rough surfaces. Because of the different wind gradients above smooth and rough surfaces, there are characteristic forms of internal boundary layers as illustrated in Fig. 3.9.

The classical model for the height of an internal boundary layer based on Eq. (3.1) was given by Elliott (1958):

$$\delta_{Elliott} = z_{02}\left(\frac{z_{01}}{z_{02}}\right)^{y_{(1-R)}} \quad with \quad R = \frac{u_{*2}}{u_{*1}} \tag{3.15}$$

Current parameterizations give a value that is multiplied by the base of the natural logarithm. *i.e.*

$$\delta \approx e\,\delta_{Elliott} \tag{3.16}$$

Parameterizations base also on Eq. (3.1) where developed by Radikevitsch (1971) and Logan and Fichtl (1975):

$$\frac{u_{*2}}{u_{*1}} = 1 - \frac{\ln\dfrac{z_{01}}{z_{02}}}{\ln\dfrac{\delta}{z_{01}}} \tag{3.17}$$

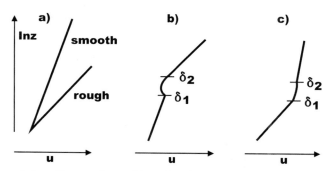

Fig. 3.9. The wind profile at an internal boundary layer for neutral stratification: (**a**) typical profile for rough and smooth surfaces, (**b**) change of the surface roughness from rough to smooth, (**c**) change of the surface roughness from smooth to rough

The height of an internal boundary layer as a function of fetch was found in hydrodynamical investigations, and is given by a 4/5-exponential law (Garratt 1990; Savelyev and Taylor 2001; Shir 1972):

$$\delta = f_1\left(z_{01}\middle/z_{02}\right) x^{\frac{4}{5}+f_2\left(z_{01}\middle/z_{02}\right)} \tag{3.18}$$

3.2.2 Experimental Findings

There are many experimental results available about internal boundary layers, but more fundamental experiments which could update our present models and parameterizations in a more physical way are missing (Garratt 1990; Raabe 1991). Consequently classical experiments, such as Bradley (1968a), are still used for model comparisons (Fig. 3.10). The first experimental investigations in nature were made by Taylor (1969) and Peterson (1969) in a coastal zone. An example of the difficulties in the evaluation of experimental data is given by measurements made with an ascending and descending probe by Hupfer *et al.* (1976) and shown in Fig. 3.11. Their high-resolution temperature and wind profile measurements showed a narrow internal boundary layer, which cannot be found in the mean wind profile.

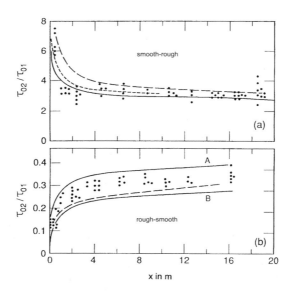

Fig. 3.10. Ratio of the friction velocities on both sides of a sudden roughness change (Bradley 1968a)

A number of experiments were done to determine the heights of either the internal boundary layer or the new equilibrium layer using an equation comparable to Eq. (3.18). At this time, the influences on the IBL height of the stratification, which increases the exponent from 0.8 to 1.4 according Rao *et al.* (1974), and the possible differences in the transitions from smooth to rough or from rough to smooth are not yet determined. Garratt (1990) assumed a greater increase of the IBL height in the case of the transition from smooth to rough than from rough to smooth. Because of the large scatter of the results from all experiments, most authors assume a simplified relation

$$\delta = a\, x^{b} \tag{3.19}$$

instead of Eq (3.18). Experimental data for this equation are given in Table 3.6. These data are related to the new equilibrium layer according to Rao *et al.* (1974).

Table 3.6. Experimental results for the coefficients in Eq. (3.19) to estimate the height of the internal boundary layer (new equilibrium layer according to Rao *et al.* 1974); for more data see Savelyev and Taylor (2005)

author	a	b	conditions
Bradley (1968a), Shir (1972)	0.11	0.8	$z_{01}/z_{02} = 125$ and 0.08 artificial roughness
Antonia and Luxton (1971; 1972)	0.28	0.79	$x \leq 10$ m, rough–smooth
	0.04	0.43	$x \leq 10$ m, smooth–rough wind tunnel
Raabe (1983)	0.30±0.05	0.50±0.05	beach, on- and off-shore winds, 5 m $< x <$ 1000 m

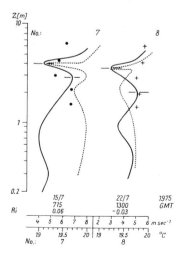

Fig. 3.11. Wind and temperature profile measurements in the costal zone at 75 m distance to the shore line according to Hupfer *et al.* (1976): bold line: wind profile measured with an ascending and descending probe within 1–2 min, dotted line: temperature profile measured with an ascending and descending probe, horizontal double line: position of the internal boundary layer, points: mean wind velocity from profile measurements

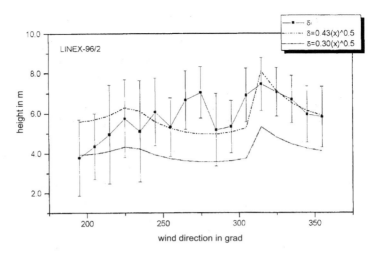

Fig. 3.12. Measured heights of the internal boundary layer from the boundary layer measuring site *Falkenberg* of the German Meteorological Service (Jegede and Foken 1999), roughness changes from smooth (z_o = 0.008 m) to rough (z_0= 0.032 m).

The relation proposed by Raabe (1983)

$$\delta = 0.3\sqrt{x} \qquad\qquad (3.20)$$

can be classified as exceedingly *robust* for the determination of the new equilibrium layer. This was experimentally shown by Jegede and Foken (1999), who determined internal boundary layers from wind profile data for roughness changes from smooth to rough and from rough to smooth, and for different wind directions which are related to different fetches (Fig. 3.12). Evaluated was the intersection of both wind profiles above and below the internal boundary layer. The height of the middle of the internal boundary layer is well represented by Eq. (3.19) with a = 0.43 and b = 0.5. The prominent disagreement for a wind direction of about 270° is caused by an additional internal boundary layer due to a group of bushes about 500 m away, which makes the actual internal boundary layer not verifiable. The remarkable scatter for wind directions < 230° is caused by changing wind directions which are partly connected with short fetches.

3.2.3 Thermal Internal Boundary Layer

Analogous to the mechanical internal boundary layer due to a step change of the surface roughness, there is also a thermal internal boundary layer due to a step change of the surface temperature. Such changes in temperature are possibly due to different land use characteristics, moisture, or gas exchange condition. Few experimental results are available because the investigations are often made in costal zones where the thermal internal boundary layer is combined with the mechanical internal boundary layer.

The height of the thermal internal boundary layer is given by Raynor *et al.* (1975):

$$\delta_T = c\left(\frac{u_*}{u}\right)\left[\frac{x\left(\theta_1 - \theta_2\right)}{\left|\partial T/\partial z\right|}\right]^{1/2} \tag{3.21}$$

The temperature gradient is measured on either the upwind side of the surface temperature change or above the internal boundary layer; all other parameters are measured at a reference level. The coefficient c depends on the reference level and is in the order of one (Arya 2001). Such parameterizations are probably as robust as Eq. (3.20) for the mechanical internal boundary layer.

A special case is the thermal internal boundary layer in the afternoon mainly due to the so-called oasis effect (see Chap. 1, Stull 1988). Shortly after noon above a strongly evaporating surface, the temperature near the ground surface decreases and the stratification becomes stable. The height of the inflection point between the stable stratification near the surface and the unstable stratification in the upper layer grows with the time. Below the inversion, the sensible heat flux is downwards, and above the inversion it is upwards. The height increases 50 to 100 m above the ground after sun set, and is then identical with the stable boundary layer. In the early morning, a similar effect can be found but with a much shorter duration.

Another special case of a thermal internal boundary layer, is a near-surface inversion layer (height about 1 m), which cannot be described by any of the phenomena already discussed. This layer was observed first by Andreev *et al.* (1969) above the Black Sea. Further results are available from the Caspian Sea (Foken and Kuznecov 1978) and from Antarctica during the FINTUREX experiment (Sodemann and Foken 2005). In all cases, the surface was relatively smooth and the stratification was either neutral or stable. Explanations for the layer include a release of heat by chemical reactions, or eddy dissipation (Chundshua and Andreev 1980) or condensation processes (Foken and Kuznecov 1978); also possible are decoupling effects due to a large wind shearing near the surface or countergradient fluxes (see Chap. 3.5.2) are possible.

3.2.4 Blending-Height Concept

An important question for modelers is the possible height of an internal boundary layer. According to the structure of internal boundary layers, it can be assumed that layers are separate only up to a certain height (Fig. 3.7). At a greater horizontal distance from the change in surface roughness, they merge. This so-called blending height is assumed to be about 30–100 m above the ground surface. Above this height, an area-averaged flux is assumed. This means that the properties near the surface are smoothed (Taylor 1987). This idea was proposed by Mason (1988) and updated by (Claussen 1991; Claussen and Walmsley 1994). The concept considers especially larger-scale changes of surface roughness with characteristic horizontal distances of $L_x > 1$ km. The blending height can be estimated by $l_b = L_x/200$ (Mahrt 1996). The blending height concept has practical applications for area averaging in numerical models (see Chap. 5.7) because it can be assumed that for the model level at approximately the height of the blending height, the fluxes above a heterogeneous surface are area averaged.

From the experimental point of view, this concept is not much accepted. In an atmospheric boundary layer with free convection, the various surfaces can be detected by aircraft measurements through the depth of the boundary layer if the single areas are large enough for convection to develop above them. This is the case for horizontal scales greater than 200 m (Shen and Leclerc 1994). This is basic knowledge for glider pilots because free convection does not start at the surface, but above a growing internal boundary layer when $\delta/L < -1$ (Andreas and Cash (1999).

Because the structure of the atmospheric boundary layer over heterogeneous surfaces is very complicated, the blending height is meaningful for models with grid sizes greater than 2–10 km. In smaller scale models, the blending height concept seems not to be the best tool.

3.2.5 Practical Relevance of Internal Boundary Layers

Internal boundary layers are disturbances, which cannot be treated in experiments and models. Because of this the *Way back to Kansas experiment* (see Chap. 2.3.2), which implies an absolutely homogeneous surface, cannot be the aim of micrometeorology. However, for measuring and modeling the properties of the near-surface layer the identification of internal boundary layers is essential.

Experimentally this can be done with wind and temperature profile measurements or measurements of turbulence characteristics, which disagree with the typical values within the internal boundary layer. The necessary measurements are made in the new equilibrium layer or above the internal boundary layer. Problematical are measuring methods that are based on gradient measurements. Extreme errors may occur if an internal boundary layer is within the profile. Measurements should be done in the new equilibrium layer because due to the logarithmic profile functions the gradients above an internal boundary layer are often very small. When thermal internal boundary layers are present during the daily cycle, the measured gradients are not proportional to a flux.

For modeling, internal boundary layers should be taken into account if the grid size is less than 1 km. Such models are presently very rare because models with high spatial resolution were developed from large-scale low-resolution models. By applying these model physics, the modeling of internal boundary layers is impossible (Herzog *et al.* 2002).

Investigations of internal boundary layers are very important in the use of wind power (WMO 1981). To realize an optimal energy output, wind power stations should be installed on the side of a roughness change that has the smoothest surface for most of the wind directions as illustrated in Fig. 3.13.

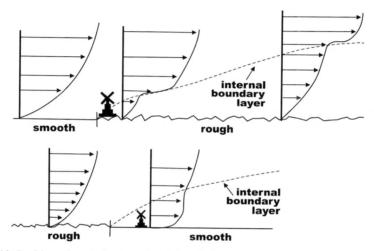

Fig. 3.13. Position of a wind power station in relation to an internal boundary layer (WMO 1981)

3.3 Obstacles

Obstacles have a remarkable influence on meteorological measurements, though the disturbance is not only at the downwind side but also on the windward side of the obstacle. In general, this range is excluded from measurements. This is impossible in forest, mountain regions, and urban environments. Often, modeling approaches are compared with wind tunnel simulations (Schatzmann *et al.* 1986). However, wind tunnel measurements are limited to neutral stratification, and the atmospheric turbulence structures can be created only provisionally.

For classical meteorological measurements and especially for micrometeorological measurements, the influences of obstacles should be excluded to the greatest possible extent. Therefore, the available rough approximations presented in this section are relevant only for mean meteorological parameters. For measurements of energy fluxes, the distances should be increased by a factor 2–5, and the footprint conditions must be applied (see Chap. 3.4).

The following approximation from the engineering technique is used for wind measurements and the use of wind power. This simple approach (WMO 1981) is illustrated in Fig. 3.14.

In Germany (VDI 2000), for meteorological standard measurements of the wind velocity at 10 m height, the necessary distance from an obstacle, A, will be a function of the height, H, and the width, B, of the obstacle. In the exceptional case of a short distance to the obstacle, D, the measuring height should be increased an amount h' where

$$h' = \frac{H}{A}(A - D), \tag{3.22}$$

see also Table 3.7.

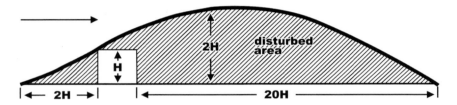

Fig. 3.14. Disturbance area at an obstacle (WMO 1981)

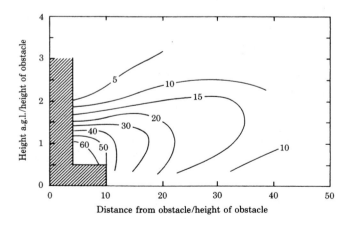

Fig. 3.15. Determination of the factor R_1 (Petersen and Troen 1990)

Table 3.7. Minimal distance of meteorological measurements, especially wind measurements, to an obstacle (VDI 2000)

height » wide	height ≈ wide	height « wide
$A \geq 0.5\,H + 10\,B$	$A = 5\,(H + B)$	$A \geq 0.5\,B + 10\,H$
for $B « H \leq 10\,B$		for $H « B \leq 10\,H$
$A \geq 15\,B$		$A \geq 15\,H$
for $H > 10\,B$		for $B > 10\,H$

The European Wind Atlas (Petersen and Troen 1990) applies a method to determine the lee side of an obstacle on the basis of a paper by Perera (1981). Accordingly, a corrected wind velocity can be calculated from the measured velocity and the porosity P of the obstacle (buildings: 0.0; trees: 0.5):

$$u_{corr} = u \left[1 - R_2\,R_1\,(1 - P)\right] \tag{3.23}$$

The factor R_1 is given in Fig. 3.15 and R_2 can be determined from

$$R_2 = \begin{cases} \left(1 + 0.2\dfrac{x}{L}\right)^{-1} & for \quad \dfrac{L}{x} \geq 0.3 \\[2ex] 2\dfrac{L}{x} & for \quad \dfrac{L}{x} < 0.3 \end{cases}, \tag{3.24}$$

where x is the distance between obstacle and the measuring point, and L is the length of the obstacle).

Results from the three methods (Eqs. (3.22), (3.23) and Fig. 4.14) are not comparable because the length of the obstacle in each case is used in a different way. However, for similar obstacles the results are not greatly different.

It should be noted that an instrument tower can also influence the measurements. Generally, measurements should not be made in the lee of a tower, and on the windward side the distance of wind and flux sensors from the mast should be large. For the installation of measuring devices, helpful rules of thumb for distances are 5-times the diameter of the mast for lattice masts, and 10-times the diameter of the mast for concrete masts.

Obstacles in connection with the generation of internal boundary layers have practical applications. For example, bushes used as windbreak lines reduce soil erosion on the lee side remarkable well, and distances less than 100 m from these break lines are optimal. Windbreaks are also used to shelter rain gauges, and snow fences are used to collect snow in the lee side.

3.4 Footprint

3.4.1 Definition

Measurements made at a particular site and at a particular height do not represent the properties or the fluxes below that sensor. Instead, the measurements represent the conditions of the underlying surface upwind of the sensor (Gash 1986). This effective area of influence is called the *footprint*. While the phenomenon of the footprint has been well known for a long time the term footprint is relatively recent. Thus, different forms of presentation and notation can be found in the literature. For example Schmid and Oke (1988) call this a source-weight function, and Leclerc and Thurtell (1989) call it a footprint function. In subsequent papers, the footprint description was widely used (Leclerc and Thurtell 1990; Schmid and Oke 1990; Schuepp *et al.* 1990), but the final mathematical form according our present knowledge was given by Horst and Weil (1992; 1994). The footprint function f describes the source area Q_η of a measuring signal η (scalar, flux) in relation to its spatial extent and its distribution of intensity, as illustrated in Fig. 3.16, and is given by:

$$\eta(x_m, y_m, z_m) = \int_{-\infty}^{\infty} \int_{-\infty}^{\infty} Q_\eta(x', y', z' = z_0) \cdot \tag{3.25}$$

$$\cdot f(x_m - x', y_m - y', z_m - z_0) \, dx' \, dy'$$

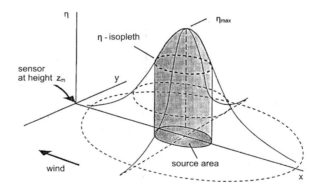

Fig. 3.16. Schematic picture of the footprint function according to Schmid (1994)

> The footprint of a special point (measuring device) is the influence of the properties of the upwind source area weighted with the footprint function.

It is interesting, that the horizontal and vertical scales of internal boundary layers and footprint concepts are also identical with the blending height concept (Horst 1999).

3.4.2 Footprint Models

The presently available footprint models (Table 3.8) are mainly diffusion models with the assumptions of homogeneous surfaces, uniform fluxes with height, and constant advection. Important input parameters are the measuring height, the roughness height, the stratification, the standard deviation of the lateral wind component, and the wind velocity. The core of analytical footprint models are diffusion models such as the often-used model developed by Gryning *et al.* (1983). More sophisticated models include Lagrangian footprint models and backward trajectory models, which can determine the footprint for heterogeneous surfaces. An overview of the present state in footprint modeling is given by Schmid (2002) and Vesala *et al.* (2004, 2008).

Table 3.8. Overview about the most important footprint models (if no remark: analytical model)

author	remarks
Pasquill (1972)	first model description
Gash (1986)	neutral stratification
Schuepp *et al.* (1990)	use of source areas, but neutral stratification and aver aged wind velocity
Leclerc and Thurtell (1990)	Lagrange' footprint model
Horst and Weil (1992)	only 2-dimensional
Schmid (1994, 1997)	separation of footprints for scalars and fluxes
Leclerc *et al.* (1997)	LES model for footprints
Baldocchi (1998)	footprint model within forests
Rannik *et al.* (2000, 2003)	Lagrange' model for forests
Kormann and Meixner (2001)	analytical model with exponential wind profile
Kljun *et al.* (2002)	three dimensional Lagrange' model for diabatic stratification with backward trajectories
Sogachev and Lloyd (2004)	boundary-layer model with 1.5 order closure
Strong *et al.* (2004)	footprint model with reactive chemical compounds

While footprint models are widely used, their experimental validation is still an outstanding issue (Finn *et al.* 1996). This validation is all the more necessary because the models are often applied under conditions that are not in agreement with the model assumptions. Currently, two methods of validation are available. The first method is the classical validation of dispersion model with tracers (Finn *et al.* 2001), and the second method tries to find the influences of obstacles in the footprint area by examination of the integral turbulence characteristics (Foken *et al.* 2000). In the second method, the differences in the integral turbulence characteristics of the footprint area due to changes of stability are investigated. This approach was validated using natural tracers (Foken and Leclerc 2004). These natural tracers are surfaces with different properties which can be verified with measurements according their footprint (Göckede *et al.* 2005). At this time, we are at the beginning of footprint model validation, and to a large extent only comparisons between different footprint models are available.

3.4.3 Application of Footprint Models

The first applications of footprint models were in the planning of field experiments (Horst and Weil 1994, 1995), *i.e.* how complex is the target surface in the footprint area of the measuring site. Changes in the underlying surface create internal boundary layers, which can change the local turbulence structure by adding mechanical or thermal-induced turbulence. An example of such a footprint calculation is shown in Fig. 3.17, where it is clearly seen that only the lowest measuring height can measure the influence of the surface on which the mast is standing. The upper measuring heights record increasing parts of the neighboring surfaces. The measurements of gradients can no longer be used for the determination of fluxes; it would more likely reflect horizontal differences.

Another important application is the so-called footprint climatology (Amiro 1998), which is shown in Fig. 3.18 for a selected measuring place. Depending on the stratification and the standard deviation of the lateral wind component, a data set over a long time period is necessary in order to calculate the climatology of the footprint areas.

For measuring places with changing roughness in the surrounding region, surface maps are used. The averaging of the roughness heights can be made according to Table 3.2, and the results used with some iterations optimal to the footprint model (Göckede *et al.* 2004). With an averaging model for fluxes (Hasager and Jensen 1999) the approach can be much better applied to a heterogeneous surface (Göckede *et al.* 2006).

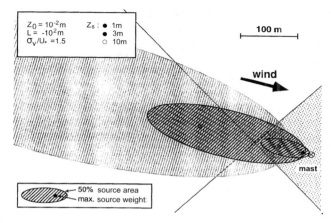

Fig. 3.17. Schematic picture of footprint areas according the model by Schmid (1997). It is seen that for different measuring heights different footprints and influences of underlying surfaces occur.

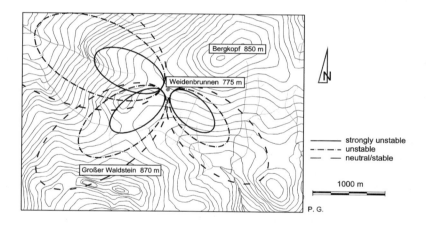

Fig. 3.18. For the measuring station Waldstein/Weidenbrunnen in the Fichtelgebirge mountains (Germany) calculated footprint areas from a two monthly data set in summer 1998 (Foken and Leclerc 2004)

Another application is the evaluation of the quality of carbon dioxide flux measuring stations with respect to the target surface in the footprint area and the spatial distribution of the estimated data quality within the footprint (see Chap. 4.1.3 and Rebmann *et al.* 2005). From Fig. 3.19, it can be seen that in the NW- and S-sectors the data quality is not sufficient for the calculation of the latent heat flux. The reason for this is that clouds and precipitation often occur in the NW-sector at the station Waldstein/Weidenbrunnen.

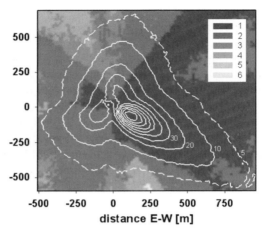

Fig. 3.19. Spatial distribution of the data quality according to Table 4.3 for the latent heat flux at the measuring station Waldstein/Weidenbrunnen for stable stratification. The relevant footprint climatology is underlined with contour lines, according to Göckede *et al.* (2006). The measuring tower is located at point (0,0).

a)

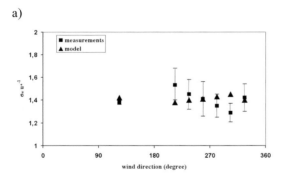

b)

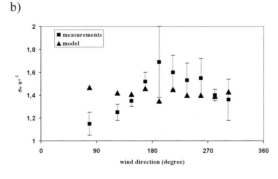

Fig. 3.20. Integral turbulence characteristics of the vertical wind velocity at the station Waldstein/Weidenbrunnen as a function of wind direction for (**a**) unstable and (**b**) stable stratification; measurements are plotted as • and model calculations according to Table 2.11 as ▲ (Foken and Leclerc 2004)

Figure 3.20 shows the measured and modeled integral turbulence characteristics of the vertical wind velocity for stable and unstable stratification at the Waldstein/Weidenbrunnen station. Significant differences can be seen between the model and experiment in the wind sector 180°–230°. However, for neutral stratification this is not the case. The reason is that the Waldstein massif is about 1 km from the measuring site, and its effects on the measurements are seen only for stable stratification. Thus, the disturbances of the vertical wind can explain the low data quality of the latent heat flux in the S-sector. Similar influences on the integral turbulence characteristics were found by DeBruin *et al.* (1991). They distinguish between heterogeneities in the surface roughness and in the thermal or other surface properties, the latter can be found with the integral characteristics of the scalars.

3.5 High Vegetation

The description of meteorological processes over high vegetation is opposite to that for dense vegetation (*e.g.* grain), which was described in Chap. 3.1.2 as a porous layer with a vertical displacement of the height (zero-plane displacement) in the profile equation. In a forest, the canopy and the understory are regions where complex micrometeorological processes occur. Therefore the crown is a layer that often decouples the understory from the atmosphere above the forest canopy. The energy and matter exchange between the atmosphere and the upper crown is often completely different than those between the lower crown and the trunk space and the soil. The stratification in both layers can be different, and the gradients in each layer can be of opposite sign. This has significant influence on measurement and modeling of exchange processes, which are very complex. Currently, these processes and their consequences are not fully understood (Finnigan 2000; Lee 2000).

3.5.1 Behavior of Meteorological Parameters in a Forest

The typical daily cycle of the temperature in a forest (Baumgartner 1956; Lee 1978) is illustrated in Fig. 3.21. The maximum air temperature occurs in the upper crown usually about 1–2 h after local noon. Below the crown, the daytime temperatures are lower. During the night, the minimum temperature occurs in the upper crown due to radiation cooling. Because the cool air in the crown drops to the ground surface, the minimum of the temperature near the ground occurs a short time later. Especially in the evening, it is warmer in the forest than in the surroundings. Fields near the forest edges may be damaged by frost in the morning due to the outflow of cold air, especially on slopes. The temperature structure in a forest has an absolutely different stratification than commonly observed above low vegetation. During day time above low vegetation and above the crown, unstable stratification is typical, and during the night stable stratification is typical. Due to the relatively cool forest ground during daytime and warm forest ground during the nighttime, inside the forest the opposite stratifications are realized.

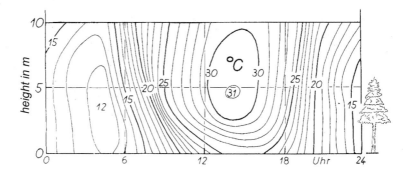

Fig. 3.21. Daily cycle of the air temperature in and above a forest according to Baumgartner (1956)

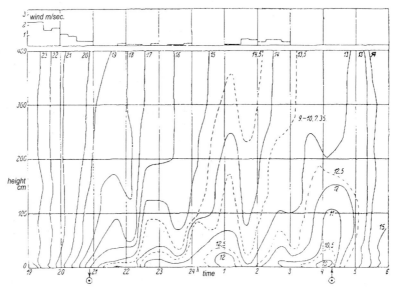

Fig. 3.22. Nigh time plot of the temperature in a forest with a sudden inflow of warm air connected with a short increase of the wind velocity at 01:40 according to Siegel (1936)

A typical phenomenon of the exchange processes in and above a forest is a de-coupling due to the crown. For example, during a calm night the upper canopy is relatively warm, but cool air is stored in the understory. A very small increase of the wind speed (gust, sweep) can cause a sudden inflow of warm air into the understory as seen in Fig. 3.22 at 01:40. Also, the opposite effect (burst, ejection) often occurs during the day, when warm or moist air is suddenly ejected from the understory. Therefore, the exchange of energy and matter within and above a forest is not a continuous process but is limited to single events.

The wind velocity in and above a forest is greatly reduced due to friction. But in a clear trunk space with nearly no understory growth a secondary maximum of the wind appears (Shaw 1977). This can change the momentum flux below the crown (Fig. 3.23). The wind profile near the top of the canopy has an inflection point, and at forest edges internal boundary layers can develop. Near the edge, as shown by measurements and modeling studies, increased turbulent fluxes occur (Klaassen *et al.* 2002).

3.5.2 Counter Gradient Fluxes

Exchange processes of energy and matter according to Chap. 2.3 are proportional to the gradient of the particular state parameter. Because of the change of the gradients in a forest site, it follows that convergence and divergence layers for fluxes must exist. However, the observed fluxes are opposite to the local gradients of the state parameters. This was found to be the case not only for the fluxes of sensible and latent heat but also for the fluxes of matter, *e.g.* carbon dioxide. This phenomenon is called *counter-gradient fluxes* (Denmead and Bradley 1985) and schematically shown in Fig. 3.24.

While the profiles are determined by the mean structures, the turbulence fluxes contain short-time transports by gusts and burst, which cause little change to the profiles. Relatively large turbulence eddies called coherent structures, which are not comparable with the turbulent elements of the gradient transport, make remarkable flux contributions (s. Chap. 3.5.4, Bergström and Högström 1989; Gao *et al.* 1989).

With these findings, the application of either the Monin-Obukhov similarity theory or the K-approach for the exchange process at a forest site gives no meaningful results. Alternatives are modeling with the transilient theory if the transilient matrix is well parameterized (Inclan *et al.* 1996), LES modeling (see Chap. 5.6), or higher-order closure approaches (Meyers and Paw U 1986; 1987).

Counter-gradient fluxes are not limited to forest sites. They also occur in connection with decoupling effects in the near-water air layer (Foken and Kuznecov 1978) or above ice shields (Sodemann and Foken 2005). It is interesting from the point of view of the history of science, that experimental findings were already known before the book chapter by Denmead and Bradley (1985) which opened the possibility to published these investigations in reviewed journals.

Fig. 3.23. Typical wind profile in a forest (Bailey *et al.* 1997); use of data by Amiro (1990)

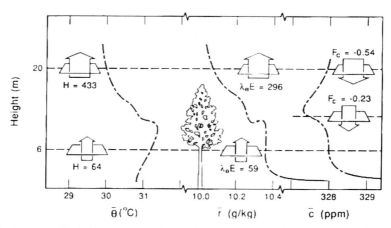

Fig. 3.24. Profil of the temperatur θ, the mixing ratio r, and the trace gas concentration c, as well as the sensible heat flux H, the latent heat flux λE, and the trace gas flux F_c in a forest with counter-gradients (Denmead and Bradley 1985)

3.5.3 Roughness Sublayer

Above a forest canopy, the profiles of state parameters are modified due to the high roughness compared with the ideal profile equations given in Chap. 2.3. This layer together with the canopy is called the roughness sublayer, which has a thickness of about three times of the canopy height. This means that above a forest within a distance of two times of the canopy height undisturbed conditions can be found. The verification of the roughness sublayer was made first in the wind tunnel (Raupach *et al.* 1980). Experiments in nature were done, for example, by Shuttleworth (1989). While above low vegetation the ratio z/z_0 for typical measurement heights has values of 100–1000, above a forest roughness sublayer this ratio is typically 5–10 (Garratt 1980). Accordingly, the profile equations (Eqs. (2.71) to (2.73)) must be changed by including functions which compensate for the effects of the increased turbulent diffusion coefficient in the roughness sublayer (Garratt 1992; Graefe 2004)

$$u_* = \sqrt{-\overline{u'w'}} = \frac{\kappa}{\varphi_*\left(\frac{z}{z_*}\right)\varphi_m(\varsigma)} \frac{\partial u}{\partial \ln z},\tag{3.26}$$

$$\overline{w'T'} = -\frac{\alpha_0 \; \kappa \; u_*}{\varphi_*\left(\frac{z}{z_*}\right)\varphi_H(\varsigma)} \frac{\partial T}{\partial \ln z},\tag{3.27}$$

$$\overline{w'q'} = -\frac{\alpha_{0E} \; \kappa \; u_*}{\varphi_*\left(\frac{z}{z_*}\right)\varphi_E(\varsigma)} \frac{\partial q}{\partial \ln z},\tag{3.28}$$

where z_* is the height of the roughness sublayer. Below the roughness sublayer the profile equations (Eqs. (2.71) to (2.73)) are no longer valid. The correction functions do not change with stratification and can be determined as empirical exponentials according to Garratt (1992):

$$\varphi_*\left(\frac{z}{z_*}\right) = \exp\left[-0.7\left(1 - \frac{z}{z_*}\right)\right] \quad for \quad z < z_*\tag{3.29}$$

For neutral stratification (3.26) has the simple form:

$$u_* = \sqrt{-\overline{u'w'}} = \frac{\kappa}{\varphi_*\left(\frac{z}{z_*}\right)} \frac{\partial u}{\partial \ln z}\tag{3.30}$$

The corrected functions must be used above the forest in all cases; otherwise the profile method (see Chap. 4.2) gives false values. The application of Bowen-ratio similarity (see Chap. 2.3.3) is in principal possible as long as the corrected functions for both fluxes are identical, which cannot be assumed from the simple form in Eq. (3.29). However, for the application of Bowen-ratio similarity the relatively small gradients above the forest are the more limiting factors.

Another possibility for the correction of the roughness sublayer was proposed by DeBruin and Moore (1985). They assumed that the mass flow inside the forest compensates the missing mass flow above the forest. The mathematical application needs either a detailed assumption about the wind profile in the forest or better measurements. From the mass flow in the forest, the mass flow above the forest and the resulting wind profile can be determined.

3.5.4 Turbulent Structures in and above a Forest

In contrast to low vegetation, above a forest ramp-like structures in the time series as seen in Fig. 3.25 typically occur. These generally are short-lived but strong turbulent events, which make remarkable contributions to the transport process. These ramps have some regularity, and persist for time periods typically ranging from tens of seconds to a few minutes; they are different from the normal fluctuations and gravity waves, which can be observed above the canopy in the night. The ramp structures are commonly called *coherent structures* (Holmes *et al.* 1996).

> Coherent structures, in contrast to stochastically distributed turbulence eddies, are well organized, relatively stable long-living eddy structures, which occur mostly with regularity in either time or space.

The structures shown in Fig. 3.22 are probably the first observations of coherent structures. More systematic investigations require the applications of statistical approaches, for example the wavelet analysis (Supplement 3.1), to clearly isolate time and frequency scales. This is not possible with spectral analysis, which gives only the frequency distribution. The first papers on this technique were presented by Collineau and Brunet (1993a; 1993b).

The turbulent fluctuations of the vertical wind are shown in the upper part of Fig. 3.26. Below this but on the same time axis is the wavelet energy plot. On the ordinate, the logarithm of the reciprocal time scale (frequency in Hz) is plotted. Long periods of minimum and maximum energy are seen, which are correlated with the ramp structures. These structures show a maximum in the variances at 300 s (lower part of the Figure). We also see in Fig. 3.26, a maximum variance in the micro-turbulence range at about 50 s, and another maximum for the very long waves. Ruppert *et al.* (2006b) have shown that not all meteorological variables have the same coherent structure, in contrast to stochastically organized eddies (Pearson jr. *et al.* 1998). The main reasons are different sources. This phenomenon is called *scalar similarity*.

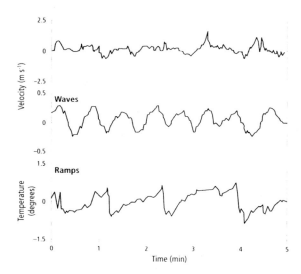

Fig. 3.25. Fluctuations, waves and ramp structures above a forest (Bailey *et al.* 1997) by using data from Gao *et al.* (1989) and Amiro and Johnson (1991)

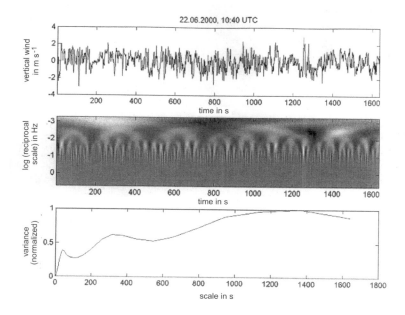

Fig. 3.26. Structure of the vertical wind velocity above a forest, above: time series, middle: wavelet analysis, below: scale of typical variances (Wichura *et al.* 2001)

Supplement 3.1.   Wavelet Analysis

The wavelet transformation $T_p(a,b)$ of a function $f(t)$ is the convolution of a time series $f(t)$ with a family of Wavelet functions $\Psi_{a,b}(t)$ (see Torrence and Compo 1998)

$$T_p(a,b) = \int_{-\infty}^{\infty} f(t)\,\overline{\Psi_{p,a,b}(t)}\,dt = \frac{1}{a^p} \int_{-\infty}^{\infty} f(t)\,\overline{\Psi\left(\frac{t-b}{a}\right)}\,dt, \qquad (3.31)$$

where $T_p(a,b)$ is the wavelet transform; $\Psi_{p,1,0}(t)$ is the mother wavelet; b is the translation parameter, a is the dilatation parameter (time scale), $1/a$ is a frequency scale, and p=0.5. If $a > 1$, then a widening of the mother wavelet occurs, and if $a < 1$ then a contracting of the mother wavelet occurs. Wavelet functions differ in their response to time and frequency, and so they must be selected according to the application. A wavelet that responds equally well in both scales is the Mexican-hat:

$$\Psi(t) = \frac{2}{\sqrt{3\sqrt{\pi}}}\left(1 - t^2\right)e^{-t^2/2} \qquad (3.32)$$

Wavelet plots show the time on the x-axis and the period or frequency on the y-axis. In the wavelet plot in Fig. 3.26, light areas correspond to high energy densities and the dark areas to low energy densities. Thus, an explicit localization of frequencies and energy densities in time is possible. This is not possible in spectral analysis.

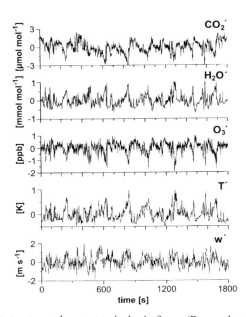

Fig. 3.27. Turbulent structures above a tropical rain forest (Rummel *et al.* 2002b)

Scalar similarity is also shown in Fig. 3.27. While the ramp structures in the vertical wind are not well developed, the ramp structures in the temperature, water vapor, ozone, and carbon dioxide fluctuations are; however, the characteristics of the ramps are not similar. For example, due to the different sources of carbon dioxide and water vapor, their ramps are mirror images.

Coherent structures are found over all surfaces. Their characteristics above highly heterogeneous surfaces are much greater than those over low uniform surfaces. Obstacles such as forest edges can generate additional structures, which generate periods that are dependent on the distance from the edge (Zhang *et al.* 2007). In general, the whole mixed layer is dominated by such structures. Also, these structures provide a good coupling between a canopy and the atmospheric boundary layer (Thomas and Foken 2007a).

3.5.5 Mixing Layer Analogy

In the case of high wind shear, a mixing layer which controls the exchange processes can develop above a forest (Raupach *et al.* 1996). A roughness sublayer alone cannot explain this observation. The development of the mixed layer is very similar to wind tunnel experiments as illustrated in Fig. 3.28. Turbulent exchange is possible if the characteristic turbulence scales of the vertical wind velocity and a scalar are similar to the scales of the mixing layer. If this is not the case, then decoupling may occur.

The characteristic length scale of the mixing layer is the shearing scale. Assuming that the inflection point of the wind profile is at the top of the canopy and that the wind velocity within the canopy is significantly lower it follows (Raupach *et al.* 1996):

$$L_s = \delta_w \Big/ 2 = \frac{u(z_B)}{\left(\partial u \Big/ \partial z \right)_{z=z_B}} \qquad (3.33)$$

The development of the mixing layer above a plant canopy was comprehensively described by Finnigan (2000) as is illustrated in Fig. 3.29. The second length scale is the characteristic distance between coherent eddies, which is the wavelength of the initial Kelvin-Helmholtz instability which is the underlying cause of the developing mixing layer. This length scale, Λ_x, can be determined, for example, by a wavelet analysis. According to Raupach *et al.* (1996) a linear relation exists between both scales

$$\Lambda_x = m \, L_s, \qquad (3.34)$$

with values for m ranging from 7 to 10 for neutral stratification.

Fig. 3.28. Developing of the mixing layer of the height δ_w (Raupach *et al.* 1996)

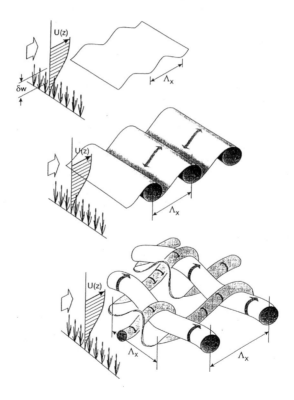

Fig. 3.29. Schematic description of the developing of the mixing layer above a plant canopy (Finnigan 2000)

The determination of energy exchange above a forest is extremely complicated. Not only the crown space but also the wind shearing above the canopy can decouple the forest from the atmosphere. Therefore, the exchange processes between the forest and the atmospheric boundary layer is occasionally intermitted.

3.5.6 Coupling Between the Atmosphere and Plant Canopies

Tall plant canopies are coupled with the atmosphere by turbulent eddies most often during the daytime. Then, the energy and matter exchange takes place between the ground, the trunk space, and the atmosphere above the canopy. Reasons for the occasional decoupling between the canopy and the atmosphere are the high roughness at the upper part of the canopy, which damps the wind field, and the stable stratification in the canopy. The latter is responsible for the enhancement of emitted gases from the ground in the trunk space and large concentrations above the canopy. This can be seen for nitrogen monoxide and carbon dioxide produced by microbiological and respiration processes in the ground. As a rule these concentrations will be removed by turbulent mixing and, in the case of nitrogen monoxide, by reactions with ozone. However, during calm conditions high concentrations may occur above low vegetation.

The reaction of nitrogen monoxide with ozone can be used as a tracer for the turbulent mixing (Rummel et al. 2002a). Also, the stable carbon isotope ^{13}C can be used as a tracer because isotope signatures in the atmosphere and the soil differ significantly, and carbon dioxide emitted from the ground can be identified in the atmosphere even above the canopy (Bowling et al. 2001; Wichura et al. 2004). Coherent structures are very effective for the characterization of the turbulent exchange process if they can be measured at different heights in the canopy (Thomas and Foken 2007b).

The dependence of the penetration depth of coherent structures into the canopy can be classified by five states of coupling (Table 3.9). In the case of gravity waves, (Wa) or no penetration of turbulence into the canopy (Dc) no coupling exist. In the case of penetration of turbulence into the crown (Ds), at least the crown is coupled with the atmosphere. A complete coupling (C) is given only for short time periods. More often is seen an occasional or partial coupling of the trunk space and the crown with the atmosphere (Cs). The typical daily cycle of the phase of coupling together with the contribution of the coherent structures to the flux is shown in Fig. 3.30.

Table 3.9. Characterization of the states of coupling between the atmosphere and tall plant canopies (Thomas and Foken 2007b). The signature with letter is according to the coupling stages in Fig. 3.30 (Figures from Göckede *et al.* 2007)

Wa	Dc	Ds	Cs	C
gravity waves above the canopy	turbulent eddies only above the canopy	turbulent mixing up to the crown	occasionally complete mixing of the canopy	complete mixing of the canopy
no coupling	no coupling	occasionally weak coupling	occasionally coupling	good coupling

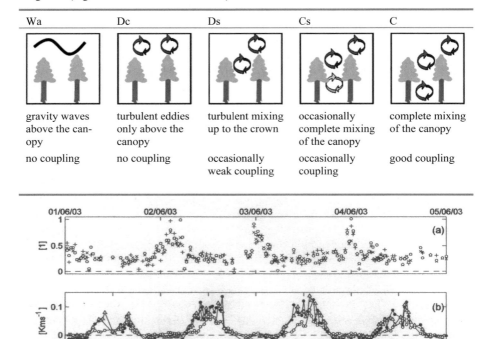

Fig. 3.30. Measurements of the characteristics of the turbulent exchange in the period from June 1 to June 4, 2003 during the experiment WALDATEM-2003 (Waldstein/Weidenbrunnen) in a 19 m high spruce forest in the Fichtelgebirge mountains: **(a)** Relative contribution of coherent structures to the carbon dioxide exchange ($\circ$), buoyancy flux ($\square$) and latent heat flux (+) at $1{,}74\ z_B$ (z_B: canopy height); **(b)** Kinematic buoyancy flux of the coherent structure at $1.74\ z_B$ (●), $0.93\ z_B$ (▲) and $0.72\ z_B$ ($\circ$); **(c)** Friction velocity (——), global radiation ($\circ$) and wind direction (●) at $1.74\ z_B$; **(d)** States of coupling between the atmosphere and the forest (abbreviations see Table 3.9) according to Thomas and Foken (2007b)

3.6 Conditions Under Stable Stratification

During the last few years, the stable or nocturnal atmospheric boundary layer has become an important research topic In contrast to the convective boundary layer, the stable boundary layer is remarkably shallow. Often it reaches a depth of only 10–100 m. It is referred to as a long-living stable (nocturnal) atmospheric boundary layer if, as in polar regions, it continues over a long time. The stable free atmosphere and the stable boundary layer are often connected by internal gravity waves (Zilitinkevich and Calanca 2000). Due to the negative buoyancy of the stable stratification, the turbulence is strongly damped, and longwave radiation processes becomes an especially strong influence. These processes have an important role in the development of ground inversions under calm and widely non-turbulent conditions. Under these situations, the phenomenon of a low level jet often occurs at heights of 10–300 m. As a general rule, intermittent turbulence and gravity waves are present, and steady-state conditions do not exist. The stability of the layer is most often greater than that at the critical Richardson number.

In Fig. 3.31 it is shown that under stable conditions scalar time series have a turbulent and wavy part. Because of gravity waves, perturbations of the vertical wind velocity are highly connected with changes of scalar quantities, and thus the correlation coefficients may be significantly greater than those for turbulent flow alone (see Chap. 4.1.3). Under this circumstances, flux measurements according to the eddy-covariance method would give an unrealistic overestimation of the turbulent fluxes if the time series were not filtered excluded to remove the wave signal (Foken and Wichura 1996).

Under stable conditions, intermittency often occurs. According to a classification of the stable atmospheric boundary layer by Holtslag and Nieuwstadt (1986), this occurs in a range beyond the z-less scaling, i.e. for local Obukhov length $z/\Lambda > 10$ (Fig. 3.32). The exact definition of intermittency is somewhat vague because the events must be characterized by relatively long periods with quasi laminar condition and interruptions by turbulent processes and their intensities. Therefore intermittency cannot be defined only in the micrometeorological scale. Intermittence factors are the ratio between the local (spatial and temporal) turbulent fluctuations and those over a longer time period or spatial area. Relatively convenient is the definition of a local intermittency factor using the wavelet coefficients (Supplement 3.1) according to Farge (1992)

$$I(a,b) = \frac{\left|T_p(a,b_0)\right|^2}{\left(\left|T_p(a,b_0)\right|^2\right)_b},$$

(3.35)

where the contribution to the energy spectrum at the moment b_0 in the scale a can be considered as the time average in this scale. For $I(a,b) = 1$, no intermittency occurs. Because the intermittency factor must be determined for all values of a, it is obvious that intermittent events develop differently at different scales. In Fig. 3.33, it is shown that a turbulent event happens at two different turbulent scales (shown as functions of frequency) while shortly after the event intermittent conditions can be detected only in the low-frequency range.

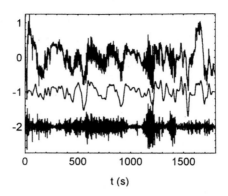

t (s)

Fig. 3.31: Examples of turbulence time series (above: temperature, experiment FINTUTREX Antarctica) with selection of the wave (middle) and the turbulent signal (below) after filtering (Handorf and Foken 1997)

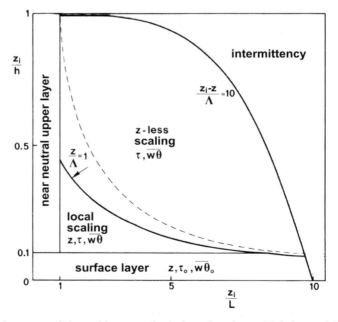

Fig. 3.32. Structure of the stable atmospheric boundary layer (Holtslag and Nieuwstadt 1986)

The frequencies of gravity waves at each height are lower than the corresponding Brunt-Väisälä frequency (Stull 1988):

$$N_{BV} = \frac{g}{\theta_v} \frac{\partial \overline{\theta_v}}{\partial z}$$

(3.36)

Using the Brunt-Väisälä frequency, the influence of the stable atmospheric boundary layer on the near surface exchange processes can be determined (Zilitinkevich et al. 2002) and this can be used in modeling (see Chap. 5.5).

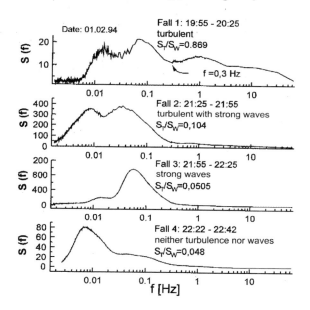

Fig. 3.33. Temperature spectrum with turbulent and wavy structures on Feb. 01, 2004 in Antarctica (experiment FINTUREX), the separation was made at a frequency of 0.3 Hz (the scale on the ordinates are different in the single figures!) according to Heinz et al. (1999)

3.7 Energy Balance Closure

According to experimental investigations using measuring, technical, and methodical reasons, the energy balance equation (Eq. 1.1) is not fulfilled. Fig. 1.14 shows that even over a homogeneous surface, a large residual of the energy balance closure exists. Thus, during the day the ground surface gets more energy by radiation than can be carried away by turbulent fluxes. Consequently, the ground heat flux and energy storage must be large. In the night, the opposite conditions exist but without a compensation for the day time energy lost (Culf *et al.* 2004; Foken 2008).

Table 3.10 shows results of experiments with personal data access by the author to guaranty the comparability. These results are in agreement with many experiments about energy balance closure in the last 20 years. Further investigations are given by Laubach (1996).

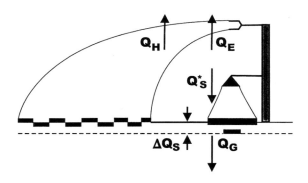

Fig. 3.34. Schematic view of the measuring area of the different terms of the energy balance equation

Table 3.10. Experimental results for the energy balance closure, *Re*: residual

experiment	author	$Re/(-Q_s^* - Q_G)$ in %	surface
Müncheberg 1983 and. 1984	Koitzsch *et al.* (1988)	14	winter wheat
KUREX-88	Tsvang *et al.* (1991)	23	different surfaces
FIFE-89	Kanemasu *et al.* (1992)	10	Step
TARTEX-90	Foken *et al.* (1993)	33	barley and bare soil
KUREX-91	Panin *et al.* (1996b)	33	different surfaces
LINEX-96/2	Foken *et al.* (1997a)	20	middle high grass
LINEX-97/1	Foken (1998b)	32	Short grass
LITFASS-98	Beyrich *et al.* (2002c)	37	nearly bare soil
EBEX-2000	Oncley *et al.* (2007)	10	irrigated cotton field
LITFASS-2003	Mauder *et al.* (2006)	30	different agricultural field

Table 3.11. Error on the closure of the energy balance near the surface and possible reasons (Foken 1998a)

term of the energy balance equation	error in %	energy in W m^{-2}	horizontal scale in m	Measuring height in m
latent heat flux (carefully corrected)	5–20	20–50	10^2	2–10
sensible heat flux	10–20	15–30	10^2	2–10
net radiation	10–20	50–100	10^1	1–2
ground heat flux	50	25	10^{-1}	−0.02– −0.10
storage term	?	?	10^{-1}–10^1	

Reasons for the imbalance are partly the relatively large errors of the single components of the energy balance equation and the use of different measuring methods (Table 3.11, Fig. 3.34). The single terms of the energy balance equation are not measured directly at the surface, but either in the air above it or in the soil. Furthermore, the information of the measured fluxes comes from different source areas. This is very critical for the turbulent fluxes where the measurement, *e.g.*, at 2 m height is connected with a much larger area according to the footprint (see Chap. 3.4), which is up to 200 m upwind from the measuring point (ratio about 1:100). Of special relevance for the energy balance closure, is energy storage in the upper soil layer (above the soil heat flux measurement). Due to changing cloudiness, this can be very variable, while all the other parameters are determined over longer time periods (Foken *et al.* 1999). The calculated fluxes are highly unsteady (Kukharets *et al.* 2000). At night, when the turbulent fluxes can be neglected, optimal measurements and an exact determination of the ground heat flux at the surface gives a very good closure of the energy balance (Heusinkveld *et al.* 2004; Mauder *et al.* 2007a). Currently, the reason for the energy balance closure problem is no longer seen in the soil. Compared with the relatively high energy storage in the soil, the storage in the plants, the storage in the air below the turbulence measurements, and the energy loss by photosynthesis in most cases are negligible (Oncley *et al.* 2007).

The energy balance equation is the basis for most atmospheric models. Therefore, the energy balance closure problem is still an ongoing issue of research to validate models with experimental data. The experiment EBEX-2000 (Oncley *et al.* 2007), which was designed for this purpose, yielded some progress regarding the sensor configuration and correction methods.

For the determination of the net radiation, a relatively high effort is necessary (see Chap. 6.2.1), because inexpensive net radiometers underestimate the net radiation, and can cause an apparent more accurate closure of the energy balance (Kohsiek *et al.* 2007). The turbulent fluxes must be carefully corrected (see Chap. 4.1.2), and their quality must be comprehensively controlled (see Chap. 4.1.3). Highly relevant, are the measurements of the moisture fluctuations and the calibration (also during experiments) of the sensors. A careful calculation of the turbulent fluxes can reduce the energy balance closure gap significantly (Mauder and Foken 2006; Mauder *et al.* 2007b) but cannot close it.

Energy balance closures cannot be used as a quality criteria for turbulent fluxes (Aubinet *et al.* 2000). This is because the influencing factors are so greatly different, wrong conclusions are possible. In comparison with similar experiments, energy balance can give only a rough criterion about the accuracy of the fluxes.

Panin *et al.* (1998) have shown that heterogeneous surfaces in the vicinity of the measurements can cause low frequency turbulent structures, which cannot be measured with the turbulence systems. A significantly longer averaging time, say over several hours, gives nearly a closure (Finnigan *et al.* 2003), while averaging times up to two hours show no significant effects (Foken *et al.* 2006b). Also, the assumption by Kanda *et al.* (2004) that organized turbulent structures in the lower boundary layer make a contribution to the energy balance closure support these findings. In this context, two experiments are of special interest; one in areas with large-scale homogeneous surfaces in the dessert (Heusinkveld *et al.* 2004), and the other in the African bush land (Mauder *et al.* 2007a). In both cases, a closed energy balance was found. Considering the results of these experiments, Foken *et al.* (2006a) have attempted to close the energy balance by investigating secondary circulations and low frequency structures using for the LITFASS-2003 experiment (Beyrich and Mengelkamp 2006). This was followed by determination of circulations using large-eddy simulation (see Chap. 5.6), and the determination of spatially-averaged fluxes using scintillometer measurements (Meijninger *et al.* 2006). Although for the absolute clarity about these phenomena, further investigations are necessary, it can be stated that the energy balance closure problem is a scale problem. A complete closure can be reached only if larger scale turbulent structures are taken into account. These secondary circulations are often stationary. The measurements in the surface layer record only smaller scales, and cannot be used to close the energy balance.

These findings have far ranging consequences. For the measurement of turbulent fluxes in heterogeneous areas, measurement methods based on the energy balance equation should not be used, *e.g.* Bowen-ratio method (see Chap. 4.2.2). Other methods measure the right flux on the small scale, but these methods cannot be transferred to larger areas and do not close the energy balance. In a first approximation, a correction can be made with the Bowen-ratio (Twine *et al.* 2000) by the distribution of the residual to both turbulent fluxes according this ratio. Because scalar similarity is not always fulfilled (Ruppert *et al.* 2006b), this method can cause large errors. The comparison of experimental data and model output remains problematic.

4 Experimental Methods for Estimating the Fluxes of Energy and Matter

In meteorology and climatology, routine observation programs generally select only state variables of the atmosphere for measurement. Because extent of these routine measurements is limited, they cannot be used for estimating fluxes of energy and matter. The growing research on the problem of climate change in the last 10–15 years has increased the demand for reliable measurements of evaporation, carbon dioxide uptake by forests, and fluxes of greenhouse gases. Nearly exclusively research institutes made such measurements. The measurements are very complex and need comprehensive micrometeorological knowledge. Most of the measuring methods are based on simplifications and special conditions, and therefore their implementation is not trivial. In the following chapter, the reader will be given overview tables listing the fields of applications and their costs.

4.1 Eddy-Covariance Method

Flux measurements using the eddy-covariance method (often also called eddy-correlation method, but this can bring some confusions, see Chap. 4.3) are a direct measuring method without any applications of empirical constants (Businger 1986; Foken *et al.* 1995; Haugen 1973; Kaimal and Finnigan 1994; Lee *et al.* 2004). However, the derivation of the mathematical algorithm is based on a number of simplifications so that the method can be applied only if these assumptions are exactly fulfilled (see Chap. 2.1.2). The quality of the measurements depends more on the application conditions and the exact use of the corrections than on the presently available highly sophisticated measuring systems. Therefore experimental experience and knowledge of the special character of atmospheric turbulence have a high relevance. The most limiting conditions are horizontally homogeneous surfaces and steady-state conditions. The exact determination of the footprint area (see Chap. 3.4), which should be over a uniform underlying surface for all stability conditions, and the exclusion of internal boundary layers and obstacle influences (see Chaps. 3.2 and 3.3) have an outstanding influence on the selection of the measuring place. This is especially relevant for forest sites, where additional specifics of tall vegetation must be taken into account (see Chap. 3.5).

The basic equations are comparatively simple (see Eqs. 2.21 and 2.22):

$$u_*^2 = -\overline{u'w'}\,,\quad \frac{Q_H}{\rho\,c_p} = \overline{T'w'}\,,\quad \frac{Q_E}{\rho\,\lambda} = \overline{q'w'}\,,\quad \frac{Q_c}{\rho} = \overline{c'w'} \tag{4.1}$$

The covariance of the vertical wind velocity, w, and either one of the horizontal wind components or of a scalar can be determined in the following way:

$$\overline{w'x'} = \frac{1}{N-1} \sum_{k=0}^{N-1} \left[\left(w_k - \overline{w_k} \right) \left(x_k - \overline{x_k} \right) \right] =$$

$$= \frac{1}{N-1} \left[\sum_{k=0}^{N-1} w_k \, x_k - \frac{1}{N} \left(\sum_{k=0}^{N-1} w_k \sum_{k=0}^{N-1} x_k \right) \right] \qquad (4.2)$$

4.1.1 Basics in Measuring Technique

According to Eq. (4.1), the turbulent fluctuations of the components of the wind vector and of scalar parameters must be measured at a high sampling frequency (see Chap. 6.1.2) so that the turbulence spectra (see Chap. 2.5) can be extended to 10–20 Hz. The measuring devices used for such purposes are sonic anemometers for the wind components and sensors that can measure scalars with the required high resolution in time. The latter are often optical measuring methods (see Chap. 6.2.3). The measuring or sampling time depends on the atmospheric stratification, the wind velocity, and the measuring height. For heights of 2–5 m, 10–20 min would be required for daytime unstable stratification (summer) and about 30–60 and sometimes as high as 120 min for nighttime stable stratification. There will not be remarkable errors if a sampling time of 30 min is used over the whole day. For short sampling times, the low frequency contributions to the fluxes are missed, and for long sampling times the steady state condition may not be fulfilled. Accordingly, the flux can be determined only after the measurements have been made. It is also possible to use filtering options. Low pass filters or trend eliminations can create errors in the fluxes (Finnigan *et al.* 2003; Rannik and Vesala 1999), thus block averaging with an averaging time of 30 min is now recommended.

Because the size of the turbulence eddies increases with distance above the ground surface, both the measuring path length and the separation between a sonic anemometer and an additional device (*e.g.* hygrometer) depend on the height of the measurements. Devices with a path length less than 12 cm should not be used below 2m, and devices with a path length more than 20 cm should be not used below 4 m. The minimum distance between a sonic anemometer and an additional device, depends on the flow distortions caused by the devices and should be determined in a wind tunnel. Typically, for fine-wire temperature sensors, the minimum distance is 5 cm, and for hygrometers is 20–30 cm. These additional instruments should be mounted downwind of the sonic anemometers and 5–10 cm below the wind measuring path (Kristensen *et al.* 1997). Therefore, to reduce the corrections of the whole system (see Chap. 4.1.2) the measuring height must be estimated not only dependent on the path length of the sonic anemometer but also dependent on the separation of the measuring devices. Also, the measuring height should be twice the canopy height in order to exclude effects of the roughness sublayer (Fig. 4.1).

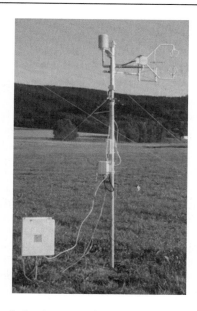

Fig. 4.1. Eddy-covariance measuring complex of the University of Bayreuth with a sonic anemometer CSAT3 and a hygrometer KH20 (Photograph: Foken)

In basic research, sonic anemometers (see Chap. 6.2.2) with a selected inflow sector to exclude flow distortion are used. For most applications, wind direction-independent omni-directional sonic anemometers are sufficient, but these have flow distortions due to mountings and sensor heads. It is recommended that only devices with few interfering parts installed below the measuring path for the vertical component of the wind should be applied.

Most of the sonic anemometers also measure the fluctuations of the sound velocity and therefore indicate the so-called sonic temperature (nearly identical with the virtual temperature). The flux calculated with this temperature is the buoyancy flux, about 10–20% greater than the sensible heat flux:

$$\frac{Q_{HB}}{\rho \, c_p} = \overline{w'T_v'} \tag{4.3}$$

The sensible heat flux can be determined by applying additional corrections, which need additional moisture measurements (see Chap. 4.1.2). More expensive are direct temperature measurements made with thin thermocouples or free spanned resistance wires (diameter < 15 μm).

Hygrometers are used for the determination of the latent heat flux (evaporation). Such devices are nowadays mostly optical devices. These have either an open path or are a closed path. The open path hygrometer is mounted near the sonic anemometer, and the closed path is mounted some meters away and the air is aspirated below the sonic anemometer. The first works in the UV and IR range, and the second works only in the IR range. UV devices should be used for low humidity conditions (water vapor pressure 0–20 hPa) and IR devices for high moistures (10–40 hPa). Closed path devices need extensive data corrections. One correction accounts for the time delay of the measuring signal in relation to the

wind measurements, and another correction accounts for the filtering of the fluctuations by the tube (Aubinet *et al.* 2000; Ibrom *et al.* 2007; Leuning and Judd
1996). The effort for maintenance and calibration for all devices is considerable.
The lifetime of UV devices is very limited (< 1000 h). Very fast optical devices
for other gaseous components (*e.g.* ozone, nitrogen oxide, sulfur dioxide) are also
available and the deposition flux can be measured in a similar way.

Unfortunately, for a long time no software was commercially available for
eddy-covariance measurements. Widely used, are versions of software developed
by McMillen (1988), which includes an internal rotation of the coordinates to
force the mean vertical wind to zero. This software should not be used in uneven
terrain because it records also advection components. The software developed by
producers of sonic anemometers should be used only if it is documented. The applications of the flow distortion correction need some care because these are determined in the wind tunnel, but these corrections have much lower values in the
atmosphere (Högström and Smedman 2004). Often these software packages offer
very short averaging intervals (1–10 min). Due to the spectral character of turbulence, these fluxes will be measured only partially. The simple summation over
longer time periods is incorrect. Special correction algorithms are necessary, and
these need further statistical parameters of the short time periods. For the covariance of a long time series of M data points, based N short-time series each with U
data points, where $N=M/U$, it follows (Foken *et al.* 1997a):

$$\overline{w'x'} = \frac{1}{M-1}\left[(U-1)\sum_{j=1}^{N}\left(\overline{w'x'}\right)_j + U\sum_{j=1}^{N}\overline{w}_j\ \overline{x}_j - \frac{U^2}{M}\sum_{j=1}^{N}\overline{w}_j\sum_{j=1}^{N}\overline{x}_j\right] \qquad (4.4)$$

For the past 5–10 years, several commercial software programs and programs
developed at universities (partly free of charge) are available (Appendix A8).
Comparison experiments, which are available for most of the programs, have
shown that these can be reliably applied (Mauder *et al.* 2008; Mauder *et al.*
2007b). Nevertheless, the programs differ slightly in the applications of correction
and quality control methods. Therefore, the user needs some basic knowledge
about these methods.

4.1.2 Correction Methods

The application of correction methods is closely connected with the data quality
control (see Chaps. 4.1.3 and 6.3). It starts with the exclusion of missing values and
outliers, which can be found by electrical and meteorological plausibility tests.

Further test should detect unfavorable meteorological influences which often
cannot be separated from measuring technical problems (Vickers and Mahrt
1997). Spikes are often electronically caused. The value is significantly above the
normal measuring value but often within the possible range. The usual test is the
determination of the standard deviation. All values greater than 3.5σ (Højstrup 1993)

will be signed as spikes. If single spikes are large, it is recommended to repeat the test 2–3 times, because otherwise erroneous data cannot be identified. Measuring series with more than 1% of spikes should not be used.

The eddy-covariance method is based on a number of assumptions (see Chap. 2.1.2), which must be controlled, and if the assumptions are not met, then necessary corrections must be applied. Because necessary corrections cannot be seen from the recorded data, extensive tests must be made (see Chap. 4.1.3). Most of these tests will be done after applying all corrections. Important are the tests for steady-state conditions and developed turbulence.

Because of the logging of several measuring signals, it may occur that the data measured at the same point of time are not stored at the same point of time. This can be partially corrected in the software of the data logger, but such time delays can change with time. A typical example is a measuring system where a gas is aspirated with a tube at the sonic anemometer. Then, depending on the stream velocity within the tube, concentration measurements can be recorded significantly later than the wind component. A cross-correlation analysis is recommended with the vertical wind and a shift of the time series according to the time delay for maximum cross-correlation. This must be done before any further calculations are made.

After the time shifting, initial covariances can be calculated. It is helpful to store these, because the first two steps use most of the calculation time. The next step is the rotation of the coordinates. Currently, the *planar-fit* method (Wilczak *et al.* 2001) is most often used. This rotation method is applied to the data over the whole measuring campaign. Therefore, only preliminary results are available up to the end of the experiment.

The following corrections are given in the order of their application. Because some values, for example the stability parameter, must be calculated from the corrected data, but they are also necessary for correction, the correction should be calculated with iterative updated values. The iteration often consists of only a few cycles, and an improvement of the fluxes by 1% is possible (Mauder *et al.* 2006).

Coordinate Rotation (Tilt Correction)

A basic condition for applying the eddy-covariance method is the assumption of a negligible mean vertical wind component (see Eq. 2.5). Otherwise advective fluxes must be corrected. This correction is called tilt correction and includes the rotation of a horizontal axis into the mean wind direction. It is based on works by Tanner and Thurtell (1969) and Hyson *et al.* (1977). The first correction is the rotation of the coordinate system around the z-axis into the mean wind. Using the measured wind components (subscript m), the new components are given by (Kaimal and Finnigan 1994)

$$u_1 = u_m \cos\theta + v_m \sin\theta,$$
$$v_1 = -u_m \sin\theta + v_m \cos\theta, \tag{4.5}$$
$$w_1 = w_m,$$

where

$$\theta = \tan^{-1}\left(\frac{\overline{v_m}}{\overline{u_m}}\right). \tag{4.6}$$

For an exact orientation of the anemometer into the mean wind and for small fluctuations of the wind direction ($< 30°$), this rotation is not necessary (Foken 1990). The friction velocity can be calculated from the horizontal wind component orientated into the mean wind direction; otherwise Eq. (2.24) must be applied. With today's high computer power, this special case has little significance.

The second rotation is around the new y-axis until the mean vertical wind disappears (Kaimal and Finnigan 1994)

$$u_2 = u_1 \cos\phi + w_1 \sin\phi,$$
$$v_2 = v_1, \tag{4.7}$$
$$w_2 = -u_1 \sin\phi + w_1 \cos\phi,$$

where

$$\phi = \tan^{-1}\left(\frac{\overline{w_1}}{\overline{u_1}}\right). \tag{4.8}$$

Both rotations are graphically shown in Fig. 4.2a (above).

With these rotations, the coordinate system of the sonic anemometer is moved into the streamlines. Over flat terrain, these rotations would correct errors in the vertical orientation of a sonic anemometer. In a sloped terrain, the streamlines are not implicitly normal to the gravity force, and at least for short averaging periods rotations may be questionable. This is especially true for short convective events or for flow distortion problems of the sensor. These effects have a remarkable influence on the observed wind components, but are not associated with the rotations. Therefore, the second rotation has recently become controversial. The main problem is how to apply these rotations. The often-used software by McMillen (1988), which does not need any storage of the raw data, determines the averaged vertical wind by low pass filtering over about 5 min periods, and rotates the coordinate system with this value. However, for longer averaging periods the method is not without doubt. For convection or periods of low wind velocities, rotation angels up to 20–40° are typical.

The third rotation (Fig. 4.2a, below) around the new x-axis was proposed by McMillen (1988) to eliminate the covariance from the vertical and the horizontal, normal to the mean wind direction, wind component (Kaimal and Finnigan 1994)

a)   b)

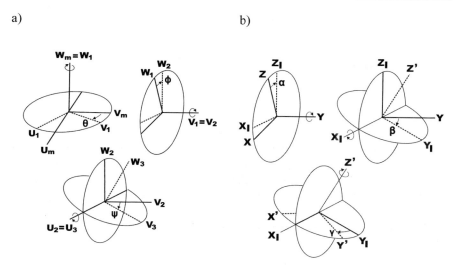

Fig. 4.2. Definition of the coordinate rotations: (**a**) double rotation (*both upper figures*) including a possible third rotation (*below*), with m the measuring parameters are described and with the numbers the rotations; (**b**) planar-fit method according to Wilczak *et al.* (2001), consist in two rotations after a linear multiple regression (*both upper figures*) followed by a rotation into the mean wind field (*below*), with index *I* subrotations are described and with apostrophe the coordinates after rotation.

$$u_3 = u_2,$$
$$v_3 = v_2 \cos \psi + w_2 \sin \psi, \tag{4.9}$$
$$w_3 = -v_2 \sin \psi + w_2 \cos \psi,$$

where

$$\psi = \tan^{-1} \left(\frac{\overline{v_2 w_2}}{\overline{v_2^2} - \overline{w_2^2}} \right). \tag{4.10}$$

This rotation does not significantly influence the fluxes, and introduces some problems. Therefore, it is recommended to use only the first two rotations (Aubinet *et al.* 2000).

A rotation into the mean stream lines was proposed by Paw U *et al.* (2000) and Wilczak *et al.* (2001). With this so called *planar-fit method*, the differences between the measuring device and the mean stream field for an unchanged mounting of the anemometer is estimated over a long time period (days to weeks) for a given measuring place. This means that the measuring device must be orientated over this period without any changes. It is therefore recommended to combine the sonic anemometer with an inclinometer and to observe the inclinometer data regularly. If necessary, the anemometer must be set in its initial orientation.

A matrix form is suitable to describe the planar-fit method (Wilczak *et al.* 2001)

$$\vec{u}_p = P(\vec{u}_m - \vec{c}),$$ (4.11)

where $\vec{u}_m$ is the vector of the measured wind velocities, $\vec{u}_p$ is the vector of the planar-fit rotated wind velocities, $\vec{c}$ is an offset vector, and P is a transformation matrix. The equations of rotations are then:

$$\overline{u}_p = p_{11}(\overline{u}_m - c_1) + p_{12}(\overline{v}_m - c_2) + p_{13}(\overline{w}_m - c_3)$$
$$\overline{v}_p = p_{21}(\overline{u}_m - c_1) + p_{22}(\overline{v}_m - c_2) + p_{23}(\overline{w}_m - c_3)$$ (4.12)
$$\overline{w}_p = p_{31}(\overline{u}_m - c_1) + p_{32}(\overline{v}_m - c_2) + p_{33}(\overline{w}_m - c_3)$$

The offset vector is necessary because, for example, the flow distortion of the sonic anemometer generates a slightly positive vertical wind velocity (Dyer 1981), and therefore a value of c_3 which differs from zero is required. The offset of the horizontal wind components can be assumed negligible. Nevertheless, before using a sonic anemometer the offset values during calm conditions (or observed in a box) should be controlled and if necessary corrected.

The planar-fit coordinate system fitted to the mean-flow streamlines is characterized by $\overline{w}_p = 0$. The tilt angels can be calculated according to Eq. (4.12) with multiple linear regressions

$$\overline{w}_m = c_3 - \frac{p_{31}}{p_{33}}\overline{u}_m - \frac{p_{32}}{p_{33}}\overline{v}_m,$$ (4.13)

where $p_{31} = \sin\alpha$, $p_{32} = -\cos\alpha \sin\beta$ and $p_{33} = \cos\alpha \cos\beta$. Knowing these angles, the coordinate system can be rotated according to Fig. 4.2b in the proposed order. This means that the rotation is first around angle α and than around β. The rotation angles for this method are only a few degrees. If the rotation angles differ for different wind directions and velocities, then this method must be applied for single wind sectors and velocity classes.

After the rotation into the mean streamline level, each single measurement must be rotated into the mean wind direction according to (Fig. 4.2b, below):

$$\gamma = \arctan\left(\frac{\overline{v}_p}{\overline{u}_p}\right)$$ (4.14)

This is an absolute analogue to other rotation methods, but is the last step. In general, for the single measurements a block averaging over 30 min is used.

Spectral Correction in the High Frequency Range

An important correction to the actual available turbulence spectra is the adjustment of the spectral resolution of the measuring system. Hence, the resolution in
time (time constant), the measuring path length, and the separation between different measuring paths must be corrected. Currently, the correction method according to Moore (1986) is usually applied. Note that the published software program has errors. The stability-dependent spectral function must be taken from the
original source (Kaimal *et al.* 1972), see Chap. 2.5. Furthermore, the published
aliasing correction should not be applied. Take in mind that the spectral functions
are based on a few measurements from the Kansas experiment, and therefore its
universal validity is limited.

The spectral correction is made using transfer functions (see Chap. 6.1.3). For
each combination of the path length for the vertical wind velocity of a sonic anemometer (w) and the sensor for the determination of the relevant flux parameter
(x), there are separate filters for the time constant (τ), the measuring path length
(d), and the senor separation (s), which must be determined. The product of these
single functions is the total transfer function:

$$T_{w,x}(f) = \sqrt{T_{\tau,w}(f)} \sqrt{T_{\tau,x}(f)} \sqrt{T_{d,w}(f)} \sqrt{T_{d,x}(f)} \, T_{s,w,x}(f) \qquad (4.15)$$

If the sensors are aligned parallel to the mean wind, the spectral correction is
partially done using the above-mentioned cross-correlation correction.

For the applied measuring system, it is also possible to use a simple analytical
correction for the observation site (Massman 2000). From Eugster and Senn
(1995), a method was proposed which is based on an electronic damping circuit.

Spectral Correction in the Low Frequency Range

Often a sampling time of 30 min is not long enough to measure the low frequency
part of the fluxes. In principle, the averaging interval can be extended by an elimi-
nation of a trend, and in most cases it is sufficient to subtract only the linear trend.
This of course involves the danger that the low frequency events, which are not
associated with turbulent fluxes, contribute to the calculated flux (Finnigan *et al.*
2003).

It is therefore advised to test if the flux has its maximum value within the usual
averaging time. This is done using the so called *ogive* test (Desjardins *et al.* 1989;
Foken *et al.* 1995; Oncley *et al.* 1990). It is calculated using the cumulative integral
of the co-spectrum of the turbulent flux beginning with the highest frequencies:

$$Og_{w,x}(f_0) = \int_{\infty}^{f_0} Co_{w,x}(f) \, df \qquad (4.16)$$

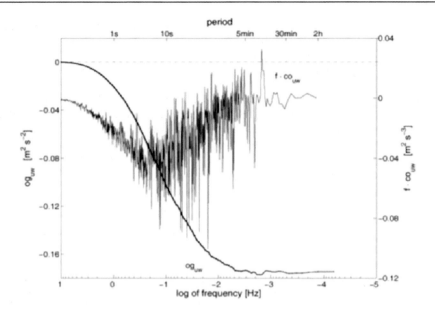

Fig. 4.3. Converging ogive (Og_{uw}) and co-spectrum ($f\,CO_{uw}$) of the momentum flux during the LITFASS-2003 experiment (09.06.2003, 12:30–16:30 UTC, Foken *et al.* 2006b)

If the value of the integral approaches a constant value (flux) for low frequencies, and if an enhancement of the averaging interval gives no significant changes, then no additional correction is necessary.

Foken *et al.* (2006b) have shown for the LITFASS-2003 experiment that in about 80% of all cases the ogive converged within an averaging time of 30 min. In the remaining cases, mainly in the transition periods of the day, the ogives did not converge or had reached a maximum value before the integration time of 30 min, and then they decrease. By applying the ogive correction for low frequency fluxes of the investigated data set, the fluxes would increase less than 5%. The ogive correction is currently not routinely applied (Fig. 4.3).

Correction of the Buoyancy Flux

The temperature measured with sonic anemometers is the so called the sonic temperature (Kaimal and Gaynor 1991):

$$T_s = T\left(1 + 0.32\,{}^{e}\!\!\big/_{p}\right) \tag{4.17}$$

This differs only slightly from the virtual temperature (see Eq. 2.69):

$$T_v = T\left(1 + 0.38\,{}^{e}\!\!\big/_{p}\right) \tag{4.18}$$

Therefore, the heat flux measured with the sonic temperature is approximately equal to the buoyancy flux. To transfer the measured buoyancy flux into the sensible heat flux, the Reynolds decomposition of Eq. (4.17) must be used in Eq. (4.1). It must be noted that the sound signal will be modified depending on the construction of the sonic anemometer and the wind velocity. Schotanus $et\ al.$ (1983) developed a correction method which is widely used. It is valid only for vertical measuring paths. For recently available sonic anemometers, Liu $et\ al.$ (2001) adapted the correction including the so called crosswind correction:

$$\rho c_p \overline{(w'T')} = \rho\, c_p\, \frac{\overline{w'T_S'} + \dfrac{2\overline{T}}{c^2}\left(\overline{u}\,\overline{u'w'}\,A + \overline{v}\,\overline{v'w'}\,B\right)}{1 + \dfrac{0.51\,\overline{T}\,c_p}{\lambda\,Bo}} \qquad (4.19)$$

Because most of the producers include the crosswind correction in the sensor software, the second term in the nominator of Eq. (4.19) can be neglected (Table 4.1).

Table 4.1. Coefficients for Eq. (4.19) according to Liu $et\ al.$ (2001), φ: angle between the measuring axis and the horizontal line for different presently used sonic anemometer types. Except for the USA-1 without turbulence module, these corrections are already included in the sensor software

factor	CSAT3	USA-1	Solent	Solent-R2
A	7/8	3/4	$1 - 1/2 \cdot \cos^2\varphi$	½
B	7/8	3/4	$1 - 1/2 \cdot \cos^2\varphi$	1

WPL-Correction

Webb $et\ al.$ (1980), discussed the necessity for a density correction (WPL-correction according to Webb, Pearman, and Leuning, formerly also called Webb-correction), which is caused by ignoring density fluctuations, a finite humidity flux at the surface, and the measurement of gas concentration per volume unit instead of per mass unit. A review of the continuing discussions over the last 20 years is given by Fuehrer and Friehe (2002). A different way is the use of the density-weighted averaging according to Hesselberg (1926), Kramm $et\ al.$ (1995), and Kramm and Meixner (2000). The application of this version would lead to inconsistencies in this book; therefore in the following, the version by Webb $et\ al.$ (1980) is used.

Webb $et\ al.$ (1980) begin their derivation assuming dry air while Bernhardt and Piazena (1988) assume moist air. The differences between both methods are negligible. Using the moist air approach, Liu (2005) showed significant differences at least for the correction of the carbon dioxide flux. The discussions of this complex correction continue. Therefore, at this time the classical WPL-correction should be used without changes.

The total flux measured per unit mass, must be represented by the specific content of the matter q_c, according to the relation

$$F_c = \overline{\rho w q_c} = \overline{\rho w}\ \overline{q_c} + \overline{(\rho w)' q_c'}\ .$$
(4.20)

Using the partial density

$$\rho_c = \rho\ q_c$$
(4.21)

the relation per unit volume is

$$F_c = \overline{\rho w q_c} = \overline{w}\ \overline{\rho_c} + \overline{w' \rho_c'}\ ,$$
(4.22)

which is be applied for measurements. The mean vertical wind velocity is included in the correction term, which is given in the following form:

$$F_c = \overline{w'\rho_c'} + \overline{q_c}\ \frac{Q_H}{c_p \overline{T}} \cdot \left[1 + 1.61 \frac{c_p \overline{T}}{\lambda} \left(1 - 0.61\,\overline{q}\right)\frac{1}{Bo} \right]$$
(4.23)

The WPL-correction is large if the turbulent fluctuations are small relative to the mean concentration. For example, this is the case for carbon dioxide where corrections up to 50% are typical. For water vapor flux, the corrections are only a few percent because the effects of the Bowen ratio and the sensible heat flux are in opposition (Liebethal and Foken 2003; 2004).

The conversion from volume into mass-related values using the WPL-correction is not applicable if the water vapor concentrations or the concentrations of other gases are transferred into mol per mol dry air before the calculation of the eddy-covariance. However, the calculation is possible depending of the sensor type and if it is offered by the manufacturers.

Correction of the Specific Heat

Due to the presentation by Stull (1988) of the correction of humidity-dependent fluctuations of the specific heat proposed by Brook (1978), which is some percentage of the flux, this correction is often used. However, shortly after the publication of this correction several authors (Leuning and Legg 1982; Nicholls and Smith 1982; Webb 1982) showed the this correction is based on incorrect conditions, and should never be used.

Advection Correction

In recent extensive measurements in heterogeneous terrain, it was found that advection couldn't be completely neglected when applying the eddy-covariance method. In Fig. 4.4 it is shown for flat and sloping terrain that in addition to the flux measured at the upper border of the volume element there also exists vertical and horizontal fluxes, which are balanced in the simplest case. Matter transformations and especially vertical fluxes in complex terrain are the reasons for the of use special corrections proposed by Lee (1998). The starting point is the balance equation for the net flux from a volume element:

$$F_n = \overline{w' \rho_c'(h)} + \int_0^h \frac{\partial \overline{\rho_c}}{\partial t} dz + \int_0^h \left(\overline{w} \frac{-\partial \overline{\rho_c}}{\partial z} + \overline{\rho_c} \frac{\partial \overline{w}}{\partial z} \right) dz \qquad (4.24)$$

The net flux is the sum of the flux above the volume element (1st term), the source and sink term due to the change in time of the partial density of the investigated substance (2nd term), and finally the advection term (3rd term). Paw U et al. (2000) showed that this advection term includes the vertical advection flux discussed by Lee (1998) as well as the effect of an additional vertical flux due to the WPL-correction (Webb et al. 1980). By choosing a streamline-adapted coordinate system, the second influencing factor can be excluded on average. The present recommendation is to not use any advection correction, but rather use the source and sink term in the net flux and apply WPL-correction (Finnigan et al. 2003; Paw U et al. 2000). However, it requires the use of the planar-fit method (Wilczak et al. 2001), and accordingly the measurements must be based on a coordinate system adapted to the mean streamlines. The problem is still a subject of recent research activities over the complex terrain (Aubinet et al. 2003b; 2005).

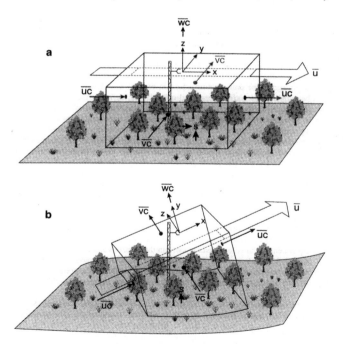

Fig. 4.4. Schematic image of the advection through a volume element in homogeneous terrain (**a**) and in a terrain with complex topography (**b**). The x–y–level is defined as an ensemble average of the wind vector, and the control volume is not parallel to the mean level of the surface (Finnigan et al. 2003).

4.1.3 Quality Assurance

Turbulence measurements with the eddy-covariance method cannot be controlled in a simple way with plausibility tests (see Chap. 6.3.1). The quality assurance of turbulence measurements is a combination of the complete application of all corrections and the exclusion of meteorological influences such as internal boundary layers, gravity waves, and intermitted turbulence. The main aim of the control of the data quality is to validate if, under given meteorological conditions, the simplifications of Eq. (4.1) are warrantable (see Chap. 2.1.2). Quality tests (Foken and Wichura 1996; Foken *et al.* 2004; Kaimal and Finnigan 1994) are used to validate the theoretical assumptions of the method such as steady-state conditions, homogeneous surfaces, and developed turbulence.

For the eddy-covariance method, steady states are required. Meteorological measurements fulfill these conditions for short time periods up to one hour only roughly. There are several tests, which can be used directly or indirectly. For example, stationarity can be determined by examining the fluxes for different averaging times (Foken and Wichura 1996; Gurjanov *et al.* 1984). In this way, the flux is determined once over M short intervals each of only about 5 min duration, and than the average over the short time intervals is calculated:

$$\left(\overline{x'y'}\right)_i = \tfrac{1}{N-1}\left[\sum_j x_j\,y_j - \tfrac{1}{N}\left(\sum_j x_j \sum_j y_j\right)\right] \tag{4.25}$$

$$\overline{x'y'} = \tfrac{1}{M}\sum_i \left(\overline{x'y'}\right)_i$$

Next, the flux is calculated over the whole averaging interval (*e.g.* 30 min):

$$\overline{x'y'} = \tfrac{1}{M\,N-1}\left[\sum_i\left(\sum_j x_j\,y_j\right)_i - \tfrac{1}{M\,N}\sum_i\left(\sum_j x_j \sum_j y_j\right)_i\right] \tag{4.26}$$

Steady-state conditions can be assumed, if both results do not differ by more than 30%. A gradation of the differences can be used as a classification of the data quality.

From Vickers and Mahrt (1997), two tests were proposed, which give similar results. They calculate the skewness and excess of the time series. For values of skewness > |2| and values of excess < 1 and > 8, the authors suggested a bad data quality, and for skewness > |1| and excess < 2 and >5 medium data quality is assumed. For wind components, the data at the beginning and end of a time series are compared. The difference of these values normalized by the mean wind velocity must fulfill the relation

$$\left|\frac{u_1 - u_N}{\overline{u}}\right| < 0.5\,, \tag{4.27}$$

if the time series can be accepted as steady-state.

The steady-state test often identifies large sudden jumps in the signal, a constant signal for a certain time, or changes of the signal level. The reason for these events is often an electronic one. Mahrt (1991) proposed a test using the Haar-wavelet for which the time of events can be well located. The test can also identify periods of intermittent turbulence.

The development of the turbulence can be investigated with the flux-variance similarity described in Chap. 2.4 (Foken and Wichura 1996). In this case, the measured integral turbulence characteristics are compared with the modeled characteristics according to Tables 2.10 and 2.11. A good data quality is assumed for differences less than about 30%. In a similar way, the comparison of the correlation coefficient between values used in the flux calculations with the mean values can be used for quality assurance testing (Kaimal and Finnigan 1994).

Table 4.2. Ranging of the data quality of the steady-state test according to Eqs. (4.21) and (4.22); the comparison of the integral turbulence characteristics with model assumptions according to Tables 2.10 and 2.11, and the wind criteria, for example for the sonic anemometer CSAT3

steady state according Eqs. (4.21) and (4.22), differences		integral turbulence characteristics, differences		horizontal inflow sector	
class	range (%)	class	range (%)	class	range
1	0–15	1	0–15	1	± 0–30°
2	16–30	2	16–30	2	± 31–60°
3	31–50	3	31–50	3	± 60–100°
4	51–75	4	51–75	4	± 101–150°
5	76–100	5	76–100	5	± 101–150°
6	101–250	6	101–250	6	± 151–170°
7	251–500	7	251–500	7	± 151–170°
8	501–1000	8	501–1000	8	± 151–170°
9	> 1000	9	> 1000	9	> ± 171°

Table 4.3. Proposal for the combination of single quality tests (Table 4.2) to an overall data quality of flux measurements (Foken *et al.* 2004)

overall quality	steady state	integral turbulence characteristics	horizontal inflow sector
1	1	1–2	1–5
2	2	1–2	1–5
3	1–2	3–4	1–5
4	3–4	1–2	1–5
5	1–4	3–5	1–5
6	5	≤ 5	1–5
7	≤ 6	≤ 6	≤ 8
8	≤ 8	≤ 8	≤ 8
	≤ 8	6–8	≤ 8
9	one ranging 9		

Additionally, the wind components should be included in a system for quality control tests. The mean vertical wind must be low such that over flat terrain and for horizontal wind velocities < 5 ms^{-1} it is < 0.15–0.20 ms^{-1}. For sonic anemometers with a limited range of acceptable wind directions (due to the support system measurements in the inflow range plus the typical standard deviation of the wind direction dependent on the location and the micrometeorological conditions) the inflow sector should be reduced by ± 30–$40°$ as a general rule. For sonic anemometers with unlimited inflow angles (omni-directional) the flow through supports and sensor heads can reduce the overall data quality.

An evaluation system for turbulent fluxes consists of two steps: The single tests should be evaluated according to the threshold values with data quality steps (Table 4.2) and the overall quality of a measurement is expressed as an appropriate combination of the single tests (Table 4.3). The highest priority should be given to the steady-state test. Note that for the test on integral turbulence characteristics for neutral stratification, the errors in the determination of the characteristics for scalars can be very high. This test should not be overly interpreted, and the test on the characteristics of the wind parameters should dominate. In any case, the ranging of the single tests should be available in cases of doubt. The classification according to Table 4.3 was done so that classes 1–3 have a high accuracy and the data can be used for basic research. Classes 1–6 can be used for long-term measurements of fluxes without limitations. Measurements of the classes 7–8 should only be used for orientation and should be if necessary deleted while class 9 is always canceled.

4.1.4 Gap Filling

The application of the eddy-covariance method in long-term measuring programs such as the international FLUXNET programme (Baldocchi *et al.* 2001) requires objective methods on how to handle missing data. This is because with such measuring programs annual sums such as the Net Ecosystem Exchange (NEE) will be determined. In substance, two types of missing data must be corrected. One type is missing data from single systems due to meteorological influences such as heavy rain or failure of the whole system. The second type is data that must be replaced. If, for example, during the night no turbulent exchange conditions exist, then the measuring method no longer works satisfactorily. For the measurements of carbon dioxide fluxes, the methods of gap filling are well developed (Falge *et al.* 2001; Gu *et al.* 2005; Hui *et al.* 2004). Nevertheless theses methods, which use an approach for the carbon assimilation at daytime and a different approach for respiration, are still under of discussion. Recently, gap filling with models (z.B. Papale and Valentini 2003) are applied.

The determination of the carbon uptake (NEE) at daytime take place with the so called Michaelis-Menton function (Falge *et al.* 2001; Michaelis and

Menton 1913), which must be evaluated for different temperature classes and the global radiation:

$$Q_{c,day} = \frac{a K \downarrow Q_{c,sat}}{a K \downarrow + Q_{c,sat}} + Q_{R,day} \tag{4.28}$$

where $Q_{c,sat}$ is the carbon flux for light saturation ($K\downarrow = \infty$), $Q_{R,\ day}$ is the respiration at daytime, and a and $Q_{c,sat}$ must be determined with multiple regression using data from an available dataset for the specific measuring site.

The respiration of an ecosystem can be determined with the Lloyd-Taylor-function (Falge *et al.* 2001; Lloyd and Taylor 1994)

$$Q_R = Q_{R,10} e^{E_0 \left[\frac{1}{283,15-T_0} - \frac{1}{T-T_0} \right]}, \tag{4.29}$$

where $Q_{R,10}$ is the respiration at 10 °C, $T_0 = 227.13$ K (Lloyd and Taylor 1994), and E_0 describes the temperature dependence of the respiration. The parameters of this equation will be determined from nighttime ($K\downarrow < 10$ Wm^{-2}) eddy-covariance measurements with the assumption that for such low radiation fluxes only the respiration can be measured. The coefficients will be determined also for temperature classes.

It must be noted that for both methods the equations for the carbon flux during the day and the respiration during the night are climatological parameterizations. These can have significant differences from the individual values.

While the Michaelis-Menton function is in general used for gap filling for the daytime, the Lloyd-Taylor function must be used at night because of low turbulence. The decision regarding the application follows from the so called u_*-criterion (Goulden *et al.* 1996). Therefore, the respiration will be normalized to exclude the temperature sensitivity using a model calculation according to Eq. (4.29), and plotted as a function of the friction velocity u_*. From a certain friction velocity, the normalized respiration will be constant. Measurements with a lower friction velocity will be gap filled. Typical threshold values are in the range $u_* = 0.3$–0.4 ms^{-1}. Not for all stations can such a threshold value be determined (Gu *et al.* 2005).

A more objective approach seems to be the application of the quality criterion for turbulent fluxes as introduced in Chap. 4.1.3 (Ruppert *et al.* 2006a). According to this, very high data quality will be used to determine the Michaelis-Menton and Lloyd-Taylor functions. Then both functions will be used to fill gaps of data with low data quality. The benefit of this method is that nighttime data with high data quality and low friction velocity can be used to parameterize the Lloyd-Taylor function. On the other hand, daytime values with low data quality must be gap filled.

4.1.5 Overall Evaluation

To give an overview of the expenses of the different methods, in the following all methods costs and quality of the measuring results are given. It is seen that the eddy-covariance method (Table 4.4) is the only direct measuring method and the most accurate with the largest time resolution. However, it needs also the most comprehensive knowledge and great experimental effort.

The very complicated algorithm of the eddy-covariance method does not allow the use for the determination of the errors according the error propagation law. Nevertheless Mauder *et al.* (2006) have tried using sensor and software comparisons during the experiments EBEX-2000 and LITFASS-2003 to determine the accuracy of the measuring method. A significant dependence was found on the type of sonic anemometer (compare Table 6.10) and on the data quality. The results are summarized in Table 4.5.

Table 4.4. Evaluation of the eddy-covariance method

criterion	evaluation
area of application	basic research and expensive continuously measuring programs
financial expense	10–50 k€ per system
personal expense	continuous scientific and technical support
education	good micrometeorological and measuring technique knowledge
error	depending on the micrometeorological conditions 5–10%
sampling	10–20 Hz
time resolution of fluxes	10–60 min
application for chemical compounds	selected inert gases (gas analyzers with high time resolution)
restrictions in the application	sufficient footprint area,
	turbulent conditions necessary,
	depending on the senor, possible precipitation

Table 4.5. Evaluation of the accuracy of the eddy-covariance method on the basis of the results of the experiments EBEX.-2000 and LITFASS-2003 (Mauder *et al.* 2006) dependent on the data quality (Chap. 4.1.3) and from the type of the sonic anemometer (Table 6.9, Foken and Oncley 1995)

sonic anemometer	data quality class	sensible heat flux	latent heat flux
type A,	1–3	5% or 10 Wm^{-2}	10% or 20 Wm^{-2}
e.g.. CSAT3	4–6	10% or 20 Wm^{-2}	15% or 30 Wm^{-2}
type B,	1–3	10% or 20 Wm^{-2}	15% or 30 Wm^{-2}
e.g.. USA-1	4–6	15% or 30 Wm^{-2}	20% or 40 Wm^{-2}

4.2 Profile Method

Under the general term of the profile method, all methods are combined which are based on the flux-gradient similarity (see Chap. 2.3). Because of the progress of the eddy-covariance method in measuring technique, the profile method with measurements at several heights has become increasingly irrelevant in the last 10–15 years. The disturbing influence of internal boundary layer makes this technique applicable only in homogeneous terrain with large uniform fetch. Conversely, simple approaches with measurements only at two levels are common; the more so since these approaches are also part of many models where because of computer technical reasons no alternative versions are possible.

4.2.1 Bulk Approaches

The most simple profile method to determine the energy exchange is the bulk approach, which is also used as a model closure approach of zero order. Bulk approach means that a uniform (linear) gradient is assumed for the given layer and only values at the upper and lower boundaries are used (Mahrt 1996). If the lower boundary of this layer is identical to the surface, than the method is strictly speaking only applicable over water bodies, because only there the gradient between surface values and measuring data at a certain measuring height (mostly 10 m) can be explicitly determined. For instance, the surface temperature and moisture for land surfaces cannot be determined exactly because of roughness elements (plant cover and others). Nevertheless, the method is partly applied by the calculation of surface information measured with satellites, even when considerable losses in the accuracy must be accepted. It is also possible to fix the lowest level a certain distance above the surface. As a general rule, this is double the canopy height, and the methods described in Chaps. 4.2.2–4.2.4 are applied.

The application of the actual bulk method above water bodies is not without problems, because normally it is not the water surface temperature that is measured but rather the temperature some decimeters below the water surface. Because of the cold film at the surface due to the cooling by evaporation, this temperature is about 0.5 K lower than the temperature in measuring level. The absolute accuracy of the determination of the surface temperature with remote sensing methods is of the same order. Instead of the turbulent diffusion coefficients, bulk coefficients are used. Then the friction velocity can be determined with the drag coefficient C_D and the wind velocity

$$u_* = \sqrt{C_D(z)}\, u(z),$$
(4.30)

where it is assumed that the wind velocity at the ground surface is zero.

The sensible heat flux is parameterized with the Stanton number C_H and the latent heat flux with the Dalton number C_E:

$$\frac{Q_H}{\rho c_p} = C_H(z)\,u(z)\,[T(z) - T(0)], \tag{4.31}$$

$$\frac{Q_E}{\rho \lambda} = C_E(z)\,u(z)\,[e(z) - e(0)] \tag{4.32}$$

The bulk coefficients (Table 4.6) are stability and wind velocity dependent. Over the ocean with mostly neutral stratification, the first sensitivity is not a problem. Over the land, the application for non-neutral stratification should be limited to the dynamical sublayer because the bulk coefficients have a remarkable dependency on stratification with differences up to 50% (Brocks and Krügermeyer 1970; Foken 1990; Panin 1983). In the literature, a number of relations for the drag coefficient are given (Geernaert 1999; Smith *et al.* 1996). Currently, the parameterization dependent on the wind velocity for the 10 m reference level is the most useful version. Coefficients as mean values from a large number of experiments are given in Table 4.6 for the following equation:

$$C_{D10} = [a + b\,(u_{10} - c)]10^{-3} \tag{4.33}$$

For greater wind velocities, the bulk coefficients increase dramatically, but are not more clearly determined. The reason for this is that there are higher energy fluxes under storm conditions than under normal conditions. The values over lakes are slightly higher and over land where nearly no data are available at least one order of magnitude higher than those over the ocean.

The Stanton and Dalton numbers over water are about 20% lower than the drag coefficient. For the same wind velocity classes as given in Table 4.6 follows according to Foken (1990) the following coefficients for Eq. (4.33): $a = 1.0$, $b = 0.054$, $c = 7$ ms^{-1}. With increasing roughness the differences in the drag coefficient increases (Garratt 1992).

By comparison of Eq. (4.33) with the profile equation for neutral stratification, a relation between the drag coefficient and the roughness height is:

$$C_D = \frac{\kappa^2}{\left[\ln\left(\frac{z}{z_0}\right)\right]^2} \tag{4.34}$$

For the determination of the Stanton number, the roughness length for temperature must be considered (for the Dalton number use the roughness length for specific humidity):

$$C_H = \frac{(\alpha_0\,\kappa)^2}{\ln\left(\frac{z}{z_0}\right)\ln\left(\frac{z}{z_{0T}}\right)} \tag{4.35}$$

In a similar way, the stability dependence of the bulk coefficients can be determined. Note that slight errors in the determination of the roughness length have a remarkable influence on the bulk coefficient. Therefore, these bulk coefficients are not really an alternative to experimental parameterizations.

Bulk approaches for the determination of the momentum and energy exchange over water bodies are widely used because the input data are routinely available or can be easily determined, *i.e.* these data are contained in models. Thus, the roughness parameter is often determined with the Charnock approach, Eq. (3.5). An overall evaluation of this approach is given in Table 4.7.

Table 4.6. Coefficients for the determination of the drag coefficient above water bodies according to Eq. (4.33) for $u_{10} < 20$ m s^{-1}

author	a	b	c in ms^{-1}	u_{10} in ms^{-1}
Foken (1990)	1.2	0	0	< 7
	1.2	0.065	7	≥ 7
Garratt (1992)	1.0	0	0	< 3.5
	0.75	0.067	0	≥ 3.5

Table 4.7. Evaluation of the bulk method

criterion	evaluation
area of application	application over water, modelling, if no other possibilities
financial expense	1–3 k€ per system
personal expense	low technical maintenance
education	introduction
error	according to the micrometeorological conditions 10–50%
sampling	1–10 s
time resolution of fluxes	10–30 min, higher accuracy for daily averages
application for chemical compounds	selected inert gases qualified possible
restrictions in the application	turbulent conditions necessary

4.2.2 Bowen-Ratio Method

The Bowen-ratio method is one of the most common methods used to determine the fluxes of sensible and latent heat. The method is based on Bowen-ratio similarity (see Chap. 2.3.3) and the energy balance equation

$$Bo = \gamma \, \frac{\Delta T}{\Delta e}, \qquad (4.36)$$

$$-Q_s^* = Q_H + Q_E + Q_G, \qquad (4.37)$$

where the psychrometric constant $\gamma = 0{,}667$ K hPa^{-1} for $p = 1000$ hPa and $t = 20\ ^\circ$C. From both equations, the sensible and latent heat fluxes can be determined:

$$Q_H = \left(-Q_s^* - Q_G\right)\frac{Bo}{1 + Bo} \tag{4.38}$$

$$Q_E = \frac{-Q_s^* - Q_G}{1 + Bo} \tag{4.39}$$

In addition to the simplifications discussed in Chap. 2.3.3, it is apparent that Eqs. (4.38) and (4.39) do not include the wind velocity and do not prescribe a certain difference between the measurement heights. To ensure that a sufficient turbulent regime exists, Foken et al. (1997b) recommend that only measurements with a wind velocity at the upper height greater than 1 ms^{-1} and a difference of the wind velocities between both heights greater than 0.3 ms^{-1} should be used. This requires additional instrumentation with anemometers. Even though the height difference of the measurements (Δz) is not included into the equations, an increase of Δz also increases the difference of the temperature and the humidity. Consequently the influence of the measuring errors decreases. It is therefore recommended to choose a ratio of the measuring heights greater than 4–8 (Foken et al. 1997b). These requirements are seldom taken into account in practice because measurements over high vegetation have ratios of the aerodynamical heights of about 1.5 (Barr et al. 1994; Bernhofer 1992). Equations (4.38) and (4.39) are singular for $Bo = -1$. Consequently, in the morning and evening hours interpretable values do not occur. Therefore, the range $-1.25 < Bo < -0.75$ should be excluded from further analysis. To determine the correct sign of the fluxes in the case $Bo < 0$ the following decision criteria are necessary (Ohmura 1982):

$$\text{If} \quad \left(-Q_s^* - Q_G\right) > 0 \quad \text{then} \quad \left(\lambda\Delta q + c_p\Delta T\right) > 0.$$

$$\text{If} \quad \left(-Q_s^* - Q_G\right) < 0 \quad \text{then} \quad \left(\lambda\Delta q + c_p\Delta T\right) < 0. \tag{4.40}$$

If the criteria are not fulfilled, the fluxes must be deleted.

The crucial disadvantage of the Bowen-ratio method is that because of the apparent unclosed energy balance (see Chap. 3.7) the residual is either added to the net radiation or distributed according to the Bowen ratio to the sensible and latent heat flux. In general, the fluxes determined with the Bowen-ratio method are larger than those determined with the eddy-covariance method. But they fulfill the energy balance equation, which is part of the method. The quantitative correctness of the fluxes may be limited.

Error analysis for the Bowen-ratio method are available to a large extent (Foken et al. 1997b; Fuchs and Tanner 1970; Sinclair et al. 1975), and in the given references, sources of further investigations are given. Many of these investigations are based on either single measurements or false assumptions. Often, only the electrical error of the sensor is used (about 0.01–0.001 K), but not the error in the adaptation of the sensor to the surrounding medium and atmosphere with radiation, ventilation and other influences. Only with much effort in measuring technique is it possible that sensors under the same meteorological conditions and mounted close together show differences of only 0.05–0.1 K or hPa. Therefore, the errors in temperature and humidity measurements in the atmosphere are significantly higher than the pure electrical error (Dugas *et al.* 1991).

The error plots given in Fig. 4.5 are taken from Foken *et al.* (1997b), and are based on an accepted measuring error of $\pm$ 0.05 K or hPa. From Fig. 4.5 we see that for 20 and 40% error in the Bowen ratio corresponds to about a 10 and 20% error in the sensible and latent heat flux respectively. Consideration of the three values of Bowen ratios in Fig. 4.5 shows that optimal conditions exist for $Bo = 0.677$; also, the Bowen ratio should not significantly differ from this value. To realize errors of < 20% (< 40%) for the Bowen ratio, the temperature and humidity differences must be > 0.6 (> 0.4) K or hPa. This underlines the requirement for large differences in the measuring heights. Limitations result from internal boundary layers and possible roughness sublayers, which must be excluded between the measuring heights. Note that in this error analysis, possible errors due to energy imbalance were not taken into account.

An overall evaluation of the method is given in Table 4.8. The evaluations of the errors of the method are based under the assumption of good net radiometer measurements (see Chap. 6.2.1) with sensor costs of at least 4 k€. Furthermore, the heat storage in the soil should be calculated very accurately to reduce the influence of the residual of the energy balance closure (Fig. 4.6).

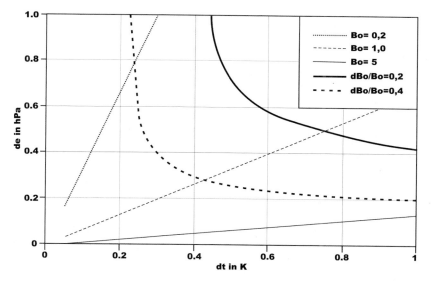

Fig. 4.5. Error of the Bowen ratio (20 und 40%) dependent on the measured temperature and water vapor differences (Foken *et al.* 1997b).

Fig. 4.6. Bowen-ratio measuring system (Photograph: Campbell Scientific, Inc. Logan UT, USA)

Table 4.8. Evaluation of the Bowen-ratio method

criterion	evaluation
area of application	applied research, partly continuously running programs
financial expense	10–15 k€ per system
personal expense	continuous scientific and technical support
education	knowledge in micrometeorology and measuring technique
error	according to the micrometeorological conditions 10–30% (assumption of a closed energy balance)
sampling	1–5 s
time resolution of fluxes	10–30 min
application for chemical compounds	principle possible (see Chap. 4.2.3)
restrictions in the application	sufficient footprint area,
	turbulent conditions necessary

4.2.3 Modified Bowen-Ratio Method

According to Businger (1986), the application of Bowen-ratio similarity (see Chap. 2.3.3) and the abandonment of the energy balance equation, which excludes the problems of energy balance (see Chap. 3.7), are the basis of the modified

Bowen-ratio method (see Eq. 2.89). On the one hand the benefits of the Bowen-ratio similarity (Eq. 2.88) can be used for the modified Bowen-ratio method, and on the other hand the sensible heat flux can be directly determined with a sonic anemometer and a correction of the buoyancy flux (Liu and Foken 2001). Such a measurement system is shown in Fig. 4.7. Usual limitations of the Bowen-ratio method are not valid when the modified Bowen-ratio is used. But the recommended ratio of the heights of $z_2/z_1 = 4$–8 is required to reduce the measuring errors. Also, the observed data are not useful when the friction velocity $u_* < 0.07 \ \mathrm{ms}^{-1}$ and the Bowen ratio $Bo \sim 0$. Because of the decreasing cost of sonic anemometers, they are now in the price range of good net radiation sensors, and so there is no overall increase measuring system costs. Furthermore, the expensive measurements of the soil heat flux and the heat storage in the soil is not necessary. The heat flux equation using the buoyancy flux with the sonic temperature T_s is

$$Q_H = \rho \, c_p \, \frac{\overline{w'T_s'} + \frac{2\overline{T}}{c^2}\left(\overline{u} \ \overline{u'w'} \, A + \overline{v} \ \overline{v'w'} \, B\right)}{1 + \frac{0{,}51 c_p \overline{T}}{\lambda \, Bo}}, \qquad (4.41)$$

where the anemometer-type dependent constants are given in Table 4.1.

If u is approximately in the direction of the mean horizontal wind, the second sum in the numerator of Eq. (4.41) is negligible. For most anemometer types, this sum is negligible, because the crosswind correction is already included in the anemometer software. The Bowen ratio can be calculated with Eq. (4.36), and the latent heat flux can be calculated with:

$$Q_E = Q_H\big/Bo. \qquad (4.42)$$

Fig. 4.7. Measuring system for the modified Bowen-ratio method (METEK GmbH Elmshorn and Th. Friedrichs & Co. Schenefeld near Hamburg; Photograph: Foken)

The modified Bowen-ratio method was originally developed for the determination of trace gas fluxes (Businger 1986; Müller *et al.* 1993). According to Eq. (2.92) a turbulent flux, *e.g.* the sensible heat flux, can be directly determined, and from the relevant difference (here temperature difference, ΔT) and the concentration difference, Δc, between two measurement levels, the trace gas flux (dry deposition) can be calculated using:

$$Q_c = Q_H \frac{\Delta c}{\Delta T} \tag{4.43}$$

In an analogues way to the Bowen-ratio method, criteria for and limitations of the method, which are dependent on the deposited or emitted matter, must be developed. The wind or friction velocity criteria can be used; also, the turbulent scales of both scalars must be similar (see Chaps. 2.5 and 4.4). An overall evaluation of the methods is given in Table 4.9.

Table 4.9. Evaluation of the modified Bowen-ratio method

criterion	evaluation
area of application	applied research, partly continuously running programs
financial expense	10–15 k€ per system
personal expense	continuous scientific and technical support
education	knowledge in micrometeorology and measuring technique
error	according to the micrometeorological conditions 10–30%
sampling	1–5 s, 10–20 Hz for turbulent flux
time resolution of fluxes	10–30 min
application for chemical compounds	for selected inert gases possible
restrictions in the application	sufficient footprint area,
	turbulent conditions necessary,
	similarity of the turbulence characteristics necessary

4.2.4 Further Parameterization Methods

The determination of turbulent fluxes with observations at only two heights is an interesting measuring approach because the fluxes can be estimated in a simple way. While the bulk method (see Chap. 4.2.1) requires a uniform (linear) gradient between both measuring heights and the Bowen-ratio method requires (see Chaps. 4.2.2 and 4.2.3) similar gradients of both parameters, there is also the possible solution of the profile equations with stability influences (Eqs. 2.71–2.73) for two heights. Corresponding proposals (Itier 1980; Lege 1981) were used by

Richter and Skeib (1984) for a method to determine the turbulent fluxes in an iterative way. They introduced a critical height, z_c, which is approximately equal to the height of the dynamical sublayer. Below this height, the equations for neutral stratification can be applied. The use of the universal functions by Skeib (1980), which allows for this method a simple layer-wise integration, gives the following equations for the flux calculation (Richter and Skeib 1991):

$$
u(z_2) - u(z_1) = \frac{u_*}{\kappa} \begin{cases} \ln\left(\frac{z_2}{z_1}\right) & z_1 < z_2 < z_c \\ \ln\left(\frac{z_c}{z_1}\right) + \frac{1}{n_u}\left[1 - \left(\frac{z_2}{z_c}\right)^{-n_u}\right] & z_1 \le z_c \le z_2 \\ \frac{1}{n_u}\left[\left(\frac{z_1}{z_c}\right)^{-n_u} - \left(\frac{z_2}{z_c}\right)^{-n_u}\right] & z_c < z_1 < z_2 \end{cases}
\tag{4.44}
$$

In an analogues way, follow the equation for the sensible and latent heat flux:

$$
T(z_2) - T(z_1) = \frac{\overline{w'T'}}{\alpha_0 \,\kappa u_*} \begin{cases} \ln\left(\frac{z_2}{z_1}\right) & z_1 < z_2 < z_c \\ \ln\left(\frac{z_c}{z_1}\right) + \frac{1}{n_T}\left[1 - \left(\frac{z_2}{z_c}\right)^{-n_T}\right] & z_1 \le z_c \le z_2 \\ \frac{1}{n_T}\left[\left(\frac{z_1}{z_c}\right)^{-n_T} - \left(\frac{z_2}{z_c}\right)^{-n_T}\right] & z_c < z_1 < z_2 \end{cases}
\tag{4.45}
$$

The coefficients in Eqs. (4.44) and (4.45) are given in Table 4.10. The critical height, which is a function of the bulk-Richardson number (Eq. 2.80) and a weighting factor R (the stability-dependent curvature of the profile according to Table 4.11), is:

$$
z_c = \frac{\varsigma_c}{\varsigma_1} z_1 = \frac{\varsigma_c \, z_1}{R \, Ri_B}
\tag{4.46}
$$

The application of Eq. (4.46) requires an iterative solution of the equation. Starting with an initial estimate of ς, the friction velocity and then the sensible heat flux are calculated. Using these values, an updated value for ς is calculated, and the process is repeated. After about 3–6 iteration steps, the method converges. As a measuring technique the Bowen-ratio system without net radiometer but with anemometers at both measuring heights can be used. The wind velocity criterion of the Bowen-ratio method, which is a test on developed turbulence, must be used in the analogues form. Also, measurements must be excluded when the differences between the levels is of the order of the measuring error. The method can be extend in principle by using humidity and/or concentration measurements to measure the latent heat and/or deposition flux. An overall evaluation is given in Table 4.12.

Table 4.10. Coefficients for Eqs. (4.44) and (4.45)

stability range	$\varsigma < 0$	$\varsigma > 0$
ς_c	−0.0625	0.125
n_u	0.25	−1
n_T	0.5	−2

Table 4.11. Weighting factor R for Eq. (4.46) according to Richter and Skeib (1984)

z_2 / z_1	2	4	8	16
$-0.0625 \leq \varsigma \leq 0.125$	0.693	0.462	0.297	0.185
$-1 < \varsigma < -0.0625$	0.691	0.456	0.290	0.178
$0.125 < \varsigma < 1$	0.667	0.400	0.222	0.118

Table 4.12. Evaluation of the parameterization approach according to Richter and Skeib (1991)

criterion	evaluation
area of application	applied research, partly continuously running programs
financial expense	10–15 k€ per system
personal expense	continuous scientific and technical support
education	knowledge in micrometeorology and measuring technique
error	according to the micrometeorological conditions 10–30%
sampling	1–5 s
time resolution of fluxes	10–30 min
application for chemical compounds	for selected inert gases possible
restrictions in the application	sufficient footprint area,
	turbulent conditions necessary

4.2.5 Profile Calculation

In Chaps. 4.2.1–4.2.4, special cases of the profile method with only two measuring heights were discussed. The classical profile method (Fig. 4.8) is based on wind, temperature and moisture measurements made at 3–6 levels, where 4–6 levels are optimum (Foken and Skeib 1980). These measurements were used for more than 10 years because to a larger extent the eddy-covariance method was too costly. Today, profile measurements are applied most often in basic research to determine parameters of the profile equation or disturbances by internal boundary layers.

The basis for the profile method in the neutral case are Eqs. (2.46)–(2.48) or in the integral form as in Eq. (2.59). The simplest case is the linear approximation, which is more or less also the basis of the bulk and Bowen-ratio method:

$$\left(\frac{\partial X}{dz} \right)_{z_a} \cong \frac{\Delta X}{\Delta z} = \frac{X_2 - X_1}{z_2 - z_1}$$

$$z_a = (z_1 - z_2)/2$$

(4.47)

The logarithmic approximation to Eq. (4.47) is a much better representation of the physical facts with a geometric average of the heights:

$$\left(\frac{\partial X}{\partial \ln z}\right)_{z_m} \cong \frac{\Delta X}{\Delta \ln z} = \frac{X_2 - X_1}{\ln(z_2/z_1)} \tag{4.48}$$

$$z_m = (z_1\, z_2)^{\frac{1}{2}}$$

The basis for the profile method in the diabatic case are Eqs. (2.71) to (2.73). For a simple graphical analysis, the integral form (Eq. 2.83) is used with $\ln z - \psi(z/L)$ on the ordinate and u or T on the abscissa according to the following equations (Arya 2001):

$$\ln z - \psi_m\left(\tfrac{z}{L}\right) = \frac{\kappa}{u_*}\, u + \ln z_0 \tag{4.49}$$

$$\ln z - \psi_H\left(\tfrac{z}{L}\right) = \frac{\alpha_0\, \kappa}{T_*}\, T - \frac{\alpha_0\, \kappa}{T_*}\, T_0 + \ln z_{0T} \tag{4.50}$$

Before both equations can be calculated, either the Richardson number (Eq. 2.79) or the Obukhov length (Eq. 2.68 and Eq. 2.70) must be determined. The method can be solved iteratively.

There are approximations that are not directly based on the profile equation. The simplest is the series expansion:

$$u(z) = a_0 + a_1 \ln z + a_2 z \tag{4.51}$$

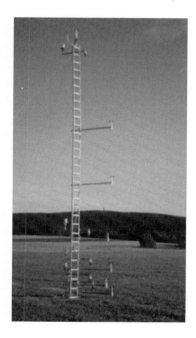

Fig. 4.8. Measuring mast for profile measurements (Photograph: Foken)

Table 4.13. Overall evaluation of the profile method with 4–6 measuring heights

criterion	evaluation
area of application	basic and applied research, partly continuously running programs
financial expense	10–15 k€ per system
personal expense	continuous scientific and technical support
education	good knowledge in micrometeorology and measuring technique
error	according to the micrometeorological conditions 5–20%
sampling	1–5 s
time resolution of fluxes	10–30 min
application for chemical compounds	for selected inert gases possible
restrictions in the application	sufficient footprint area,
	turbulent conditions necessary

A more commonly used expansion is one developed by Kader and Perepelkin (1984):

$$u(z) = a_0 + \frac{a_1 \ln z + a_3 + a_4 z^{2/3}}{5 + z} \tag{4.52}$$

For the interpolation of profiles, several approaches are used, for example, spline methods. To insure that these methods do not overcorrect measuring errors and falsify the result, special cubic interpolation methods between neighboring grid points should be used, *e.g.*, Akima (1970).

Because of the availability of powerful personal computers, lavish interpolation methods are used such as the Nieuwstadt-Marquardt approach. In this approach, a quadratic cost function is calculated as a measure of the tolerance between the measured data and the model based on the profile equations (Nieuwstadt 1978). The nonlinear system of equations to minimize the cost function can be solved using the method described by Marquardt (1983).

As with the Bowen ratio method, the profile method requires high accuracies in the measuring data. High ratios of the upper to the lower measuring height, and negligible influences from the surface are also necessary. The wind criteria for the Bowen ratio method to exclude non-turbulent cases can be also used. A complete error analysis was made by Foken and Skeib (1980). An overall evaluation is given in Table 4.13.

4.2.6 Quality Assurance

For all profile measurements, it should be demonstrated that the measuring accuracy is at least 10-fold greater than the expected difference between the two measuring heights so that the flux can at least be determined with an accuracy of 20% (positive and negative measuring errors are assumed). It is further assumed that wind and temperature measurements can be made with a sufficient accuracy. According to a method proposed by Foken (1998), the profile equations Eqs, (2.71) to (2.73) can be divided into a turbulence-caused part and in the difference of the state parameter between both heights:

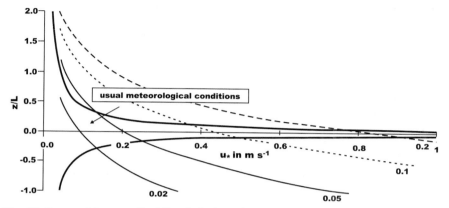

Fig. 4.9. Course of the normalized flux Q_N in dependency of the stratification and the friction velocity for $m = z_2 / z_1 = 8$ (UBA 1996)

Table 4.14. Minimal determinable flux (20% error) for energy and different matter fluxes above low ($m = z_2 / z_1 = 8$) and tall ($m = 1.25$) vegetation for neutral stratification and $u_* = 0.2$ m s^{-1} (units µg m^{-3} for concentrations and µg s^{-1} m^{-2} for fluxes), the italic fluxes are larger than the typical fluxes (Foken 1998, completed)

energy or matter flux	c_{min}	$\Delta_{c.min}$	flux m = 8	flux m = 1.25
sensible heat	0.05 K	0.5 K	0.025 m K s^{-1} 30 W m^{-2}	0.05 m K s^{-1} 60 W m^{-2}
latent heat	0.05 hPa	0.5 hPa	0.025 hPa ms^{-1} 45 W m^{-2}	0.05 hPa ms^{-1} 90 W m^{-2}
nitrate particles	0.01	0.1	0.005	0.01
ammonium particles	0.02	0.2	0.01	*0.02*
CO_2	100	1000	50	100
NO	0.06	0.6	*0.03*	*0.06*
NO_2	0.1	1.0	0.05	0.1
O_3	1.0	10.0	0.5	*1.0*
NH_3	0.014	0.14	0.007	0.014
HNO_3	0.2	2.0	*0.1*	0.2
HNO_2	0.25	2.5	*0.125*	0.25

$$Q_c = Q_N \left[u_*, \varphi(z/L), \ln(z - d) \right] \Delta_c \tag{4.53}$$

The normalized flux Q_N is plotted Fig. 4.9. The minimal measurable flux with 20% accuracy is the 10-fold resolution of the measuring system c_{min}:

$$Q_{c,min} = Q_N \, 10 \, c_{min} \tag{4.54}$$

Typical values of still-measurable fluxes over low and high vegetation are given in Table 4.14.

4.2.7 Power Approaches

For many applied methods, the power approach is widely used for the determination of the wind distribution near the ground surface. This is not only done in the surface layer but also in the lower part of the atmospheric boundary layer (Doran and Verholek 1978; Hsu et al. 1994; Joffre 1984; Sedefian 1980; Wieringa 1989):

$$\frac{u_1}{u_2} = \left(\frac{z_1}{z_2} \right)^p \tag{4.55}$$

For wind power applications, an exponent of $p = {}^1/_7$ is applied often (Peterson and Hennessey jr. 1978). Detailed approaches use a stability and roughness dependency on the exponent. After differentiation of Eq. (4.55), it follows according to Huang (1979):

$$p = \frac{z}{u} \frac{\partial u}{\partial z} \tag{4.56}$$

This version allows complicated approaches including the roughness of the surface and the stratification expressed with universal functions. Irvin (1978) proposed the approach

$$p = \frac{u_*}{u \, \kappa} \varphi_m(\varsigma). \tag{4.57}$$

Similar is the approach by Sedefian (1980):

$$p = \frac{\varphi_m\left(\dfrac{z}{L} \right)}{\left[\ln\left(\dfrac{z}{z_0} \right) - \psi_m\left(\dfrac{z}{L} \right) \right]} \tag{4.58}$$

Huang (1979) used this form of the exponent with the universal function given by Webb (1970) and Dyer (1974) and a special integration for large roughness elements. For unstable stratification, it follows that

$$p = \frac{\left(1-19.3\frac{z}{L}\right)^{-\frac{1}{4}}}{\ln\frac{(\eta-1)(\eta_0+1)}{(\eta+1)(\eta_0-1)} + 2\tan^{-1}\eta - 2\tan^{-1}\eta_0}$$

$$\eta = \left(1-19.3\frac{z}{L}\right)^{\frac{1}{4}} \qquad \eta_0 = \left(1-19.3\frac{z_0}{L}\right)^{\frac{1}{4}} \tag{4.59}$$

and for stable stratification:

$$p = \frac{1+6\frac{z}{L}}{\ln\frac{z}{z_0} + 6\frac{z}{L}} \tag{4.60}$$

The numbers in both equations differ from those in the original reference according to the investigations by Högström (1988). This approach is for instance used in the footprint model by Kormann and Meixner (2001).

The application of this method is not without problems. In the morning daylight hours, these approaches are in a good agreement with measurements, but in the early afternoon – with the start of the thermal internal boundary layer (see Chap. 3.2.3) – the values of the stratification measured in the surface layer cannot be applied to the whole profile. An overall evaluation of the method is given in Table 4.15.

Table 4.15. Evaluation of exponential approaches for the determination of the wind profile in the lowest 100 m

criterion	evaluation
area of application	engineer–technical applications, continuously running programs
financial expense	1–3 k€ per system
personal expense	continuous technical support
education	experiences in measuring technique
error	according to the micrometeorological conditions 5–20% (unstable stratification), otherwise significant larger errors
sampling	1–5 s
time resolution of the gradients	10–30 min
restrictions in the application	only for unstable stratification sufficient accuracy, significant influences by internal boundary layers band thermal internal boundary layers at the afternoon for near surface measurements,
	turbulent conditions necessary

An interesting approach was given with the definition of the radix layer (Santoso and Stull 1998), which is approximately one fifth of the atmospheric boundary layer (surface layer and lower part of the upper layer). In the uniform layer above the radix layer, no increase of the wind velocity occurs and a constant wind velocity u_{RS} predominates:

$$\frac{u(z)}{u_{RS}} = \begin{cases} \left(\frac{z}{z_{RS}}\right)^A \exp\left[A\left(1 - \frac{z}{z_{RS}}\right)\right] & z \le z_{RS} \\ 1 & z > z_{RS} \end{cases} \qquad (4.61)$$

The application of this method is limited to convective conditions. Furthermore there are difficulties to determine the height z_{RS} of the radix layer.

4.3 Flux-Variance Methods

The flux-variance relation (variance method) according to Chap. 2.4 allows the calculation of fluxes using the measured variance of a meteorological parameter and the integral turbulence characteristics. Many investigations of this method have been made. However, the variance method has never reached a practical relevance even though the method has an accuracy comparable to the eddy-covariance method (Tsvang *et al.* 1985). Thus, the application of the variance method according to Eqs. (2.98) and (2.99) as well as Tables 2.11 and 2.12 is possible.

The equations can also be derived from the definition of the correlation coefficient or the covariance. Therefore, the method is often called eddy-correlation method, which should not be confused with the eddy-covariance method:

$$r_{wX} = \frac{\overline{w' X'}}{\sigma_w \sigma_X} = \frac{F_X}{\sigma_w \sigma_X} \qquad (4.62)$$

$$r_{wY} = \frac{\overline{w' Y'}}{\sigma_w \sigma_Y} = \frac{F_Y}{\sigma_w \sigma_Y} \qquad (4.63)$$

By using assumed values of correlation coefficients (Table 4.16), well-known standard deviations, and one well-known flux, a second flux can be determined. The sign of the flux must be determined by additional measurements, *e.g.* by measurements of the temperature gradient:

$$|F_X| = |F_Y| \frac{\sigma_X}{\sigma_Y} \qquad (4.64)$$

The standard deviations can be determined only for a low-frequency spectral range, which differs from the necessary spectral range of the eddy-covariance method for the reference flux. This method can be applied for example, only if one of the two gas fluxes can be measured with high time resolution using the eddy-covariance method.

Furthermore the flux can also be determined according to Eq. (4.65), if the correlation coefficient is well known:

$$F_X = r_{wX}\, \sigma_w\, \sigma_X \tag{4.65}$$

The correlation coefficient can be roughly parameterized according to the stratification (Table 4.16)

$$r_{wX} = \tilde{\varphi}(\varsigma). \tag{4.66}$$

But the table gives no sufficient data basis.

An evaluation of the practically rarely used flux-variance method is given in Table 4.17.

Table 4.16. Typical values of the correlation coefficient

author	r_{uw}	r_{wT}
Kaimal and. Finnigan (1994)	−0.35	0.5 (unstable)
		−0.4 (stabile)
Arya (2001)	−0.15	0.6 (unstable)

Table 4.17. Evaluation of the flux-variance method

criterion	evaluation
area of application	basic research
financial expense	2–10 k€ per system
personal expense	continuous scientific and technical support
education	good micrometeorological and measuring technique knowledge
error	depending on the micrometeorological conditions 10–30%
sampling	10–20 Hz (probably lower)
time resolution of fluxes	10–30 min
application for chemical compounds	selected inert gases (gas analyzers with high time resolution)
restrictions in the application	sufficient footprint area,
	turbulent conditions necessary,
	if necessary similarity of the scalars

4.4 Accumulation Methods

4.4.1 Eddy-Accumulations-Method (EA)

The basic idea of the eddy-accumulation method (conditional sampling) originates in the work of Desjardins beginning in 1972 (Desjardins 1977). He assumed that the covariance of the turbulent flux could be averaged separately for positive and negative vertical wind velocities:

$$\overline{w'c'} = \overline{w^+c} + \overline{w^-c} = \left(\overline{w^+} + \overline{w^-}\right)\overline{c} + \overline{w^+c'} + \overline{w^-c'}$$

$$\left(\overline{w^+} + \overline{w^-}\right) = \overline{w} = 0$$

(4.67)

Realization of this direct measuring technique should be done by concentration measurements in two separate reservoirs for positive and negative vertical wind velocities weighted with the actual vertical wind velocity (Fig. 4.10). However, the therefore necessary valve control technology did not yet exist.

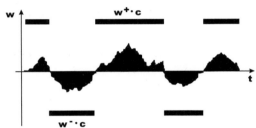

Fig. 4.10. Schematical view of the eddy-accumulation method (Foken *et al.* 1995)

4.4.2 Relaxed Eddy-Accumulation Method (REA)

The accumulation method comes to the fore by the work of Businger und Oncley (1990). They combined the eddy-accumulation method (EA) according to Eq. (4.67) with the flux-variance similarity according to Eq. (2.97). Their method becomes an indirect method:

$$\overline{w'c'} = b\sigma_w\left(\overline{c^+} - \overline{c^-}\right)$$

(4.68)

where the coefficient $b = 0.627$ for an ideal Gaussian frequency distribution (Wyngaard and Moeng 1992), otherwise low variations occur which are also probably different for different gases (Businger and Oncley 1990; Oncley *et al.* 1993; Pattey *et al.* 1993):

$$b = \frac{r_{wc}\sigma_c}{\left(c^+ - c^-\right)} = 0.6 \pm 0.06 \tag{4.69}$$

The coefficient b is to a large extent independent of the stratification. This is probably the opposite stability dependency of the integral turbulence characteristics for the vertical wind and matter (Foken et al. 1995). The relaxed eddy-accumulation (REA) method is schematically shown in Fig. 4.11. The weighting of the concentrations is no longer necessary, but the high switching frequency of the valves for zero passages of the vertical wind is still necessary.

A further improvement was realized with the modified relaxed eddy-accumulation method (MREA) according to Businger and Oncley (1990), which is of practical use and generally called REA. In this method, the air in the case of positive and negative maximal values of the vertical wind velocity are collected into two separate reservoirs; for fluctuations around zero the air is rejected or collected in a control volume (Fig. 4.12).

Equations (4.68) and (4.69) get the following modifications

$$\overline{w'c'} = b\sigma_w \left(c^+(w > w_0) - c^-(w < -w_0)\right) \tag{4.70}$$

with

$$\frac{b\left(w_0/\sigma_w\right)}{b(0)} = e^{-\frac{3}{4}\frac{w_0}{\sigma_w}} \pm 0.012. \tag{4.71}$$

The threshold value w_0 is dependent on the experimental conditions and the gas. This requires that with a parallel simulation using a proxy parameter, e.g. the temperature or the water vapor, b must be actually determined:

$$b = \frac{\overline{w'c_{proxy}'}}{\sigma_w\left(c_{proxy}^+(w > w_0) - c_{proxy}^-(w < -w_0)\right)} \tag{4.72}$$

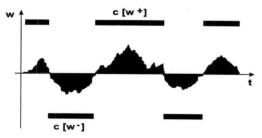

Fig. 4.11. Schematical view of the relaxed eddy-accumulation method (Foken et al. 1995)

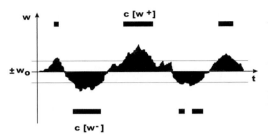

Fig. 4.12. Schematical view of the modified relaxed eddy-accumulation method (Foken *et al.* 1995)

Simulation experiments have shown (Ruppert *et al.* 2006b) that the optimal value for *b* is approximately 0.6. The accuracy of the method is up to one order lower if a constant *b*-value is used with respect to the determination with a proxy value. It must be taken into consideration that only for small eddies are scalars similar (Pearson jr. *et al.* 1998). This is not the case for the application of larger eddies (Ruppert *et al.* 2006b). This so-called scalar similarity can also change during the daily cycle (see also Chap. 3.5.5). Therefore the choice of the proxy scalar must be made very carefully.

For the implementation of the method, the measuring system must be adapted to the stream field by a coordinate transformation, and the planar-fit method (Wilczak *et al.* 2001) is recommended (see Chap. 4.1.2). Furthermore attention must be paid so that the integral turbulence characteristics of the vertical wind velocity, which must follow the known dependencies, are not specifically modified by the measuring place. An overall evaluation is given in Table 4.18.

Often, the accuracy of gas analyzers is not high enough to measure sufficient concentration differences between both reservoirs. This can be overcome by the hyperbolic relaxed eddy-accumulation method (HREA), which is based on an idea by von Shaw (1985). Bowling *et al.* (1999) and Wichura *et al.* (2000) developed this method for practical use for carbon dioxide isotope fluxes. In this method, only air is collected outside a hyperbolic curve, which must be defined for the special measuring case (Fig. 4.13)

$$\left| \frac{\overline{w'\,c'}}{\sigma_w\,\sigma_c} \right| > D \text{ for } w > 0, \tag{4.73}$$

$$\left| \frac{\overline{w'\,c'}}{\sigma_w\,\sigma_c} \right| > D \text{ for } w < 0. \tag{4.74}$$

The hyperbolic relaxed eddy-accumulation method needs a considerable expense to determine exactly the dead band. The value calculated by Bowling *et al.* (1999) $D = 1.1$ is probably too large so that only extreme events are collected which are not adequately distributed among the quadrants. A simulation study by Ruppert *et al.* (2006b) showed that an optimal value is $D \sim 0.8$. Very important is the correct choice of the proxy scalar and the control of the scalar similarity of the measuring and proxy parameters. Deviations have a much larger effect than on the simple REA method.

The benefit of the hyperbolic relaxed eddy-accumulation method is the significant increase of the concentration differences in both reservoirs, which may be compensated by the high efforts to control turbulent similarity relations. An overall evaluation is given in Table 4.19.

The relaxed eddy-accumulation method is still under development. Further research is desirable because the method opens perspectives for matter fluxes that cannot be measured with other methods. This indirect method can be successfully used only if the underlying flux-variance similarity is fulfilled. Therefore methods are also necessary to control these basics.

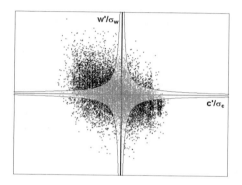

Fig. 4.13. Schematical view of the hyperbolic relaxed eddy-accumulation method (Ruppert *et al.* 2002)

Table 4.18. Evaluation of the relaxed eddy accumulation method (REA)

criterion	evaluation
area of application	basic research and extravagant continuous measuring programs
financial expense	10–50 k€ per system
personal expense	continuous scientific and technical support
education	good micrometeorological and measuring technique and probably also chemical knowledge
error	depending on the micrometeorological conditions 5–20%
sampling	10–20 Hz
time resolution of fluxes	30–60 min
application for chemical compounds	selected inert gases (gas analyzers with high time resolution)
restrictions in the application	sufficient footprint area,
	turbulent conditions necessary,
	similarity of turbulent scales of the scalars,
	no local influences on integral turbulence characteristics

Table 4.19. Evaluation of the hyperbolic relaxed eddy accumulation method (HREA)

criterion	evaluation
area of application	basic research
financial expense	10–50 k€ per system
personal expense	intensive scientific and technical support
education	good micrometeorological and measuring technique and probably also chemical knowledge
error	depending on the micrometeorological conditions 5–20%
sampling	10–20 Hz
time resolution of fluxes	30–60 min
application for chemical compounds	selected inert gases (gas analyzers with high time resolution)
restrictions in the application	sufficient footprint area,
	turbulent conditions necessary,
	similarity of turbulent scales of the scalars,
	no local influences on integral turbulence characteristics

4.4.3 Disjunct Eddy-Covariance Method (DEC)

Up to now, only methods have been discussed which can be used for inert gases or gases that do not react during their stay in the reservoirs. Gas analyzers with sampling rate about 10–20 Hz are necessary for the measurement of turbulent fluxes with the eddy-covariance method. Such instruments are available for only a few gases such as ozone. The basic idea of the disjunct eddy-covariance method borrows from the eddy-covariance method for aircraft measurements and is therefore a direct measuring method. Due to the velocity of the aircraft and the nearly identical sampling frequencies as for surface measurements, there is no optimal adaptation of the sampling frequency on the resolvable turbulent eddies according to the sampling theorem (see Chap. 6.1.2). According to investigations by Lenschow *et al.* (1994), it is possible to estimate fluxes for a fully-developed turbulent regime even when the sampling frequency is low in comparison to the eddy size. This means a larger separation in time (*disjunct*) of the single samples.

This is the benefit of the disjunct eddy-covariance method where samples are taken only in certain time intervals (Rinne *et al.* 2000). Although the direct sampling is taken over a time interval < 0.1 s, the gas analyzer can process the data for several second due to its high inertia if no remarkable reactions occur during this time.

To reduce the error < 10% relative to the eddy-covariance method, based on simulations the time difference between two samplings should be in the range of 1–5 s (Ruppert *et al.* 2002). The method is still in development, and it is anticipated that it will be an applicable method for fluxes of reactive gases, particles, *etc.* An initial evaluation of the method is given in Table 4.20.

Table 4.20. Evaluation of the disjunct eddy-covariance method (DEC)

criterion	evaluation
area of application	In the moment only basic research
financial expense	10–20 k€ per system
personal expense	Intensive continuous scientific and technical support
education	good micrometeorological, measuring technique and chemical knowledge
error	depending on the micrometeorological conditions 5–20%
sampling	1–10 s, sampling duration < 0.1 s
time resolution of fluxes	30–60 min
application for chemical compounds	selected inert gases (gas analyzers with time resolution < 10 s)
restrictions in the application	sufficient footprint area,
	turbulent conditions necessary,
	influences of turbulent scales up to now not investigated

4.4.4 Surface Renewal Method

The surface renewal method (Paw U *et al.* 1995) is based on the concept of eddy accumulation (conditional sampling). Turbulence measurement are not used, instead concentration measurements of ramp structures (see Chap. 3.5.4), which occur over the vegetation, are used. Because these structures *renewal* permanent the flux can be defined with the emptying of the storage. The derivation was made using the sensible heat flux but it can also be applied for matter fluxes. The sensible heat flux is defined as the temperature change in time in a volume V above an area A, where V/A can be used as canopy height z_B:

$$Q_H = \rho \, c_p \, \frac{dT}{dt} \left(\frac{V}{A} \right) \tag{4.75}$$

A practical equation was developed by Snyder *et al.* (1996) based ramp structures such as those shown in Fig. 4.14 for stable and unstable stratification. Accordingly, the temperature change in time dT/dt in Eq. (4.75) can be substituted by a/l and is multiplied with the relative duration of the heating or cooling $l/(l+s)$:

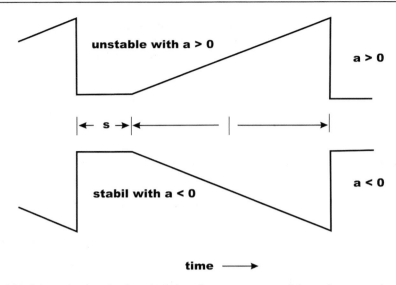

Fig. 4.14. Schematic view fort he calculation of ramp structures of the surface renewal method (Snyder *et al.* 1996)

Table 4.21. Evaluation of the surface renewal method

criterion	evaluation
area of application	basic research and extravagant continuous measuring programs for special cases
financial expense	5–10 k€ per system
personal expense	intensive continuous scientific and technical support
education	good micrometeorological and measuring technique and probably also chemical knowledge
error	depending on the micrometeorological conditions 10–30%
sampling	10–20 Hz
time resolution of fluxes	30–60 min
application for chemical compounds	selected inert gases (gas analyzers with high time resolution)
restrictions in the application	only over vegetation with ramp structures,
	influences of turbulent scales up to now not investigated

$$Q_H = \rho\, c_p \frac{a}{l+s} z_B \tag{4.76}$$

The method needs careful analysis of the turbulence structure and promises possibilities for the application of fluxes above high vegetation, especially for predominant stabile stratification with significant ramp structures. An initial evaluation is given in Table 4.21.

4.5 Fluxes of Chemical Substances

The deposition of chemical substances occurs in three different ways (Finlayson-Pitts and Pitts 2000; Foken *et al.* 1995):
− wet deposition of solute gases and substances in rain water,
− moist deposition of solute gases and substances in fog water,
− dry deposition by turbulent transport of gases and particles (aerosols).

Only the dry deposition can be measured with the previously described methods of energy and matter exchange. The wet deposition is a precipitation measurement made with so-called wet-only collectors, which open only during precipitation to avoid the collection of dry deposition and sedimentation. The problems with this measurement are similar to those for precipitation measurement (see Chap. 6.2.4). The moist deposition is important only in areas with frequent fog conditions such as hilly regions (Wrzesinsky and Klemm 2000).

In ecological studies, bulk deposition is measured with open collectors located in the trunk space of a forest. These measurements are the sums of the wet deposition, the sedimentation, the wash up of deposited substances from leafs, and partly dry deposition. The crown-eaves method attempts to compare the bulk deposition with the wet deposition outside the forest, and to estimate the dry deposition as a difference term (Guderian 2000). This method cannot measure the dry deposition quantitative, because substances are also directly absorbed by the plant surfaces or deposited in the understory or the soil.

Table 4.22. Ratio of the dry deposition (DD) to the wet deposition (WD) resp. dry particle deposition (DPD) over rural areas (Foken *et al.* 1995)

matter	high roughness (forest)		low roughness (meadow)	
SO_2/SO_4^-	3–4 : 1	DD/WD	1–1.5 : 1	DD/WD
	1:1	DD/DPD	3–10 : 1	DD/DPD
$NO_2 + HNO_3$	1.8–4 : 1	DD/WD	1–2 : 1	DD/WD
NO_3^-	1.2–2 : 1	DD/DPD	2–10 : 1	DD/DPD
NH_3/NH_4^+	0.2–5 : 1	DD/WD	?	DD/WD
	0.2–5 : 1	DD/DPD	1 : 1	DD/DPD
metal	~ 1 : 1	DD/WD	1 : 20	DD/WD

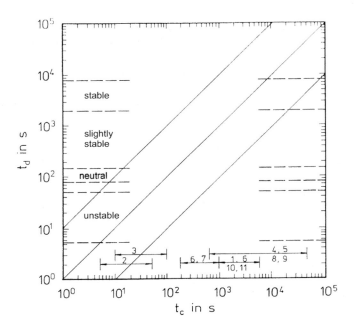

Fig. 4.15. Relation between the characteristic transport time t_d and characteristic reaction time t_c for different chemical reactions for a layer of 10 m thickness and different thermal stratifications according to Dlugi (1993), 1: $HO_2 + HO_2 \rightarrow H_2O_2 + O_2$, 2: $HNO_3 + NH_3 \leftrightarrow NH_4NO_3$, 3: $O_3 + NO \rightarrow NO_2 + O_2$, 4: $O_3 + $ isoprene $\rightarrow$ reaction products (R), 5: $O_3 + $ monoisoprene $\rightarrow$ R, 6: $NO_3 + $ monoisoprene $\rightarrow$ R, 7: $NO_3 + $ isoprene $\rightarrow$ R, 8: $OH + $ isoprene $\rightarrow$ R, 9: $OH + $ monoisoprene $\rightarrow$ R, 10: $O_3 + $ olefins $\rightarrow$ R, 11: $O_3 + NO_2 \rightarrow NO_3 + O_2$

The distribution of the total deposition of the three deposition paths depends significantly on the properties of the surface. High vegetation with a large surface roughness is much better able to absorb matter by dry deposition from the air (comb out) than by wet or moist deposition. Depending on the substances, dry deposition makes a contribution to the total deposition of about ¾ above high vegetation and ⅔ above low vegetation (Foken *et al.* 1995). The substance-dependent differences are shown in Table 4.22.

The dry deposition can be determined, in principle, with all the methods described thus far. One of the difficulties is the limited number of gases for which high-frequency gas analyzers are available for the eddy-covariance method. For profile measurements, the absolute accuracy of the gas analyzers is often not good enough for the necessary resolution or else the sampling times are larger than the typical meteorological averaging times. Often, modified Bowen-ratio methods, flux-variance similarities, or accumulation methods are utilized. All these methods are not suitable for continuously running measuring programs. For recording of the dry deposition estimates, the so-called deposition velocity is used. For climate research worldwide, carbon dioxide exchange measurements with the eddy-covariance method are used (Valentini *et al.* 2000). Therefore, extensive technical guides are available (Aubinet *et al.* 2000; Aubinet *et al.* 2003a; Moncrieff *et al.* 1997). The existence worldwide of more than 200 measuring stations (Baldocchi *et al.* 2001) does not ignore the fact that technical and methodical problems still exist.

The previous remarks are only valid for inert gases as carbon dioxide or sulfur dioxide. For reactive gases, the methods can be used only if one gas is significantly in surplus and reactions do not play an important role, for example ozone deposition during daytime in a rural area. It is also possible to combine complexes of matter, for instance the NO_x-triade consisting of nitrogen monoxide, nitrogen dioxide, and ozone, or the ammonium-tirade consisting of ammoniac, ammonium ions and ammonium nitrate. With the application of the profile method, it must be taken into account that due to reactions the gradient do not always present the direction of the flux (Kramm *et al.* 1996a). For the application of the eddy-covariance and eddy-accumulation methods, it must be considered that the reaction times of typical chemical reactions are in the order of 10^1–10^4 s and therefore just in the rang of the turbulence fluctuations (Fig. 4.15). Therefore, these methods more or less cannot be used except for multipoint measurements under the assumption of isotropy turbulence (Foken *et al.* 1995).

A measure for the ratio of the transport time to the reaction time is the Damköhler number (Molemaker and Vilà-Guerau de Arellano 1998)

$$Da_t = \frac{t_d}{t_c} k \langle c * \rangle, \tag{4.77}$$

where k is the kinematical reaction constant and $\langle c * \rangle$ is a dimensionless volume-averaged concentration of one of the reaction partners in the equilibrium. Other definition scale with the emission flux (Schumann 1989). Because the reactions often occur within the smallest eddies, the Kolmogorov-Damköhler number is applied

$$Da_k = \sqrt{\frac{v}{\varepsilon}} \, k \langle c * \rangle \tag{4.78}$$

with the kinematic viscosity, v, and the energy dissipation ε. An estimate of the dependence of chemical reactions on the Damköhler number was given by Bilger (1980):

$$Da_t \langle\langle 1 \qquad\qquad\qquad \text{slow chemistry}$$

$$Da_k < 1 < Da_t \qquad\qquad \text{moderate chemistry} \quad (4.79)$$

$$Da_k \rangle\rangle 1 \qquad\qquad\qquad \text{fast chemistry}$$

A simple and routinely applicable method, but physically unrealistic (see below), is based on the deposition velocity defined by Chamberlain (1961):

$$v_D(z) = -\frac{Q_c}{c(z)} \tag{4.80}$$

This definition is inconsistent with the gradient approach (Roth 1975). But the deposition velocity can be assumed to be a reciprocal transport resistance (see

Chap. 5.3), in which the concentration can be replaced by the concentration difference between the measuring height and a second reference height (in the ground), where the concentration is constant or very low. In this way, the physical incorrectness can be overcome. This assumption has the disadvantage that short time changes of the dry deposition (daily cycle or shorter) cannot be exactly reproduced. The method is therefore applicable for long-term measurements.

Deposition velocities are on the order of 10^{-3} m s^{-1} and are highly variable depending on the surface and the meteorological conditions (Table 4.23). For countrywide investigations, the deposition velocity listed by the relevant agencies is dependent on the time of the year in which the averaged dry deposition was measured.

The proportionality given in Eq. (4.80) between the flux and the concentration is generally fulfilled only for unstable and neutral stratification. Figure 4.16 shows that this proportionality can be realized only with a large scatter. For stable stratification, changes of the concentration are nearly independent of the flux.

Due to the non-physical definition of the deposition velocity, there are also problems with its experimental determination. The exact definition would be the transport or transfer velocity (Arya 1999; Roth 1975):

$$v_D(z) = -\frac{Q_c}{c(z) - c(0)} \qquad (4.81)$$

Only for the case $c(0) = 0$, $i.e.$ for matter which disappears nearly completely by reactions at the surface, the deposition velocity would be identical with the transfer velocity. Otherwise one obtains unrealistic values (Businger 1986). The determination of the transfer velocity can be made with all the methods described in this Chapter; however, the profile method (see Chap. 4.2) is the simplest method (Table 4.24).

The usual way of the determination of the deposition velocity is the resistance approach (Seinfeld and Pandis 1998), which is described in detail in Chap. 5.3. Thus, the calculation using simple models is possible (Hicks et $al.$ 1987).

Table 4.23. Examples for the deposition velocity (Helbig et $al.$ 1999)

gas	surface	v_D in 10^{-2} ms^{-1}	conditions
SO$_2$	grass	0.5	neutral stratification
	needle forest	0.3–0.6	averaged value
O$_3$	grass	0.55	neutral, 5 ms^{-1}
	spruce forest	0.4	averaged value
NO	grass	0.05	neutral, 5 ms^{-1}
	spruce forest	0.1–0.4	averaged value
NO$_2$	grass	0.6	neutral, 5 ms^{-1}
	spruce forest	1.2	in spring

Table 4.24. Evaluation of the determination of the dry deposition with the deposition velocity

criterion	evaluation
area of application	routinely measuring programs
financial expense	2–15 k€ per system
personal expense	technical support
education	knowledge in measuring technique
error	according to the micrometeorological conditions 20–50%, partly > 50%
sampling	1–5 s, determination of averages over 10–30 min
time resolution of fluxes	decade and monthly averages
application for chemical compounds	selected inert gases
restrictions in the application	only for rough estimates usable

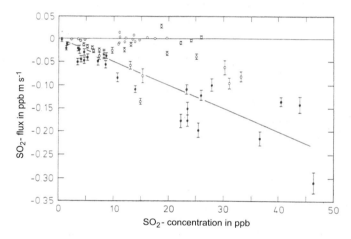

Fig. 4.16. Dependency of the sulfur dioxide deposition on the concentration under the assumption of a constant deposition velocity; filled symbols are values at noon time with unstable stratification, and open symbols values with stable stratification (Hicks and Matt 1988)

5 Modeling of the Energy and Matter Exchange

Within micrometeorology the term *modeling* is not defined especially well, and covers a range of complexity extending from simple regressions up to complicated numerical models. In applied meteorology (agro meteorology and hydro meteorology) simple analytical models are widely distributed over this range. An important issue is the modeling of evaporation. Sophisticated numerical models made nearly no entrance into the field. The following chapter describes different types of models beginning with simple analytical methods up to the consideration of energy and matter transport in the surface layer in numerical models for specific problems. The application of models in heterogeneous terrain and flux averaging is addressed in a special subchapter.

5.1 Energy Balance Methods

Energy exchange measurements and modeling form the bases for many applied investigations. Regardless of the related methodical problems of measurement (see Chap. 3.7), the energy balance equation is addressed using methods such as the Bowen-ratio method (Chap. 4.2.2). Many models in the field are based on similar theoretical backgrounds. Often, the general problems of application are unknown to the user. Most applications can be used only for hourly values at noon with unstable stratification or for daily and weekly averages (see below). Various models distinguish between potential evaporation from free water bodies or water saturated surfaces and actual evapotranspiration, which is the sum of the evaporation of the soil and transpiration from the plants.

   In most of the models, parameters are used which were empirically determined by experiments. Strictly speaking, only the climatological conditions of the experimental area are valid. The values of parameters can vary from place to place, and more importantly they are valid only for special mean climatological conditions. Thus, the parameters are functions of the place, the time of the year, and the weather or climate of the period when the experiment was done. They may not be any more valid than those determined recently or in the near future (Houghton *et al.* 2001). Therefore, hourly and daily values have only a low representativness. Depending on the sensitivity of the parameterization, decade or monthly averages have an acceptable accuracy. Therefore, it is necessary to associate the time scale of the model with the intended use and the geographically valid region.

5.1.1 Determination of the Potential Evaporation

Dalton Approach

The simplest way to determine the potential evaporation over open water is the Dalton approach, which is comparable to the bulk approach discussed in Chap. 4.2.1, and cannot be assigned to the energy balance approach. Instead of the Dalton number alone, simple correction functions are used which include the dependency of the wind velocity,

$$Q_E = f(u) \left[E(T_0) - e(z) \right],$$
$$f(u) = a + b\ u^c,$$

(5.1)

where $a = 0.16$; $b = 0.2$; $c = 0.5$ for lakes in Northern Germany (Richter 1977) (Table 5.1).

Table 5.1. Application possibilities of the Dalton approach

criterion	evaluation
defining quantity	potential evaporation of free water bodies
area of application	according to the validity of the area specific constants (DVWK 1996)
resolution of ten input parameters	10–60 min averages
representiveness of the results	(daily-), decade and monthly averages
error	20–40%

Turc Approach

Many approaches are based on radiation measurements, *e.g.* the evaporation calculation, where only the air temperature (t in °C) and the global radiation (in Wm^{-2}) are used as input parameters (Turc 1961):

$$Q_{E\text{-}TURK} = k\ (K\downarrow + 209)\frac{0.0933\,t}{t+15}$$

(5.2)

The method was developed for the Mediterranean Sea and for the application in Germany Eq. (5.2) needs a correction factor of about $k = 1.1$ (DVWK 1996) (Table 5.2).

Table 5.2. Applications of the Turc approach

criterion	evaluation
defining quantity	potential evaporation of free water bodies, possible for well saturated meadows
area of application	Mediterranean Sea, Germany (lowlands) with correction factor k=1.1
resolution of ten input parameters	10–60 min averages
representiveness of the results	decade and monthly averages
error	20–40%

Priestley-Taylor Approach

The Bowen ratio is the starting point for the derivation of several methods for the determination of sensible and latent heat flux. The Priestley-Taylor approach starts with Eq. (2.91), which can be written with the potential temperature and the dry adiabatic temperature gradient as:

$$Bo = \gamma \frac{\partial \bar{\theta} / \partial z}{\partial \bar{q} / \partial z} = \frac{\gamma \left[\left(\partial \bar{T} / \partial z \right) + \Gamma_d \right]}{\partial \bar{q} / \partial z} \tag{5.3}$$

With the temperature dependence of water vapor pressure for saturation according to the Clausius-Clapeyron's equation

$$\frac{d\,q_s}{dT} = s_c \left(\bar{T} \right) \tag{5.4}$$

it follows that

$$Bo = \frac{\gamma \left[\left(\partial \bar{T} / \partial z \right) + \Gamma_d \right]}{s_c \left(\partial \bar{T} / \partial z \right)} = \frac{\gamma}{s_c} + \frac{\gamma \, \Gamma_d}{s_c \left(\partial \bar{T} / \partial z \right)}. \tag{5.5}$$

For the further derivation, the second term on the right-hand-side of Eq. (5.5) will be ignored; however, this is valid only if the gradient in the surface layer is significantly greater than the dry adiabatic gradient of 0.0098 Km^{-1}. This is not possible in the neutral case, but the fluxes then are generally small. After introduction of the Priestley-Taylor coefficient $\alpha_{PT} \sim 1.25$ for water saturated surfaces and of the energy balance equation Eq. (1.1), follows the Priestley-Taylor approach (Priestley and Taylor 1972):

$$Q_H = \frac{\left[(1 - \alpha_{PT}) s_c + \gamma \right] \left(-Q_s^* - Q_G \right)}{s_c + \gamma} \tag{5.6}$$

$$Q_E = \alpha_{PT} \, s_c \, \frac{-Q_s^* - Q_G}{s_c + \gamma} \tag{5.7}$$

Typical values of the ratios $c_p/\lambda = \gamma$ and $de_s/dT = s_c$ are given in Table 5.3 and can be calculated approximately with the following relation: (Table 5.4)

$$\frac{s_c}{\gamma} = -0.40 + 1.042\, e^{0.0443 \cdot t} \tag{5.8}$$

Table 5.3. Values for the temperature dependent parameters γ und s_c based on the specific moisture (Stull 1988)

temperature in K	γ in K^{-1}	s_c in K^{-1}
270	0.00040	0.00022
280	0.00040	0.00042
290	0.00040	0.00078
300	0.00041	0.00132

Table 5.4. Applications of the Priestley-Taylor approach

criterion	evaluation
defining quantity	potential evaporation of free water bodies
area of application	universally
resolution of ten input parameters	10–60 min averages
representiveness of the results	(daily-), decade and monthly averages
error	10–20%

Penman Approach

A commonly used method for the determination of the potential evaporation is that proposed by Penman (1948). This method was developed for Southern England and underestimates the evaporation for arid regions. The derivation is based on the Dalton approach and the Bowen ratio, whereas the equation of the Priestley-Taylor type is an intermediate stage (DVWK 1996). The evaporation in mm d^{-1} is

$$Q_E = \frac{s_c\left(-Q_s^* - Q_G\right) + \gamma\, E_a}{s_c + \gamma}, \tag{5.9}$$

where the available energy must be used in mm d^{-1}. The conversion factor from mm d^{-1} to Wm^{-2} is 0.0347. The second term in the numerator of Eq. (5.9) is called the ventilation term E_a (also in mm d^{-1}) and contains the influence of turbulence according to the Dalton approach. It is significantly smaller than the first term and is often ignored in the simplified Penman approach (Arya 2001). The Priestley-Taylor approach follows when $\alpha_{PT} = 1.0$.

The ventilation term is a function of the wind velocity and the saturation deficit:

$$E_a = (E - e)(f_1 + f_2\, u) \qquad (5.10)$$

While one can use daily averages in Eq. (5.10), the use of 10–60 min averages is considerably more meaningful; however, in this case the units must be converted. Typical values for both wind factors f are given in Table 5.5. These values are valid for water surfaces, but they can also be used for well-saturated grass surfaces, which to a large degree is the actual evaporation. To include the effects of larger roughness, the ventilation term according to the approach by van Bavel (1986) can be applied in hPa m s^{-1} (Table 5.6):

$$E_a = \frac{314\,K}{T} \frac{u}{\left[\ln\left(\frac{z}{z_0}\right)\right]^2}(E - e) \qquad (5.11)$$

Table 5.5. Wind factors in the ventilation term of Eq. (5.10)

surface and reference	f_1 in mm d^{-1} hPa^{-1}	f_2 in mm d^{-1} hPa^{-1} m^{-1} s
original approach for water bodies (Hillel 1980)	0.131	0.141
small water bodies (DVWK 1996)	0.136	0.105
water bodies (Dommermuth and Trampf 1990)	0.0	0.182
grass surfaces (Schrödter 1985)	0.27	0.233

Table 5.6. Applications of the Penman approach

criterion	evaluation
defining quantity	potential evaporation of free water bodies
area of application	universally
resolution of ten input parameters	10–60 min averages
representiveness of the results	(daily-), decade and monthly averages
error	10–20%

Overall Evaluation of Approaches for the Determination of the Potential Evaporation

All approaches presented thus far are valid only for long averaging intervals. Less sophisticated approaches have only low accuracy for short averaging intervals (Table 5.7).

Even when the calculation must be done with 30–60 min averages, which is necessary due to the non-linearity of the approaches, no statements can be made for these short time periods.

Table 5.7. Methods for the determination of the potential evaporation of water bodies. The underlying grey scale corresponds to the accuracies given in the last line

minute				
hour				
day				
decade	Dalton approach	Turc approach	Priestley-Taylor approach	Penman approach
month				

very good	good	satisfactory	rough estimate	inadequately
5–10%	10–20%	20–40%	40–100%	> 100%

5.1.2 Determination of the Actual Evaporation

Empirical methods for the determination of evaporation are widely used, but are applicable only in the areas of their development. Therefore, the following chapter is written without such approaches. Thus, in Germany the commonly used methods developed by Haude (1955) and Sponagel (1980) or the modified Turc approach according Wendling *et al.* (1991) are described extensive in the German version of this book (Foken 2006b).

Penman-Monteith Approach

The transition from the Penman to the Penman Monteith approach (DeBruin and Holtslag 1982; Monteith 1965; Penman 1948) follows by consideration of non-saturated surfaces and cooling due to evaporation, which reduce the energy of the sensible heat flux. Including both aspects lead to the Penman-Monteith method for the determination of the actual evaporation (evapotranspiration)

$$Q_H = \frac{\gamma\left(-Q_s^* - Q_G\right) - F_w}{R_G \, s_c + \gamma}, \tag{5.12}$$

$$Q_E = \frac{R_G \, s_c \left(-Q_s^* - Q_G\right) + F_w}{R_G \, s_c + \gamma}, \tag{5.13}$$

with the so-called ventilation term

$$F_w = C_E \, \overline{u} \, \left(R_G - R_s\right) q_{sat}, \tag{5.14}$$

where R_G is the relative humidity of the surface; R_s is the relative humidity close to the surface, and q_{sat} is the specific humidity for saturation. Eq. (5.14) can also be formulated according the resistance concept (see Chap. 5.3) without the molecular-turbulent resistance:

$$F_w = \frac{q_{sat} - q_a}{r_a + r_c} \qquad (5.15)$$

In the simplest case, the canopy resistance, r_c, will be replaced by the stomatal resistance r_s. The stomatal resistance can be calculated from the stomatal resistance of a single leaf r_{si} and the leaf-area index (LAI, leaf surface of the upper side per area element of the underlying surface)

$$r_s = \frac{r_{si}}{LAI_{aktiv}}, \qquad (5.16)$$

where LAI_{aktiv} is the LAI of the active sunlight leafs. Generally, this is only the upper part of the canopy, and therefore is $LAI_{aktiv} = 0.5\ LAI$ (Allen et al. 2004). In the simplest case, the turbulent resistance is given (Stull 1988) as:

$$r_a = \frac{1}{C_E \overline{u}}. \qquad (5.17)$$

But usually r_a is calculated from Eqs. (2.61) and (2.63):

$$r_a = \frac{\ln\left(\dfrac{z-d}{z_0}\right) \ln\left(\dfrac{z-d}{z_{oq}}\right)}{u(z)} \qquad (5.18)$$

Table 5.8. Typical values of the LAI (Kaimal and Finnigan 1994) and the stomata resistance of single leafs (Garratt 1992)

surface	height in m	LAI in m^2 m^{-2}	r_{si} in s m^{-1}
seat (begun to grow)	0.05	0.5	
cereal	2	3.0	50–320
forest	12–20	1–4	120–2700

Table 5.9. Application of the Penman-Monteith approach

criterion	evaluation
defining quantity	actual evaporation
area of application	universally
resolution of ten input parameters	10–60 min averages
representiveness of the results	hourly and daily averages
error	10–40%

In the non-neutral case, universal functions can be used in Eq. (5.18). Typical values of the parameters are given in Table 5.8.

The Food and Agriculture Organization of the United Nations (FAO) has put much effort into the development of a uniform method to determine the evaporation, and recommended an applicable equation with a limited input data set (Allen et al. 2004)

$$Q_E = \frac{s_c \left(-Q_s^* - Q_G\right) + \rho\, c_p\, \dfrac{0{,}622}{p}\, \dfrac{E - e}{r_a}}{s_c + \gamma \left(1 + \dfrac{r_s}{r_a}\right)},$$ (5.19)

where r_s and r_a are given by Eqs. (5.16) and (5.18), respectively. The factor $0.622/p$ was included contrary to the original reference, and is necessary for consistency; the constants s_c and γ are used in the dimension K^{-1} and not as in the original reference, i.e. hPa K^{-1} (Table 5.9).

To compare worldwide evaporation rates and to use input parameters, which are available everywhere, the FAO has formulated a (grass) reference evaporation (Allen et al. 2004). This is based in principle on Eq. (5.19), but includes the estimated input parameters given in Table 5.10.

The Penman-Monteith approach is widely used in diverse applications, for example, in the atmospheric boundary conditions of many hydrological and ecological models. With satisfactory accuracy, it allows at least during the daytime, the calculation of hourly data, generally the determination of daily sums of the evaporation, and the sensible heat flux.

The available energy is the main forcing, but because the atmospheric turbulence and the control by the plants influence the ventilation term, the method is inaccurate if the turbulent conditions differ from an average stage. Therefore, this approach is often not used in meteorological models that have several layers in the surface layer (see Chap. 5.3).

Table 5.10. Fixing of the input parameters for the FAO-(grass)-reference evaporation (Allen et al. 2004)

parameter	value	remark
r_a	$d = 2/3\ z_B$; $z_0 = 0.123\ z_B$; $z_{0q} = 0.1\ z_0$ with $z_B = 0.12$ m and $z = 2$ m follows $r_a = 208 / u\ (2\text{m})$	it is $\kappa = 0.41$ applied
r_s	$LAI_{aktiv} = 0.5\ LAI$; $LAI = 24\ z_B$ with $r_{si} = 100$ s m^{-1} and $z_B = 0.12$ m follows $r_s = 70$ s m^{-1}	
$-Q_s^* - Q_G$	various simplifications possible with an albedo of 0.23	Allen et al. (2004)

Other approaches not discussed here require long averaging intervals and thus have low accuracy for short averaging periods. Water balance methods use the water balance equation Eq. (1.23) while runoff and precipitation are measuring parameters.

5.1.3 Determination from Routine Weather Observations

The equations previously presented are generally unsuitable to determine the energy exchange with routine available data. Holtslag and van Ulden (1983) developed a method to determine the sensible heat flux under application of the Priestley-Taylor approach. They included an advection factor $\beta = 20$ W m^{-2} according to DeBruin and Holtslag (1982), and varied α_{PT} with the soil moisture between 0.95 and 0.65; however, for summer conditions with good water supply $\alpha_{PT} = 1$ can be assumed. Eq. (5.6) then has the following form with temperature-dependent constants according to Table 5.3:

$$Q_H = \frac{[(1-\alpha_{PT})s_c + \gamma](-Q_s^* - Q_G)}{s_c + \gamma} + \beta \tag{5.20}$$

To estimate the available energy an empirical equations is used

$$(-Q_s^* - Q_G) = 0.9\frac{(1-\alpha)K\downarrow + c_1 T^6 - \sigma T^4 + c_2 N}{1 + c_3}, \tag{5.21}$$

where T is the air temperature; N is the cloud cover; $K\downarrow$ is the downward radiation; α is the surface albedo, and the constants $c_1 = 5.3 \cdot 10^{-13}$ Wm^{-2}K^{-6}, $c_2 = 60$ Wm^{-2} and $c_3 = 0.12$. The disadvantage of this method is that the cloud cover from routine weather observations is often not available. The application is limited to daylight hours with either neutral or unstable stratification and neither rain nor fog.

Göckede and Foken (2001) have tried to use, instead of the cloud cover, the often-measured global radiation as input parameter. They applied the possibility to parameterize the radiation fluxes from cloud observation according to Burridge and Gadd (1974), see Stull (1988), to determine a general formulation for the transmission in the atmosphere, see Eqs. (1.5) and (1.7). After that follows for the available energy

$$(-Q_s^* - Q_G) = 0{,}9\ K\downarrow\left(1 - \alpha - \frac{0.08\ K\ m\ s^{-1}}{K\downarrow_G}\right), \tag{5.22}$$

where $K\downarrow$ is the measured global radiation, and $K\downarrow_G$ is the global radiation near the surface which can be calculated from the extraterrestrial radiation and the angle of incidence:

$$K\downarrow_G = K\downarrow_{extr}\ (0.6 + 0.2\sin\Psi) \tag{5.23}$$

With Eq. (5.25) it is necessary only to calculate the angle of incidence for hourly data with astronomical relations (Appendix A4). The method can be applied as well on Eq. (5.20) and on the Penman-Monteith approach in Eq. (5.12). Possibilities for application are similar to those by the method of Holtslag and van Ulden (1983) as shown in Table 5.11.

Table 5.11. Application of the Holtslag-van-Ulden approach

criterion	Evaluation
defining quantity	sensible heat flux and actual evaporation
important input parameters	cloud cover (original method)
	global radiation (modified method),
area of application	universally
resolution of ten input parameters	10–60 min averages
representiveness of the results	hourly and daily averages
error	10–30%

5.2 Hydrodynamical Multilayer Models

The development of multilayer models began soon after the start of hydrodynamic investigations (see Chap. 1.3). In these models, the energy exchange in the molecular boundary layer, the viscous buffer layer, and the turbulent layer of the surface layer (Fig. 1.4) was separately parameterized according to the special exchange conditions. The exchange of sensible heat can be shown to be dependent on the temperature profile in dimensionless coordinates (Fig. 5.1) with the dimensionless height $z^+ = z\,u_*/v$ and the dimensionless temperature $T^+ = T/T_*$ (T_*: dynamical temperature) analogous to the wind profile with dimensionless velocity $u^+ = u/u_*$ (Csanady 2001; Landau and Lifschitz 1987; Schlichting and Gersten 2003). For the molecular boundary layer, $T^+ \sim z^+$, and for the laminar boundary layer $u^+ \sim z^+$. Above the viscous buffer layer, the flow is turbulent. Therefore, the typical logarithmic profile equations $T^+ \sim \ln z^+$ and $u^+ \sim \ln z^+$ are valid. The greatest problem for the parameterization is the formulation for the buffer layer, where empirical approaches must be applied. According to Fig. 5.1, the similarity of the profiles in the nature can be applied in hydrodynamic investigations (Foken 2002).

The hydrodynamic multilayer models based on bulk approaches, where instead of the bulk coefficients the so-called profile coefficient Γ is included, can be determined by integration over all layers

$$Q_H = -\Gamma\left[T(z) - T_0\right],\qquad(5.24)$$

$$\Gamma = \left(\int_0^z \frac{dz}{K_T + v_{Tt} + v_T}\right)^{-1},\qquad(5.25)$$

where K_T is the turbulent diffusion coefficient for heat, v_{Tt} is the molecular-turbulent diffusion coefficient in the buffer layer, and v_T is the molecular diffusion coefficient. The first integrations were done by Sverdrup (1937/38) and Montgomery (1940) using a single viscous sublayer consisting of the buffer layer and the molecular boundary layer. For this combined layer, a dimensionless height $\delta_{vT}^+ \approx 27.5$ assumed, and for the turbulent layer a logarithmic wind profile with roughness length z_0 was applied. For smooth surfaces an integration constant instead of the roughness length was used (von Kármán 1934).

An integral approach for all layers including the turbulent layer was presented by Reichard (1951), who parameterized the ratio of the diffusion coefficient and the kinematic viscosity

$$\frac{K_m}{v} = \kappa\left(z^+ - z_T^+ \tanh\frac{z^+}{z_T^+}\right). \tag{5.26}$$

This approach is in good agreement with experimental data as shown in Fig. 5.1, and can be used for the parameterization of exchange processes between the atmosphere and the surface (Kramm et al. 1996b).

In the 1960s and 1970s, several papers were published with an integration of the profile coefficient over all three layers (Bjutner 1974; Kitajgorodskij and Volkov 1965; Mangarella et al. 1972; 1973). These models were based on new hydrodynamic data sets and took into account the wavy structure of the water surface (Foken et al. 1978)

$$\delta_T^+ = 7.5\frac{u_*}{v}\left[2 + \sin(\zeta - \pi/2)\right], \tag{5.27}$$

where $\varsigma=0$ is for the windward and $\varsigma=\pi$ for the lee site valid.

From measurements of the dimensionless temperature profile near the sea surface, it was possible to determine the dimensionless temperature difference in the buffer layer as $\Delta T^+ \approx 4$ (Foken et al. 1978; Foken 1984); also compare with Fig. 5.1. Based on this, the profile coefficient by application of Eq. (5.27) with $\varsigma=0$ (also valid for low friction velocities $u_* < 0.23$ m s^{-1}) is:

$$\Gamma = \frac{\kappa\, u_*}{\left(\kappa\, \mathrm{Pr} - \frac{1}{6}\right)\delta_T^+ + 5 + \ln\dfrac{u_*\, z}{30\, v}} \tag{5.28}$$

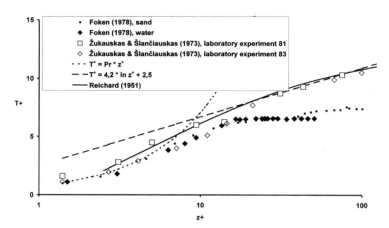

Fig. 5.1. Dimensionless temperature profile (T^+: dimensionless temperature, z^+: dimensionless height) according to laboratory measurements (Shukauskas and Schlantschiauskas 1973), and outdoor measurements (Foken 1978) and balanced profiles for the molecular layer (dotted line) and the turbulent layer (broken line) as well as profiles according to Reichardt (1951), from Foken (2002)

This model shows good results in comparison with experimental data (Foken 1984; 1986), and can be used for the calculation of the surface temperature for known sensible heat flux. However, such approaches found no practical application, and studies of the last 20 years are not available. The reason is not based in the approach but in the absolute different approaches for the energy exchanges in the surface layer in presently used models (Geernaert 1999), as it is shown in Chap. 5.5.

5.3 Resistance Approach

Recent models for the determination of the turbulent exchange are layer models. These models generally use the resistance approach for the energy and matter exchange between the atmosphere and the ground surface. They can be classified in three types:

One-layer-models consider only soil, plants and atmosphere at a close range. The plants are not separated into different layers. Instead, it is assumed that a big leaf covers the soil (*big leaf model*). Many of the so-called Soil-Vegetation-Atmosphere-Transfer (SVAT) models can be called big-leaf models, but sometimes they are also called multilayer models. These models consist manly of surface layer physics (partly several layers) and are schematically illustrated in Fig. 5.2 (Braden 1995; Hicks *et al.* 1987; Kramm *et al.* 1996a; Schädler *et al.* 1990; Sellers and Dorman 1987). A special case is the hybrid model according to Baldocchi *et al.* (1987).

Multilayer models simulate the atmosphere in several layers. The simplest models have no coupling with the atmospheric boundary layer, and only the surface layer is solved in detail. These models are available with simple closure (1st and 1.5th order), and higher order closure (Baldocchi 1988; Meyers and Paw U 1986; 1987).

Multilayer models with boundary layer coupling best represent the current state of model development. In these models, the lower layers deal with balance equations, and the upper layers use assumptions based on mixing length approaches (Fig. 5.3, see Chap. 2.1.3). These models are also available with simple closure (1st and 1.5th order) and higher order closure. The most widely used models are of 1st order closure, which use a local mixing length approach (Mix *et al.* 1994) or a non-local transilient approach (Inclan *et al.* 1996).

The resistance concept is based on the assumption that in the turbulent layer the turbulence resistance counteracts the turbulent flux, and in the viscous and molecular layer a molecular-turbulence resistance counteracts the flux, and in the plant and soil all resistances can be combined into a total resistance (canopy resistance). The canopy resistance can be divided into different transfer pathways, where the main transport paths are stomata–mesophyll, cuticula, or direct transfer to the soil, which are schematically illustrated in Fig. 5.4. The simplest picture is the comparison with Ohm's law:

$$I = \frac{U}{R} \tag{5.29}$$

Here, the flux is analogous with the current, I, and the gradient with the voltage, U. The resistance, R, can be described as a network of individual resistances (Fig. 5.4) in the following simple form:

$$r_g = r_a + r_{mt} + r_c \tag{5.30}$$

The consideration of the resistance concept in the profile equations (Eqs. 2.46 to 2.48) is illustrated in the example of the sensible heat flux:

$$Q_H = -K(z)\frac{\partial T}{\partial z} = -\frac{\int_0^z dT}{\int_0^z \frac{dz}{K(z)}} = -\frac{T(z) - T(0)}{\int_0^z \frac{dz}{K(z)}} \tag{5.31}$$

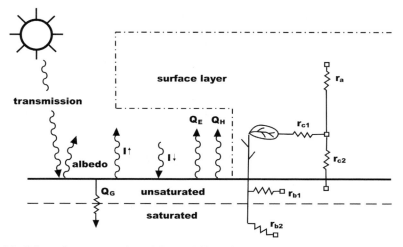

Fig. 5.2. Schematic representation of the modeling of the atmospheric surface layer including plants and soil (Blackadar 1997)

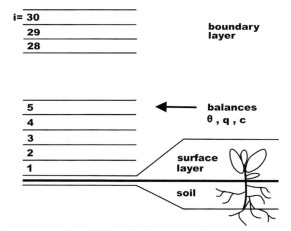

Fig. 5.3. Schematic representation of a boundary layer model (Blackadar 1997)

For the bulk approaches (Eqs. 4.30 to 4.32) follows

$$Q_H = -C_H\, u(z)[T(z) - T(0)] = -\Gamma_H\, [T(z) - T(0)], \qquad (5.32)$$

where the total resistance is:

$$r_{g(0,z)} = \int_0^z \frac{dz}{K(z)} \qquad (5.33)$$

$$r_{g(0,z)} = \frac{1}{\Gamma_{H(0,z)}} \qquad (5.34)$$

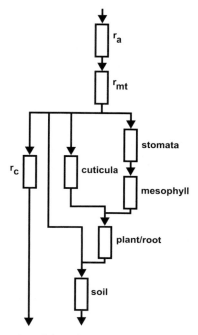

Fig. 5.4. Schematic representation of the resistance concept

The individual parts of the total resistance must be parameterized. For the turbulent resistance Eq. (2.83) under consideration of Eq. (3.7) is applied (Foken *et al.* 1995)

$$r_a = \int_{\delta}^{z_R} \frac{dz}{K(z)} = \frac{1}{\kappa u_*} \left[\ln \frac{z_R - d}{\delta - d} - \psi_H \left(\varsigma_{z_R}, \varsigma_\delta \right) \right], \tag{5.35}$$

where z_R is the reference level of the model, *i.e.* the upper layer of the model in the surface layer. The lower boundary, δ, is identical with the upper level of the molecular-turbulent layer. The resistance in the molecular-turbulent range is

$$r_{mt} = \int_{z_0}^{\delta} \frac{dz}{(D + K_H)} = (u_* B)^{-1} \tag{5.36}$$

with the so-called sublayer-Stanton number B (Kramm *et al.* 1996b; Kramm *et al.* 2002)

$$B^{-1} = Sc \int_0^{\eta} \frac{d\eta}{1 + Sc\, K_m/v}, \tag{5.37}$$

$$\eta = u_* (z - z_0)/v, \tag{5.38}$$

and the Schmidt number in the case of the exchange of gases including water vapor

$$Sc = \frac{v}{D}.$$

(5.39)

For the exchange of sensible heat, the Schmidt number in Eq. (5.37) is replaced by the Prandtl number.

Instead to a reasonable parameterization with multilayer models (see Chap. 5.2) presently the usual parameterization is done with the roughness length (Jacobson 2005)

$$r_{mt} = \ln\left(\frac{z_0}{z_{0q}}\right) \frac{\left(Sc/Pr\right)^{2/3}}{\kappa u_*},$$

(5.40)

with $Pr = 0.71$ and $Sc = 0.60$ valid in the temperature range of 0–40 °C. For the sensible heat flux follows for the sublayer -Stanton number according to Eqs. (5.36) and (5.40), (Owen and Thomson 1963):

$$\kappa B^{-1} = \ln\frac{z_0}{z_{0T}}$$

(5.41)

This equation is defined only for $z_0 > z_{0T}$, otherwise negative molecular-turbulent resistances would be calculated which are non-physical (Kramm et al. 1996b; Kramm et al. 2002). Nevertheless, in the literature negative values can be found (Brutsaert 1982; Garratt 1992), which can compensate for too large resistances in the turbulent layer and in the canopy, but in reality these negative values are due to inaccurate concepts of roughness lengths for scalars. For κB^{-1} values of 2–4 are typical. The application of the Reichardt (1951) approach would lead to a value of 4 (Kramm and Foken 1998).

The canopy resistance is often approximated as a stomatal resistance:

$$r_c \approx r_{st} = \frac{r_{st,min}\left(1+\frac{b_{st}}{PAR}\right)}{g_\delta(\delta_e)\, g_\Psi(\Psi)\, g_T(T_f)\, g_C(c_{CO_2})\, g_D}$$

(5.42)

With this the parameterization, the minimal stomatal resistance (Table 5.8) and the photosynthetically active radiation (PAR) with empirical constant b_{st} are used. The correction functions in the denominator of Eq. (5.44) have values from 0 to 1, and include the saturation deficit between the atmosphere and leafs (δ_e), the water stress (Ψ), the leaf temperature (T_f), and the local carbon dioxide concentration (c_{CO_2}). Furthermore, g_D is a correction factor for the molecular diffusivity of different gases. The parameterizations are very complicated, and require separate models (Blümel 1998; Falge et al. 1997; Jacobson 2005; Müller 1999). Therefore further literature should be consulted.

5.4 Modeling of Water Surfaces

The modeling of energy and matter exchange over water surfaces is generally simpler than over land surfaces (Csanady 2001; Geernaert 1999; Smith et al. 1996). However, methods for high wind velocities (> 20 m s^{-1}) are very inaccurate. The usual approaches are comparable with bulk approaches (see Chap. 4.2.1). Thus, Eqs. (4.30)–(4.32) can be directly used for the open ocean where water and air temperatures became more similar, and a nearly-neutral stratification occurs. Otherwise it is recommended to use the profile equations with universal functions. Furthermore, the influence of surface waves should be included with the roughness-Reynolds number (Rutgersson and Sullivan 2005), see Eq. 3.4. A well-verified approach was given by Panin (1985), and accordingly Eqs. (4.31) and (4.32) are multiplied by the following factors respectively:

$$\cdot \begin{cases} (1 - z/L)\left[1 + 10^{-2}(z_0\, u_* / \nu)^{3/4}\right] & z/L < 0 \\ \left[1/(1 + 3.5\, z/L)\right]\left[1 + 10^{-2}(z_0\, u_* / \nu)^{3/4}\right] & z/L < 0 \end{cases} \tag{5.43}$$

The Stanton and Dalton numbers are those for neutral stratification.

The approaches used in hydrodynamics can be applied in an analogues way over water bodies. For example Eq. (5.28) can be inserted in Eqs. (4.31) instead of the product $u\, C_H$ and in Eq.(4.32) instead of the product $u\, C_E$.

These approaches fail for shallow water bodies. For shallow lakes the fluxes can be calculated from the temperature regime (Jacobs et al. 1998). The above given approaches are usable under the assumption that over flat water the exchange is increased by steep waves and better mixing of the water body. According to Panin et al. (1996a) a correction function dependent on the water depth and the wave height should be included to determine fluxes in shallow water areas (depth lower than 20 m)

$$Q_H{}^{FIW} \approx Q_H(1 + 2h/H),$$
$$Q_E{}^{FIW} \approx Q_E(1 + 2h/H), \tag{5.44}$$

where H is the water depth and h the wave height, which can be calculated using

$$h \approx \frac{0.07\, u_{10}{}^2 \left(gH / u_{10}{}^2\right)^{3/5}}{g}, \tag{5.45}$$

where u_{10} is the wind velocity measured in 10 m (Davidan et al. 1985). This approach is also well verified for German lakes (Panin et al. 2006).

5.5 Modeling in Large-Scale Models

The modeling of the momentum, energy, and matter exchange in global circulation models is very simple in comparison to resistance models (Beljaars and Viterbo 1998; Brutsaert 1982; Jacobson 2005; Zilitinkevich et al. 2002). The limited computer time does not allow the application of complicated and iterative methods. The determinations of the momentum and energy exchange use bulk approaches (see Chap. 4.2.1), which are calculated for the layer between the surface (index s) and the first model layer (index 1). The surface fluxes are then given by:

$$u_*^2 = C_m \left| \overrightarrow{u_1} \right|^2 \tag{5.46}$$

$$\left(\overline{w'\theta'} \right)_0 = C_h \left| \overrightarrow{u_1} \right| \left(\theta_s - \theta_1 \right) \tag{5.47}$$

$$\left(\overline{w'q'} \right)_0 = C_q \left| \overrightarrow{u_1} \right| \left(q_s - q_1 \right) \tag{5.48}$$

This assumes constant fluxes between the surface and the first model level (i.e. 30 m). In the case of stable stratification, this assumption is questionable. The transfer coefficients $C_{m,h,q}$ can be calculated according to an approach by Louis (1979) or in a modified form by Louis et al. (1982) from the coefficient for neutral stratification $C_{mn,hn,qn}$ and a correction factors that are functions of the stratification and the roughness of the underlying surface:

$$C_m = C_{mn} F_m \left(Ri_B, z_1/z_0 \right) \tag{5.49}$$

$$C_h = C_{hn} F_h \left(Ri_B, z_1/z_0, z_1/z_{0T} \right) \tag{5.50}$$

$$C_q = C_{qn} F_q \left(Ri_B, z_1/z_0, z_1/z_{0q} \right) \tag{5.51}$$

In the neutral case, the transfer coefficients are dependent only on the roughness length:

$$C_{mn} = \left(\frac{\kappa}{\ln \frac{z_1 + z_0}{z_0}} \right)^2 \tag{5.52}$$

C_{hn} and C_{qn} will be determined according to Eq. (4.31) with z_{0T} and Eq. (4.32) with z_{0q}.

The bulk-Richardson number, Ri_B is:

$$Ri_B = \frac{g}{\theta_v} \frac{\theta_{v1} - \theta_{vs}}{\left| \overrightarrow{u_1} \right|^2} \tag{5.53}$$

The original approaches for the correction functions based on a limited number of experimental data (Louis 1979; Louis et al. 1982) are

$$F_m = \left(\frac{1 + 2\,b\,Ri_B}{\sqrt{1 + d\,Ri_B}} \right)^{-1}, \tag{5.54}$$

$$F_h = \left(\frac{1 + 3\,b\,Ri_B}{\sqrt{1 + d\,Ri_B}} \right)^{-1}, \tag{5.55}$$

where the adaptation parameters $b = 5$ and $d = 5$. Although these methods have been questioned (Beljaars and Holtslag 1991), they remain in use. Some corrections regarding the stability functions (Högström 1988) are sometimes used. But the potential of modern micrometeorology is to a large extend not exhausted. An important criticism on the use of the Louis-(1979)-scheme is the application of roughness lengths for scalars. There physical sense is controversial, and they are nearly identical with the aerodynamic roughness length. Above the ocean, the roughness is determined according to either the Charnock equation or better yet to a combination approach (see Chap. 3.1.1 and Table 3.3). The roughness lengths for scalars are parameterized according to the Roll (1948) approach for smooth surfaces (Beljaars 1995):

$$z_{0T} = 0.40 \frac{v}{u_*}, \qquad z_{0q} = 0.62 \frac{v}{u_*} \tag{5.56}$$

For a better consideration of convective cases (Beljaars 1995) the wind vector can be enhanced by a gustiness component

$$\left| \overrightarrow{u_1} \right| = \left(u_1{}^2 + v_1{}^2 + \beta\,w_*{}^2 \right)^{\frac{1}{2}} \tag{5.57}$$

with $\beta = 1$. The Deardorff velocity scale w_* (Eq. 2.42) can be simplified with the use of a mixed layer height of $z_i = 1$ km. This approach is in good agreement with experimental data, and is a good representation of the moisture exchange.

The parameterization of the stable stratification is especially difficult, because these cannot be assumed to hold for the first model layer and the universal functions are dependent on external parameters (Handorf et al. 1999; Zilitinkevich and Mironov 1996). In the simplest case, modified correction functions Eqs. (5.54) and (5.55) can be assumed (Louis et al. 1982)

$$F_m = \frac{1}{1 + 2\,b\,Ri_B \left(1 + d\,Ri_B \right)^{-\frac{1}{2}}}, \tag{5.58}$$

$$F_h = \frac{1}{1 + 3\,b\,Ri_B \left(1 + d\,Ri_B \right)^{\frac{1}{2}}} \tag{5.59}$$

with $b = 5$ and $d = 5$.

A parameterization applying external parameters was presented by Zilitinke-vich and Calanca (2000)

$$F_m = \left(\frac{1 - \alpha_u \, Fi_0}{1 + \frac{C_u}{\ln z/z_0} \frac{z}{L}} \right)^2 , \tag{5.60}$$

$$F_h = \left(\frac{1 - \alpha_\theta \, Fi \frac{Fi_0^2}{Ri_B}}{1 + \frac{C_\theta}{\ln z/z_0} \frac{z}{L}} \right) , \tag{5.61}$$

where Fi is the inverse Froude-number and Fi_0 the inverse external Froude-number

$$Fi_0 = \frac{N \, z}{u} \tag{5.62}$$

where N is the Brunt-Väisälä frequency (Eq. 3.36), $C_u = \alpha_u \, \kappa/C_{uN}$, and $C_\theta = \alpha_\theta \, \alpha_0 \, \kappa/C_{\theta N}$. The first experimental assessments of the coefficients are given in Table 5.12.

Table 5.12. Constants of the parameterization according to Zilitinkevich and Calanca (2000) in Eqs. (5.60) and (5.61)

author	experiment	C_{uN}	$C_{\theta N}$
Zilitinkevich and Calanca (2000)	Greenland experiment (Ohmura *et al.* 1992)	0.2...0.5	
Zilitinkevich *et al.* (2002)	Greenland experiment (Ohmura *et al.* 1992)	0.3	0.3
Zilitinkevich *et al.* (2002)	Cabauw tower, The Netherlands	0.04...0.9	
Sodemann and Foken (2004)	FINTUREX, Antarctica (Foken 1996), *Golden days*	0.51±0.03	0.040± 0.001
Sodemann and Foken (2004)	FINTUREX, Antarctica (Foken 1996)	2.26±0.08	0.022± 0.002

5.6 Large-Eddy Simulation

Thus far the model approaches have been based mainly on mean relations and av-eraged input parameters. They do not allow a spectral-dependent view, where the effects of single eddies can be shown. The reasons for this are the significant diffi-culties in spectral modeling and the large-scale differences in atmospheric bound-ary layers. The spatial scale extends from the mixed layer height of about 10^3 m down to the Kolmogorov micro scale

$$\eta = \left(\frac{v^3}{\varepsilon} \right)^{1\!/\!4} \tag{5.63}$$

of about 10^{-3} m. The energy dissipation, ε, is identified with the energy input from the energy conserving scale $l \sim z_i$ and the relevant characteristic velocity:

$$\varepsilon = \frac{u^3}{l} \tag{5.64}$$

For the convective boundary layer, the energy dissipation is approximately 10^{-3} m^{2}s^{-3}. The turbulent eddies in the atmospheric boundary layer cover a range from kilometers to millimeters. Thus, a numerical solution of the Navier-Stokes equation would need 10^{18} grid points. Because the large eddies, which are easily resolvable in a numerical model, are responsible for the transports of momentum, heat and moisture, it is necessary to estimate the effects of the small dissipative eddies which are not easily resolvable. The simulation technique for large eddies (Large-Eddy-Simulation: LES) consist in the modeling of the important contributions of the turbulent flow and to parameterize integral effects of small eddies (Moeng 1998). For technical applications with low Reynolds-numbers, almost all eddy sizes can be include. This method is called Direct Numerical Simulation (DNS).

The basic equations for LES are the Navier-Stokes equations, where single terms must be transferred into volume averages

$$\widetilde{u}_i = \iiint (u_i \, G) dx \, dy \, dz, \tag{5.65}$$

where G is a filter function which filters out small eddies and regards only large eddies. The total contribution of the small eddies is taken into account in an additional term of the volume averaged Navier-Stokes equations and parameterized with a special model. The widely used approach is the parameterization according to Smagorinsky-Lilly (Lilly 1967; Smagorinsky 1963), where the diffusion coefficient is described in terms of the wind and temperature gradients. For small eddies in the inertial subrange, the $-5/3$ law is assumed so that the relevant constants can be determined (Moeng and Wyngaard 1989). If small-scale phenomena have an important influence, the application of LES modeling requires a lot of care, for example, near the surface or when including chemical reactions.

The LES modeling is currently only a research instrument, which is able to investigate simple situations with high resolution in space and time, but at the cost of large computation times. It has made remarkable contributions to the understanding of the atmospheric boundary layer. Beginning with the first simulations by (Deardorff 1972), LES has been mainly applied to the convective case (Schmidt and Schumann 1989; Schumann 1989). In most cases, the ground surface is assumed homogeneous or only simply structured. Recently, the stably stratified boundary layer has become a topic of investigations. Currently, LES is a rapidly developing research field with many publications (Garratt 1992; Kantha and Clayson 2000; Moeng 1998; Moeng et al. 2004; Raasch and Schröter 2001, and others).

5.7 Area Averaging

All methods to determine the turbulent momentum and energy fluxes are related to that surface above which the fluxes are measured. But in most cases, the problem is to determine for example the evapotranspiration within a catchment, or over a large agricultural area or even over entire landscapes. Area-averaged fluxes are necessary in numerical weather and climate forecast models as input or validation parameters. It is impossible to calculate them by a simple averaging of the input parameters because complicated non-linear relations could cause large errors. Nevertheless this method has recently been used for weather and climate models, which use simple parameterizations of the interaction of the atmosphere with the surface. An overview of different methods of area averaging is given in Table 5.13. In this table, the statistical dynamical approaches are not listed, and thus only rough approximations of the land use types are given.

Using the resistance concept in the form of Eq. (5.30), the total resistance of the area is a *parallel connection* of the total resistances of areas with different land use:

$$\frac{1}{r_g} = \frac{1}{r_{g1}} + \frac{1}{r_{g2}} + \frac{1}{r_{g3}} + \dots \tag{5.66}$$

Applying the method of parameter aggregation, it follows from Eq. (5.66) for an averaging of individual resistances as in Eq. (5.30):

Table 5.13. Methods of area averaging

averaging method	procedure	example/reference
parameter aggregation	averaging for example of the roughness length	$\overline{z_0} = \frac{1}{N} \sum_i^N z_{0i}$
	averaging of *effective* parameters	*i.e.* Troen and Peterson (1989), see Chap. 3.1.1
flux aggregation	averaging for example of the roughness length with Fourier analysis	Hasager and Jensen (1999), Hasager *et al.* (2003)
	mixing method for resistances	Mölders *et al.* (1996)
	flux determination for dominant areas	
	mosaic approach	Avissar and Pielke (1989), Mölders *et al.* (1996)
	- *tile* approach	
	- subgrid approach	

$$\frac{1}{r_g} = \frac{1}{\frac{1}{N}\sum_i r_{a_i}} + \frac{1}{\frac{1}{N}\sum_i r_{mt_i}} + \frac{1}{\frac{1}{N}\sum_i r_{c_i}} \tag{5.67}$$

It is immediately clear that Eq. (5.67) is physical incorrect. Nevertheless this version is extremely practicable because, for example, the mean resistance of the turbulent layer can be determined by averaging the roughness lengths of individual areas as is done in most of weather prediction and climate models. However, one must be aware that because of non-linear relation remarkable misinterpretations of the turbulent fluxes can occur (Stull and Santoso 2000).

In contrast, for the flux aggregation of each single area, the total resistance must be calculated, which means different boundary conditions for the different areas:

$$\frac{1}{r_g} = \sum_i \frac{1}{r_{a_i} + r_{mt_i} + r_{c_i}} \tag{5.68}$$

The more simple methods of flux averaging differ in the ways Eq. (5.68) is used for the individual areas.

5.7.1 Simple Area Averaging Methods

A very simple but still-used method is the calculation of fluxes for dominant areas. For each grid element of a numerical model, the dominate land use must be determined over which the fluxes are calculated. It is assumed that over all grid elements the different land uses are statistically balanced. Therefore, each grid element has only one land use type. The averaging within a grid element is a quasi parameter-averaging process because the individual estimations of the parameters of the grid elements are widely intuitive and therefore parameter averaged.

The *blending-height* concept (see Chap. 3.2.4) can also be used for area averaging. It is assumed that at a certain height above the ground (for example 50 m) the fluxes above the heterogeneities of the surface do not differ and can be presented as an averaged flux. The fluxes for this height can be parameterized using effective parameters. A typical case is the application of effective roughness length, where the friction velocities are averaged instead of the roughness lengths (Blyth 1995; Hasager and Jensen 1999; Hasager *et al.* 2003; Mahrt 1996; Schmid and Bünzli 1995a; 1995b; Taylor 1987). From Eq. (2.61), it follows that by averaging the friction velocities of the individual areas an effective roughness length is given by:

$$z_{0eff} = \frac{\overline{u_* \ln z_0}}{\overline{u_*}} \tag{5.69}$$

A more empirical averaging of roughness lengths as presented by Petersen and Troen (1990) for the European Wind Atlas (Table 3.2) can be classified as an early stage of the above given method. A good effective averaging is of increasing importance for many practical reasons, for example micrometeorological processes in the urban boundary layer (Grimmond *et al.* 1998).

The procedure of roughness averaging with effective roughness length is widely applied for the mixing method. Here only those resistances will be averaged which are obviously different for the different individual areas. Generally, the turbulent and the molecular-turbulent resistances are assumed, and only the canopy resistance is averaged according to the land use (Fig. 5.5). This method uses the fact that in most cases meteorological information is not available for different land uses within a grid element. However, for the determination of the turbulent and molecular-turbulent resistances a parameter averaging is used because these are often not parameterized for a specific underlying surface.

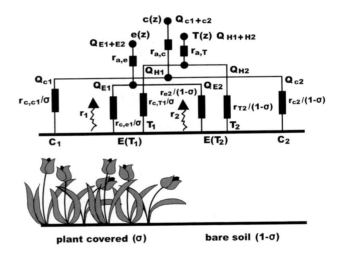

Fig. 5.5. Schematic representation of the mixing method (UBA 1996)

5.7.2 Complex Area-Averaging Methods

The mosaic approach belongs to the complex methods (Avissar and Pielke 1989; Mölders *et al.* 1996). This description is currently often used as a generic term for different applications. In the simplest case (*tile*-approach), for each grid cell contributions of similar land use types are combined and the parameterization of all resistances and the fluxes for each type will be separately calculated. The mean flux is the weighted average according to the contribution of the single land uses

(Fig. 5.6). This method is widely used for high-resolution models in space (100 m grid size), but it does not allow horizontal fluxes (advection) between the areas.

This disadvantage is overcome with the subgrid method (Fig. 5.7), where for each land use a small multi-layer model is used, which takes advection into account. For a certain height according to the blending-height concept, an average of the fluxes for a grid element is assumed. Such models correspond well with the reality, but they require very large computer capacity. Therefore, the subgrid method has been applied only for single process studies.

Model calculations with subgrid models and experiments (Klaassen *et al.* 2002; Panin *et al.* 1996b) show that fluxes above one surface are not independent of the neighborhood surfaces. Accordingly, these model studies for highly heterogeneous surfaces show an increase of the flux for the total area (Friedrich *et al.* 2000). According to numerical studies by von Schmid and Bünzli (1995b) this increase of the fluxes occurs on the lee side of boundaries between the single surfaces (Fig. 5.8).

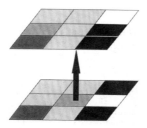

Fig. 5.6. Schematic representation of the mosaic approach (Mölders *et al.* 1996). The initial distribution of the surface structures will be combined according their contributions for further calculations

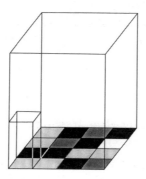

Fig. 5.7. Schematic representation of the subgrid method (Mölders *et al.* 1996). The surface structure will be used further on for selected model calculations

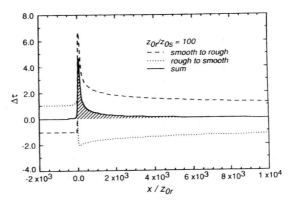

Fig. 5.8. Ratio of the friction velocities ($\Delta\tau=\tau_r/\tau_s$) above both surfaces for the flow over the roughness change. Immediately after the roughness change an increase of the friction velocity is observed. (Schmid and Bünzli 1995b)

5.7.3 Model Coupling

Averaging concepts are often used in the coupling of models. For model coupling, different approaches have been tested and are promising (Mölders 2001). The simplest versions are the direct data transfer and the one-way coupling, where the plant, soil, or hydrological model (SVAT and others) get their forcing from a meteorological model. In a two-way coupling, the SVAT model for example, gives its calculated fluxes back to the meteorological model. If sufficient computer time is available, then complete couplings are possible.

Instead of a coupling with effective parameters, a coupling with fluxes is preferred (Best *et al.* 2004), because effective parameters cannot adequately describe the high non-linearity of the fluxes. Instead of a coupling with area-averaged fluxes (Herzog *et al.* 2002), modules between the models should be included (Mölders 2001), which allow suitable coupling of the heterogeneous surface in different models, *e.g.* an averaging according to the mosaic or subgrid approach (Albertson and Parlange 1999; Mölders *et al.* 1996).

For model coupling, an unsolved problem is the usage sufficient grid structures. Meteorological models are based on rectangular grids, while land use models are based on polygons. A promising development is the use of adaptive grids (Behrens *et al.* 2005) that fit themselves to the respective surface conditions of each model, and have a high resolution in regions where the modeling is very critical or where the largest heterogeneities occur.

6 Measuring Technique

Because meteorological measurements are made mainly in the near surface layer they are located in the micrometeorological scale. The lack of literature in meteorological measuring technique is obvious, and after the classic by Kleinschmidt (1935) no further basic textbooks have been published; however, some overview textbooks are available (Brock and Richardson 2001; DeFelice 1998). For this reason, a special chapter is dedicated to micrometeorological measurements. Unlike other books with extensive descriptions of the measuring devices, only general principles of the micrometeorological measuring technique are described. Of special importance are techniques for the optimal adaptation of the sensors to the surrounding environment – the turbulent atmosphere. A special focus is taken on the quality assurance of measuring data.

6.1 Data Collection

Digital data collection systems have recently replaced analogue systems. Due to their inertia, analogue systems often required only simple filter functions; however, for digital recording systems these functions are not simple. This is a topic for the users but not the producers, who developed their loggers mostly for universal applications. For the development of micrometeorological measuring systems, the following basic considerations must be made.

6.1.1 Principles of Digital Data Collection

Modern data collection systems use data loggers that collect quasi-parallel signals and create a serial digital signal for transmission to a computer or a storage unit. Thus, each signal is sampled regularly in time, which is the sampling frequency. This sampling frequency must be adapted very precisely to the frequency of the measuring signal. For sensors with a high time resolution (*e.g.* sonic anemometers), the measuring sensor itself is no longer a low pass filter (LP). An additional low-pass filtering (see Chap. 6.1.1) must be provided so that at the input of the logger no frequencies occur that are greater than half of the sampling frequency. Often, the low pass filtering is done with the software of the sensor. Thus, the signal of the sensor is sampled with a frequency, which is much greater than necessary. The signal is then averaged over several samplings (over sampling). This low pass filtering has the benefit that often-present noise of 60 (50) Hz from the power supply has no influence on the input signal of the logger.

With a *multiplexer* (MUX), the measuring signals of the single sensors are sampled one after the other. Then, the sampling process starts again from the beginning. For turbulence measurements, covariances are often calculated between different logger channels, and the logger must be programmed in such a way that signals for covariance calculations are sampled at neighboring channels. It is often recommended that unused logger channels be grounded to prevent noise impulses from influencing the system. The logger submits the measuring signals one after the other over a *sample and hold circuit* (SH) to an analogue-to-digital converter (A/D), whose configuration influences the accuracy of the system. While often in the past only 11 or 12 bits were resolved, recent converters have mostly 16 bit resolution of the signal. For the configuration of the logging system, attention must be paid so that there is enough time available for converting the signals, which must be tuned with the sampling frequency. The high sampling frequencies required for turbulence measurements are close to the limits of the loggers. A basic circuit of a data sampling system is shown in Fig. 6.1.

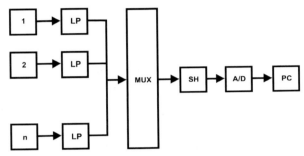

Fig. 6.1. Basic circuit of a data sampling system with 1 to n signals, low passes (LP), multiplexer (MUX), sample and hold circuit (SH), analogue-digital converter (A/D), all integrated in a modern logger and a data collection computer (PC) or storage medium

Supplement 6.1   Specification of amplification and damping in decibel

The amplification or damping of a measuring system is given by a logarithmic proportion (Bentley 2005):

$$X_{dB} = 20 \lg \frac{X_2}{X_1} \qquad (6.1)$$

This ratio is applied for amplifiers with the output voltage X_2 and the input voltage X_1 respectively for the damping of filters. The measuring unit is the logarithmic proportion expressed in decibel (dB). For power (*e.g.* product of voltage and current) one uses:

$$P_{dB} = 10 \lg \frac{P_2}{P_1} \qquad (6.2)$$

This measure is also used to express the noise level. If the signal differs from the noise level by factor 10^6, then the distance to the noise level is 60 dB.

The analogue-to-digital converter is responsible for the discretization of the signal amplitude. The specification must be dependent on the noise level, S, between the signal level and the mean noise level, where the logarithmic measure S_{dB} = 10 lg S is used. According to

$$S_{dB} = 1.76 + 6.02\,n \qquad (6.3)$$

the resolution of the A/D converter bit, n, the minimal distance to the noise level in decibel (dB, see Supplement 6.1) can be determined, or the separation from the noise level (Profos and Pfeifer 1993).

6.1.2 Signal Sampling

Micrometeorological data must be partly sampled with a high time resolution, *e.g.* for eddy covariance measurements, the sampling frequency is about 20 Hz. Even for standard meteorological data the sampling rate is 1 Hz. The sampling rate determines the discretization in time. Accordingly, the measuring signal must be on the A/D converter a long enough time so that the signal level can be adjusted and digitalized. The time difference between two samples depends on the converting time and the number of measuring channels. The sampling of a periodic function $g(t)$ must be done in such a way that the measuring data can be reconstructed for the given sampling time Δt. The necessary number of samples is given by the sampling theorem (Lexikon 1998):

> According to the sampling theorem a function $g(t)$ with sampling values $g(x_i)$ in a time period Δt can be exactly reconstructed, if their Fourier spectra $S(k)$ for $k > \pi/\Delta t$ disappears. The sampling period Δt must be chosen so that $\Delta t < 1/(2\,f_g)$ where f_g is the highest resolvable frequency.

This means that periodic oscillations must be sampled more than two times per period or the sampling frequency must be double of the measuring frequency (Bentley 2005). The frequency

$$f_N = \frac{1}{2\Delta t} > f_g \qquad (6.4)$$

is called Nyquist frequency. Therefore, the highest appearing frequency, f_g, must be limited by low pass filtering, and is equal to the frequency where a damping of 3 dB occurs. If higher frequencies nevertheless occur, *e.g.* by noise of the power frequency, the so-called aliasing effect will appear (Bentley 2005):

> The aliasing effect is a false reconstruction of a continuous function $g(t)$ made from discreetly-sampled values $g(x_i)$ with a time resolution of Δt, resulting in higher frequencies appearing as lower ones.

In Fig. 6.2, it is shown that for accurate sampling the function (f1) can be correctly reconstructed while functions with higher frequencies (f2) cannot be reconstructed. In spectra of signals with aliasing the energy of lower frequencies increases ($< f_g$). The increase is equal to the energy of the not-included frequencies above f_g. Therefore, no energy lose happens, but the frequencies are not correctly reconstructed. For frequencies that are not related to the measuring process (electrical power frequency), false measurements occur. If the power frequency cannot filtered, then the sampling frequency should be chosen in such a way that the power frequencies compensate each other, *e.g.* for 60 Hz power frequency a sampling frequency of 30 Hz and for 50 Hz one of 25 Hz (Kaimal and Finnigan 1994).

Meteorological measuring systems are already low pass filters (see Chap. 4.1.2). Due to the finite extension of the sensors, the only turbulence elements that can be measured are larger than or equal to the measuring path d. Furthermore, due to the turbulence spectrum the filter frequency depends on the wind velocity and the measuring height. As a simple approximation for the upper frequency limit with 10% damping and a height of 5–10 m, one can use according to Mitsuta (1966):

$$f_{g10\%} = \frac{u}{d} \tag{6.5}$$

Another simple approximation with the wind velocity and the measuring height (Kaimal and Finnigan 1994) is:

$$f_g > 8\frac{\overline{u}}{z} \tag{6.6}$$

The length of a measuring series depends on the lowest frequency of the turbulence spectrum to be constructed. Attention must be paid so that for an acceptable accuracy of the standard deviations and covariances a minimum of samples is necessary. According to Haugen (1978), this is at least 1000, and is illustrated for different quantities in Fig. 6.3. In the case of statistical independence, the error is approximately $N^{-1/2}$ (N: umber of samples). Because the independence of time series is generally uncertain a higher error must be assumed (Bartels 1935; Taubenheim 1969).

For spectral analysis, the length of the necessary time series is determined by the accuracy of the low frequency part of the spectrum. Assuming an error of 10%, the length, T, must be chosen so that either it is the 10 times the longest period to reconstruct or the lower frequency limit given by $f_{gl} = 10/T$ (Taubenheim 1969). This measuring length is generally significantly longer than those for the error calculation for standard deviations and covariances.

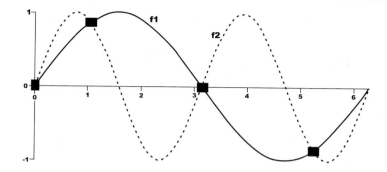

Fig. 6.2. Sampling of periodic signals: f1 is correctly represented in contrast to f2

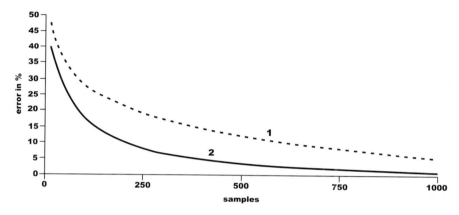

Fig. 6.3. Error in the measurements of (1) the friction velocity or the sensible heat flux; (2) error in the standard deviation of the vertical wind component as functions of the number of samples according to graph from von Haugen (1978)

6.1.3 Transfer Function

Not only can an inadequate sampling of the measuring signal be a reason for errors but also an inadequate resolution in time and space of the sensor. A transfer function of a measuring system describes the time delay between the output signal and the input signal (phase shift) and the damping of the amplitude. A transfer function can be defined in the space of the complex Laplace operator $s = \delta + i\omega$, with δ as a damping parameter (see Supplement 6.2):

$$T(s) = \frac{L\{X_a(T)\}}{L\{X_e(T)\}} \tag{6.7}$$

The determination of the Laplace transform of the output signal from the Laplace transform of the input signal is given by:

$$X_a(s) = X_e(s)T(s) \tag{6.8}$$

This description is very simple for many practical applications. For example, the product of the single functions can determine the transfer function of a measuring system. For a turbulence complex, the transfer function can be determined as the product of the transfer functions of the response in time, of the averaging by the sensor in space, and by the averaging in space of sensors that are used for eddy-covariance measurements in a certain distance (Eq. 4.15).

Supplement 6.2   Laplace transformation

In contrast to the Fourier transformation (see Supplement 2.3), which is a function of circular frequency, the Laplace transform transformation of an aperiodic signal, which disappears for $t < 0$, is a function of a complex operator $s = \delta + i\omega$ (Bentley 2005). The Laplace transformed L is defined as:

$$X(s) = L\{X(t)\} = \int_0^\infty X(t)\, e^{-st}\, dt \tag{6.9}$$

For the backward transformation is:

$$X(t) = L^{-1}\{X(s)\} = \frac{1}{2\pi i} \int_{s=\delta-i\omega}^{\delta+i\omega} X(s)\, e^{st}\, ds \tag{6.10}$$

A benefit of the Laplace transformation is that extensive tables and software for the determination of the transformation are available (Doetsch 1985; Graf 2004).

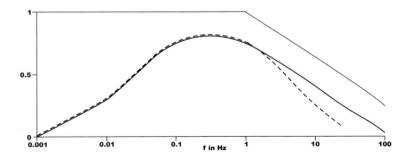

Fig. 6.4. Schematic graph of the transfer function T (*thin line*) and the turbulence spectra S (*thick line*) with the resulting spectra $T\,S$ (*dotted line*)

The error of a standard deviation or of a flux is given by the ratio of the spectrum multiplied with the transfer function to the undamped spectrum:

$$\frac{\Delta F}{F} = 1 - \frac{\int\limits_0^\infty T_{x(y)}(f) S_{x(y)}(f) df}{\int\limits_0^\infty S_{x(y)}(f) df} \tag{6.11}$$

Figure 6.4 shows schematically that the turbulence spectrum is reduced if the transfer function is less than one. Because in most of the cases the damping starts in the inertial subrange, the correction of turbulence measurements for well-known turbulence spectra is relatively simple. The spectrum in the inertial subrange is extrapolated or corrected relative to the modeled spectra, *e.g.* Moore (1986).

6.1.4 Inertia of a Measuring System

A special case of a transfer function is the sharp change of the signal from $X = X_o$ for $t \leq t_0$ to $X = X_\infty$ for $t > t_0$. For a measuring system of the first order, *e.g.* temperature measurements, follows the differential equation (Brock and Richardson 2001):

$$X_e(t) = X_a(t) + \tau \frac{dX_a}{dt} \tag{6.12}$$

The dependency of the input X_a and output X_e signals can be presented with the time constant τ. To have an indicator number, which is dependent only on the sensor and not on the size of the quantity for different meteorological systems, different indicator numbers are used.

Time Constant

For first order measuring systems the differential equation Eq. (6.12) has an exponential solution

$$X(t) = X_\infty \left(1 - e^{-\frac{t}{\tau}} \right), \tag{6.13}$$

where X_∞ is the final value after complete adjustment to the surrounding conditions and τ is the time constant which is a measure of the inertia of the measuring system:

The time constant of a measuring system is the time required for the measuring system to reach 63% of its final or equilibrium value. The value of 63% is equal to (1-1/e).

To determine the real final value due to a sharp change of the input signal, it is necessary to measure significantly longer than the time constant. This value depends on the required accuracy, and should be at least five times the time constant. The response of the measuring signal after a sharp change is schematically shown in Fig. 6.5.

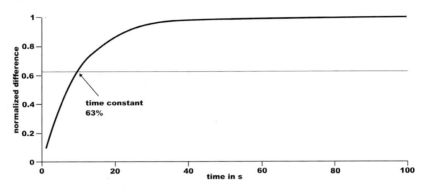

Fig. 6.5. Schematic graph for the documentation of the time constant for a change of the measuring signal by a normalized signal difference

Distance Constant

Anemometers have a wind-speed dependent time constant. For application to all anemometers and wind speeds, a comparable measure, the distance constant, was defined as:

The distance constant is the length of the wind path necessary for the anemometer to reach 63% of the final velocity.

The relation between the time constant and the distance constant L can be calculated with the final velocity V_∞:

$$L = V_\infty \cdot \tau \tag{6.14}$$

Dynamic Error

The time constant describes the dynamic error of a measuring system. The typical case in meteorology of dynamic errors is, when a nearly linear change of the meteorological element occurs over a given time span. Instead of a change in time, a change in space is also possible which is typical for moving measuring systems such radio sondes and tethersondes. For this case, Eq. (6.12) can be written in the form (Brock and Richardson 2001):

$$a \cdot t = X_a(t) + \tau \frac{dX_a}{dt} \qquad (6.15)$$

The solution of this equation is:

$$X(t) = a\,t - a\,\tau\left(1 - e^{-\frac{t}{\tau}}\right) \qquad (6.16)$$

The second term on the right-hand-side in Eq. (6.16) is responsible for the lag in the measured signal relative to the input signal and is called dynamical error:

$$\Delta X_d(t) = a\,\tau\left(1 - e^{-\frac{t}{\tau}}\right) \qquad (6.17)$$

For steady state conditions, the output signal is shifted in relation to the input signal by the time difference $\Delta t = \tau$. The dynamic error is schematically shown in Fig. 6.6. For known input signal functions, the dynamic error can be easily mathematically corrected or by correction networks. This may be relevant for the exact determination of the temperature gradient of an inversion layer of small vertical thickness. Often dynamic errors are a reason for hysteresis.

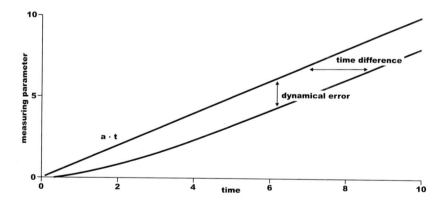

Fig. 6.6. Schematically graph of the dynamical error and of the time difference by a linear change of the input signal

6.2 Measurement of Meteorological Elements

Standard measurements in meteorological networks are made according to the international guide of the World Meteorological Organization (WMO 1996). Often, such regulations are extensive in order to guaranty a high and constant data quality for climate and environmental monitoring stations (see Appendix A6). In contrast, weather stations, which supply only data for weather forecast or for the current weather state (nowcasting), qualitative simpler sensors can be used. The research and special practical applications need specialized measuring stations (VDI 2006a), which are listed in Table 6.1. The typical measuring parameters are given in Appendix A6.

Micrometeorological measurements are related to the meteorological scales according to Orlanski (1975) (see Fig. 1.2). These relations are relevant as well for direct and for remote sensing measuring techniques (Table 6.2). For direct techniques the mast *etc.* is responsible for the scale.

Table 6.1. Classification of meteorological measuring stations (VDI 2006a), except stations of meteorological services which are classified by WMO (1996)

number	indication	feature
1	agrometeorological station	agrometeorological basic parameters *e.g.* for the determination of evaporation
2	microclimatological or micrometeorological station	miscellaneous application with different instrumentations, *e.g.* biometeorological measurements
3	micrometeorological station with turbulence measurements	including eddy-covariance measurements for research purposes
4	air pollution measuring stations	determination of parameters for air pollution calculations
5	immission measuring station	additional meteorological measurements to immission measurements
6	disposal site measuring stations	mainly for the determination of the water balance
7	noise measuring station	determination of parameters (mainly wind) fort he meteorological influence (ISO 1996)
8	road measuring station	measurement of meteorological elements which may affect the traffic
9	hydrological station	mainly precipitation measurements
10	forest climate station	measurement of meteorological elements in clearing and below trees
11	nowcasting station	measurement of the actual weather state, *e.g.* with present weather sensors
12	"hobby station"	measurement of meteorological elements with simple sensors but according the relevant installation instructions

Table 6.2. Assignment of direct and remote measuring systems to meteorological scales (the grey shading shows the degree of scale assignment)

measuring system	macro	meso			micro		
	β	α	β	γ	α	β	γ
radio sonde	▓	▓					
boundary layer sonde		▓					
tower > 100 m			░				
mast < 50 m					▓		
turbulence measuring technique					▓	▓	
satellite (vertical resolved)	▓						
wind profiler		▓	░				
Sodar		▓	▓				
RASS		▓	▓				
LIDAR			▓	▓			

Table 6.3. Assignment of scales of models to the necessary resolution of input parameters (the grey shading shows the degree of scale assignment)

resolving structures	macro	meso			micro		
	β	α	β	γ	α	β	γ
horizontal fields	▓	░					
vertical distributions			░	▓	█	█	█
boundary layer parameters		░	░		█	█	█
specified surface layer parameters				░	▓	█	█

Knowing the scale of the measuring systems is very important if the data are used as input parameters of models. The smaller the scales of the models are, the greater is the importance of vertical structures, boundary layer structures, and the surface layer structures (Table 6.3).

6.2.1 Radiation Measurements

Radiation measuring sensors are based on the principle of a radiation-caused heating and therefore an increase of the surface temperature of a receptor. For absolute devices (used for calibration) the temperature is directly measured on a black receiver surface, which is irradiated without any filters directly by the sun. Due to the selective measurement of the direct sun radiation, the longwave radiation can be neglected, and therefore absolute devices measure only the shortwave radiation. Relative devices measure the temperature difference between two different irradiated areas one black and one white. Furthermore radiation-measuring devices are classified if they measure radiation from the half space or directed (from the sun) and if they measure short or long wave radiation. For measurement of the short wave diffuse radiation the sun is shadowed. Net radiometers measure the upper and lower half space. The possible spectral range can be selected by the material, which is used for the dome (receiver protection). Domes made of quartz glass are only permeable for short wave radiation (0.29–3.0 µm). For measuring short and longwave radiation (0.29–100 µm) domes with Lupolen (special polyethylene) are used, and for longwave radiation (4.0... 5.0–100 µm) domes made from silicon are used. The gap between short and longwave radiation has no significant influence on the measurements due to the low amount of energy in this range. Using a special set of filters or spectrally sensitive photoelectric cells, the photosynthetic active radiation (PAR) can be measured. PAR has dimensions of mmol $m^{-2}s^{-1}$, and the numerical value is approximately twice the numerical value of the global radiation in Wm^{-2}. Radiation instruments working in the small or large atmospheric window of the longwave range and a relatively small dihedral angle are used for measuring surface temperature. An overview is given in Table 6.4.

In the last 10–15 years, remarkable progress was achieved regarding the accuracy of radiation sensors (Table 6.5). This success is based on the classification of radiation sensors by the World Meteorological Organization (Brock and Richardson 2001; Kasten 1985) as clear guidelines for the error limits (Table 6.6) and the building up of the *Basic Surface Radiation Network* (BSRN) of the World Climate Research Program with a well-formulated quality control scheme (Gilgen *et al.* 1994). Therefore, the world radiation centre in Davos (Schwitzerland) maintains a collection of sensor called the *World Radiation Reference*. The other world centers regularly compare their *Primary Standard* devices with this reference. Regional and national radiation centers should compare their *Secondary Standard* devices with the world centers at least every five years. Currently, the widely used pyranometers PSP (Eppley Lab. Inc., USA), CM11, and especially CM21 (producer: Kipp & Zonen, The Netherlands) conform to the Secondary Standard.

For radiation sensors, a cosine correction is of special importance because the radiation power does not exactly follow the cosine of the solar angle of incidence. The reasons for the corrections are the unequal thickness of the domes and the changing thickness of the atmosphere with changes of the angle of incidence. Therefore, the exact horizontal leveling of the senor and the clearness of the domes should be controlled.

In any case, the radiation instruments should be permanent (1–3 years) compared with national sensors for comparison (etalon). Because Secondary Standard devises are widely used, the comparisons with the sensors of different users are possible.

In contrast to shortwave radiation sensors, longwave radiation sensors are insufficiently calibrated. Serious differences were pointed out by Halldin and Lindroth (1992). While shortwave radiation sensors can be calibrated relatively simply *against sun* or with a lamp of defined radiation, calibration of longwave radiation sensors requires that radiation of the sensors must be taken into account. The measuring signal is simply the difference between the incoming longwave radiation and the longwave radiation of the housing in the form

$$I \downarrow = \frac{U_{rec}}{C} + k\sigma_{SB} T_G^4 ,$$
(6.18)

where U_{rec} is the voltage measured at the receiver, C and k are calibration coefficients, and T_G is housing temperature which must be measured.

Progress in sensor development was made by Philipona et al. (1995), who installed thermistors in the silicone domes of the longwave radiation sensor to measure the temperature of the dome to correct the effects of dome temperature and local heating by shortwave radiation. The correction requires the temperatures of the housing and the dome (T_D):

Table 6.4. Classification of radiation measuring devices

measuring device	sensor type		wave length		opening angle	
	absolute	relative	short wave	long wave	half space	directed
pyrheliometer	x		x			x
aktinometer		x	x			x
pyranometer		x	x		x	
albedometer		x	x		x*	
ayrgeometer		x		x	x	
net pyrgeometer		x		x	x*	
radiometer		x	x	x	x	
net radiometer		x	x	x	x*	
IR radiation thermometer		x		x		x

* upper and lower half space

Table 6.5. Accuracy of radiation measuring devices for BSRN stations (Ohmura et al. 1998)

parameter	device	accuracy in 1990 in Wm^{-2}	accuracy in 1995 in Wm^{-2}
global radiation	pyranometer	15	5
direct solar radiation	aktinometer, sun photometer	3	2
diffuse radiation	pyranometer with shaddow ring	10	5
longwave down welling radiation	pyrgeometer	30	10

$$I \downarrow = \frac{U_{emf}}{C}\left(1 + k_1 \sigma_{SB} T_G^3\right) + k_2 \sigma_{SB} T_G^4 - k_3 \sigma_{SB}\left(T_D^4 - T_G^4\right) \qquad (6.19)$$

Table 6.6. Quality requirements for pyranometers (Brock and Richardson 2001), percents are related to the full measuring range

property	secondary standard	first class	second class
time constant (99%)	< 25 s	< 60 s	< 4 min
offset (200 W m^{-2})	± 10 W m^{-2}	± 15 W m^{-2}	± 40 W m^{-2}
Resolution	± 1 W m^{-2}	± 5 W m^{-2}	± 10 W m^{-2}
Stability	± 1%	± 2%	± 5%
non-linearity	± 0,5%	± 2%	± 5%
cosine response			
10° solar elevation, clear	± 3%	± 7%	± 15%
spectral sensitivity	± 2%	± 5%	± 10%
temperature response	± 1%	± 2%	± 5%

Table 6.7. Accuracy and costs of net radiometers (Foken 1998a, modified)

classification	measuring principle	device (for example)	error, costs
basic research, BSRN-recommendation since 1996 (Gilgen *et al.* 1994)	short and long wave radiation (above and below) separately measured	sw: Kipp & Zonen CM21(2x), ventilated, 1x temperature; lw: Eppley PIR[*], ventilated, 4 x temp..	< 3% ~ 15.000 €
basic research, recommendation up to 1996 (Halldin and Lindroth 1992)	shortwave and whole radiation (above and below) separately measured	sw: Kipp & Zonen CM21(2x), ventilated sw+lw: Schulze-Däke 1x temp., ventilated	< 5% ~ 12.000 €
agrometeorological measurements	short and longwave radiation (above and below) separately measured	sw/lw: Kipp & Zonen CNR1, 1 x temp., unventilated	5–10% ~ 4.000 €
simple methods	whole radiation (above and below) together measured	Funk, Fritschen (Q7), NR-lite	> 10% ~ 1.000 €

[*] Probably better accuracy with CG 4 (Kipp & Zonen), but increasing costs up to larger than 20.000 €

To take into account the influence of the shortwave radiation, the so-called f-correction (Philipona *et al.* 1995) is added in Eq. (6.19)

$$ f \, \Delta T_{S-N} = f \left[\left(T_{SE} - T_N \right) + \left(T_{SW} - T_N \right) \right], \qquad (6.20) $$

where the temperature differences between the North and South, or North and Southeast and North and Southwest sides of the dome must be considered. The f-factor is calculated using the difference between the shaded and un-shaded sides of the dome:

$$ f = \frac{I\!\downarrow_{shade} - I\!\downarrow_{sun}}{\left(\Delta T_{S-N} \right)_{sun}} \qquad (6.21) $$

The f-correction is a function of location and time of year, and should be regularly checked. This can be done by shading (shadow stick) for five minutes at the highest level of the sun, and performing an actual calculation of the f-factor.

In micrometeorology, the net radiation relevancies are highly important, and thus a high accuracy is required. Instead of the formerly used radiometers, which measured the short and longwave radiation together, recently all four radiation components are measured separately. As shown in Table 6.7, this is in principle possible, but at relatively high costs. Very accurately determined values of net radiation are normally greater than those determined with radiometers.

6.2.2 Wind Measurements

Classical wind measuring devices are based on a mechanical principle, where the wind path is transformed into a rotation movement (Fig. 6.7). Other wind measuring techniques, such as the application of sound or remote sensing methods with radar, are becoming increasingly important principles. A further classification can be made if the sensors are used for turbulence measurements using the eddy-covariance method. An overview is given in Table 6.8.

For mechanical anemometers, the knowledge of the transfer function between the velocity in a wind tunnel and the velocity measured with the anemometer is of high importance:

Table 6.8. Division of wind measuring sensors (anemometer)

measuring device	measuring principle				application	
	mech.	sound	therm.	other	mean	turb.
cup anemometer	x				x	
propeller anemometer	x				x	(x)
hot wire anemometer			x			x
sonic anemometer		x			x	x
laser anemometer				x	(x)	x

The transfer function is the linear dependence between the wind velocity and the rotation velocity of the anemometer within a defined working range.

This calibration relation is linear over a wide range of velocities, but for low velocities ($<$2–4 m s^{-1}) is exponential. This is shown in Fig. 6.8 where c is the threshold velocity, which should not be confused with the intersection of the linearly extrapolated transfer function to the point with zero revolutions, a:

The threshold velocity of a rotating anemometer is the lowest wind speed that transfers the rotating anemometer into a continuous movement.

Beyond the threshold velocity (about 0.1–0.3 m s^{-1}), the so-called distance constant is important. It is a measure of inertia, and gives the necessary wind path required for an anemometer to register 63% of a wind velocity difference (see Chap. 6.1.4). The distance constant is an important parameter for mechanical anemometers; for sensitive propeller anemometers it is about 1 m; for small cup anemometers it is about 2–3 m, and for larger anemometers it is about 5 m. Note that sonic and hot wire anemometers in the usual application range are free of inertia. The distance constant should be used for the assessment of the measurement quality instead of the threshold velocity, because the starting velocity is generally within a range where the turbulent wind field is not fully developed.

In a turbulent flow, additional problems for mechanical anemometers exist (Kristensen 1998). With increasing distance constant, the wind velocity will be overestimated due to the mechanical inertia (over speeding):

Fig. 6.7. Cup anemometer (Photograph: Th. Friedrichs & Co., Schenefeld near Hamburg)

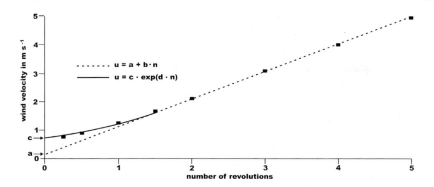

Fig. 6.8. Transfer function of a cup anemometer (VDI 2000) with n: number of revolutions of the anemometer, u: wind velocity and a, b, c and d: constants

> Overspeeding is the turbulence-caused over estimation of the measured wind velocity relative to the true wind velocity.

Wind gusts will brig a mechanical anemometer into high rotation; however, after the gust the anemometer will require some time to adjust to the lower wind speeds. There is no compensation of these additional rotations in the case of low wind speeds. Overspeeding can be as large as 10% of the wind velocity, and is particularly large for low wind velocities and high distance constants. The value is proportional to $(\sigma_u/u)^2$. For micrometeorological measurements it is especially important that the overspeeding is small because otherwise wind gradients near the ground will be inaccurate.

The cosine response of an anemometer is of importance:

> The cosine response is the ratio of the measured wind velocity for a special angle of incidence to the wind velocity of the horizontal wind field multiplied by the cosine of the angle.

$$F(\alpha) = \frac{u(\alpha)}{u(0) \cdot \cos \alpha} \tag{6.22}$$

An ideal cosine response is given by $F(\alpha) = 1$. For propeller anemometers, deviations up to 15% occur for incidence angles of about 45° (Figs. 6.9 and 6.10). These deviations can be relatively simply corrected with the relation:

$$u_{korr}(\alpha) = u_{mess}(\alpha) \cdot [\cos \alpha - a \cdot \sin(2\alpha)] \tag{6.23}$$

where $a = 0.085$ (Drinkov 1972) or $a = 0.140 - 0.009 \, u$ (Foken *et al.* 1983). For crosswinds, a dead zone of approximately ± 2° exists where the propeller does not rotate. For measurements of the vertical wind, the dead zone is overcome with two inclined sensors. A shank extension is recommended for flow from the front

(Bowen and Teunissen 1986) so that the dynamical conditions of a propeller for flow from the front and behind are nearly identical.

Sonic anemometers have a nearly cosine response except for angles with flow distortion. For cup anemometers, a cosine response cannot be assumed. If the inclination of the flow is not very great, than the same wind velocity is always measured. This means that for an inclined flow the cup anemometers overestimate the wind velocity.

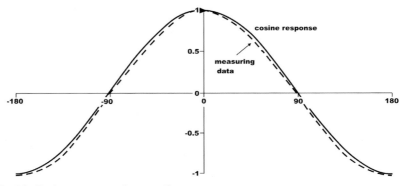

Fig. 6.9. Cosine response of a propeller anemometer

Fig. 6.10. UVW three-component propeller anemometer by Gill (Photograph: R. M. YOUNG Company/GWU-Umwelttechnik GmbH)

The first sonic anemometers used the phase shift method (Bovscheverov and Voronov 1960; Kaimal and Businger 1963). In this method, the ultrasonic signal of the transmitter is received at several points, and the phase difference between the transmitted and received signals is a function of wind velocity. The disadvantage of this method is that for different wind velocities ($5–10$ ms^{-1} – steps) equal measuring values will be observed, but the sound wave is shifted several times of 2π.

Modern sonic anemometers use the travel time principle and a direct time determination (Hanafusa et $al.$ 1982). In this method, a sonic signal (about 100 kHz) is transmitted from both sides of a measuring path and received on the opposite sides. Due to the wind velocity, one signal is faster than the other. The exact travel times of the sonic signals are used for the determination of the wind velocity:

$$t_{1,2} = \frac{\sqrt{c^2 - u_n^2} \pm u_d}{c^2 - u_n^2} d \qquad (6.24)$$

where d is the path length, u_d is the wind component along the path, u_n is the normal component of the wind, and c is the sound speed. This relation is based on the assumption that the flow in the sonic anemometer is generally slightly shifted by an angle γ from the measuring path (Fig. 6.11), and for the travel times follows (Brock and Richardson 2001; Kaimal and Finnigan 1994):

$$t_{1,2} = \frac{d}{c \cos \gamma \pm u_d} \qquad (6.25)$$

The difference of the reciprocal travel times gives the wind velocity and the sum of the reciprocal travel times gives the sound velocity:

$$\frac{1}{t_1} - \frac{1}{t_2} = \frac{2}{d} u_d \qquad (6.26)$$

$$\frac{1}{t_1} + \frac{1}{t_2} = \frac{2}{d} c \sqrt{1 - \frac{u_n^2}{c^2}} \approx \frac{2}{d} c \qquad (6.27)$$

The first sonic anemometers with the travel time principle (Mitsuta 1966) could only measure the time difference. Measuring the wind velocity depended on the sound velocity and therefore on the temperature and the moisture:

$$c^2 = 403T \left(1 + 0.32 \frac{e}{p}\right) \qquad (6.28)$$

From this equation the so-called sonic temperature can be calculated (Kaimal and Gaynor 1991), which is similar to the virtual temperature:

$$T_s = T \left(1 + 0.32 \frac{e}{p}\right) = \frac{d^2}{1612} \left(\frac{1}{t_1} + \frac{1}{t_2}\right)^2 \qquad (6.29)$$

A recalculation into the true temperature is possible by applying the geometric parameters of the sonic anemometer (see Chap. 4.1.2).

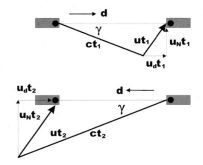

Fig. 6.11. Vector graph of the sound paths of a sonic anemometer

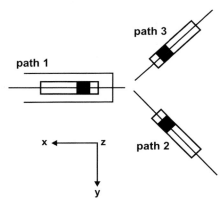

Fig. 6.12. Inclined positions of the measuring paths of the sonic anemometer CSAT3 (Liu *et al.* 2001)

Initially, the measuring paths of the sonic anemometers were predominantly Cartesian orientated (Hanafusa *et al.* 1982; Mitsuta 1966). Modern sonic anemometers have increased angles of the measuring paths, *e.g.* 120° (Fig. 6.12) so as to get lower flow distortion. Also, sensors were developed, which can be used for all flow directions (omni-directional). A problem yet to be investigated, is the significant self-correlation of the Cartesian wind components from the inclined wind components, which influences the statistical independence for the calculation of the covariances.

Disturbances of the wind field are the most important sources of errors for sonic anemometers. The reasons are the installations of the sensors and the size of the transmitters/receivers. For new sensors, a large ratio of the path length, d, to the transmitter/receiver diameter, a, of up to $d/a = 50$ is required, because under these circumstances the influences of flow distortion are small. Furthermore, the angle, θ, between the wind vector and the transmitter-receiver path should be large. For a ratio $d/a = 15$, the deviations follow (Kaimal and Finnigan 1994):

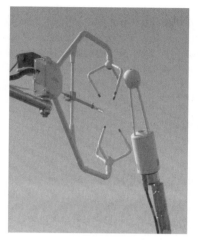

Fig. 6.13. Sonic anemometer CSAT3 with inclined measuring paths and large inflow sector; additionally an IR gas analyzer LI7500 and a sensitive temperature sensor are mounted (Photograph: Foken)

Fig. 6.14. Sonic anemometer USA-1 with inclined paths and omni-directional construction (Photograph: Foken)

Table 6.9. Classification of sonic anemometers (based on a classification by Foken and Oncley 1995; Mauder *et al.* 2006)

anemometer class		sensor type
A	basic research	Kaijo-Denki typ A,
		Campbell CSAT3, Solent HS
B	general use for flux measurements	Kaijo-Denki typ B,
		Solent Wind Master, R2, R3 ,
		METEK USA-1, Young 81000
C	general use for wind measurements	sensors of class B,
		2D-anemometer of different producers

$$(u_d)_{mess} = \begin{cases} u_d(0{,}82 + 0{,}18 \cdot \theta/75), & 0° \le \theta \le 75° \\ u_d, & 75° \le \theta \le 90° \end{cases} \qquad (6.30)$$

Devices that are predominantly used in basic research have only a limited open angle to reduce the influence from the sensor (Fig. 6.13). For routine applications, omni-directional sonic anemometers are usually used (Fig. 6.14). These must be calibrated in the wind tunnel, or the manufacturer must deliver calibration curves for correction.

In a turbulent flow, the errors due to flow distortion are often smaller than those for the laminar flow in the wind tunnel. Therefore, the application of the wind tunnel correction may lead to an overcorrection. For flux measurements, it is important that below the measuring path for the vertical wind component there are no sensor mountings, otherwise flow corrections up to 10% are possible.

The number of different sonic anemometer types has increased in the last few years. Thus, a type of classification is necessary, which is given in Table 6.9.

6.2.3 Temperature and Humidity Measurements

General Statements

The classical mercury thermometer, bimetal thermometer, and hair hygrometer are being replaced increasingly by electrical measuring principles. The measurement of electrical resistance is used most frequently for temperature measurements. Thermocouples and thermistors are reserved mostly for special measurements with small sensors. Due to the temperature sensitivity of the sound velocity, sonic anemometers are often used for temperature measurements (remark: the sonic temperature is not absolutely equal to the temperature, see Chap. 4.1.2). For humidity measurements, ceramic material is used to measure the relative humidity, and optical techniques are used to measure the absolute humidity. The dew point temperature is measured by detecting condensation on a chilled mirror. An overview is given in Table 6.10 where it is also indicated if the sensor can be used for the observation of mean or turbulent quantities.

Table 6.10. Classification of temperature (T) and humidity sensors (H)

measuring device	measuring principle				application	
	therm.	electr.	optical	other	mean	turb.
psychrometer (T,H)	x	x			x	
mercury thermometer (T)	x				x	
resistance thermometer (T)		x			x	(x)
thermistor (T)		x			x	
thermocouple (T)		x			x	(x)
sonic thermometer (T)				x	x	x
hair hygrometer (H)				x	x	
capacity hygrometer (H)		x			x	
dew point hygrometer (H)		x		x	x	
infrared hygrometer ((H)			x		(x)	x
ultraviolet hygrometer (H)			x			x

Table 6.11. Time constant of temperature and humidity measuring systems

measuring device	time constant in s
sonic thermometer	< 0.01
optical humidity measuring system	< 0.01
thin resistance wires (< 20 μm)	< 0.01
thermocouples (< 20 μm)	< 0.01
thermistors	0.1–1
mercury and resistance thermometers (3–5 mm diameter)	10–30

An important parameter for temperature and moisture sensors is the time constant (see Chap. 6.1.4). Typical time constants are given in Table 6.11. It is seen that the wet sensors (*e.g.* at the psychrometer) are less inert than dry sensors.

Temperature Measurement

The measurement of the *true* air temperature is one of the most difficult measuring problems in meteorology. Even today, it is worth reading the paper by Albrecht (1927). While improved technical possibilities are available, the measurement problems are often significantly underestimated. Because temperature sensors warm if they are exposed to radiation, they work better as a radiation sensor than a temperature sensor. Therefore, at least the incidence of direct solar radiation on the sensor must be eliminated. Furthermore, the thermometer must be ventilated to prevent heating and the turbulent exchange of heat. Both of these requirements are realized (ideally) with Assmann's aspirated psychrometer (Assmann 1887; 1888; Sonntag 1966–1968; 1994). The device consists of two thermometers ventilated at a rate > 2.5 ms^{-1} and equipped with a double radiation shield. There are several duplicated sensors available, which do not have the quality of the original sensor. Even electrical versions of the sensor have the best results if Assmann's dimensions of the double radiation shield and ventilation velocity > 2.5 ms^{-1} are fulfilled (Frankenberger 1951).

The reasons for the difficulties of temperature measurement are that outside of closed rooms measuring accuracy is about 0.1 K, and only for very well maintained devices can the accuracy be as great as 0.05 K. Therefore, the errors due to radiation and turbulence conditions are much greater than the possibilities of the recent electrical measuring techniques (< 0.001 K) (Fig. 6.15).

The radiation influence on the temperature measurement is called radiation error and can be estimated as an additional heating by the absorption of radiation by the sensor. This heating depends on the Prandtl number,

$$\mathrm{Pr} = \frac{\nu}{a_T} \tag{6.31}$$

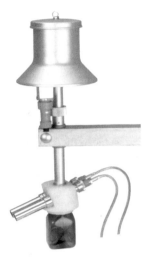

Fig. 6.15. Electrical aspiration psychrometer according to Frankenberger (1951) (Photograph: Th. Friedrichs & Co., Schenefeld near Hamburg)

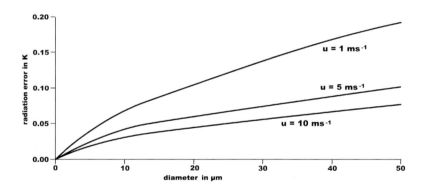

Fig. 6.16. Radiation error of thin platinum wires for $K{\downarrow} = 800$ W m^{-2} and $a = 0.5$ (Foken 1979)

which is the ratio of the kinematic viscosity to the molecular thermal conductivity, which is for air 0.71; the Reynolds number (Eq. 2.19), which is the ratio of inertial forces to frictional forces (see Chap. 2.1.2); and the Nusselt number,

$$Nu = f(\mathrm{Re}, \mathrm{Pr}) \qquad (6.32)$$

which is a function of the heat conductance and the flow characteristics. The radiation error is therefore a function of the radiation balance at the sensor surface Q_s, the sensor surface F, and the heat transfer properties, α,

$$Sf = \frac{Q_s}{\alpha\, F} \tag{6.33}$$

where

$$\alpha = \frac{Nu\,\lambda}{d}, \tag{6.34}$$

$$Q_s = a\,K \downarrow F_R, \tag{6.35}$$

and a is the absorption capacity of the surface, λ is the molecular heat transfer number, d is the sensor length, and F_R is the area affected directly by the radiation. For forced convection ($0.01 < Re < 10000$) if follows (van der Hegge Zijnen 1956) that:

$$Nu = 0.42\,\mathrm{Pr}^{0.2} + 0.57\,\mathrm{Pr}^{0.33}\,\mathrm{Re}^{0.5}$$

$$Nu_{air} = 0.39 + 0.51\,\mathrm{Re}^{0.5} \tag{6.36}$$

For the absorption capacity of platinum ($a=0.5$), the radiation errors are given in Fig. 6.16. Thus, radiation errors below 0.1 K are realized only for wire diameters < 20 µm.

The radiation error can be excluded only for very thin and spread-out resistance wires and for thermocouples of the same dimension. The use of thin resistance wires for turbulence measurements is based on Kretschmer (1954). Extensive investigations of the use of resistance wires were made, for example, by Tsvang (1960), Foken (1979), and Jacobs and McNaughton (1994). For a long time, the 12 µm Pt-wire sensor "AIR-150" was commercially available; today only the thermocouples made by Campbell Scientific have the same diameter. For thin platinum wires, it must be taken into consideration that the specific resistance in Ω·m for diameters < 50 µm, but especially for diameters < 10 µm, increases because the diameter is in the order of the free path length of the electrons.

For the measurement of the mean temperature, a suitable radiation shield, *e.g.* double protection tube or weather hut with double Venetian blind, is necessary to reduce the radiation error. Huts have the disadvantage that they heat up, and the so-called hut error of up to 1 K can occur (WMO 1996). For micrometeorological measurements, small cylindrical huts designed by Gill (Fig. 6.17) are often used. These huts were described regarding their dynamic and radiation properties by (Richardson *et al.* 1999). The hut error can be removed by sufficient ventilation. Therefore, the Reynolds number should be above the critical Reynolds number to achieve a turbulent flow:

$$\mathrm{Re} = \frac{L\,V}{\nu} \geq \mathrm{Re}_{krit} = \begin{pmatrix} 2300 & tube \\ 2800 & plate \end{pmatrix} \tag{6.37}$$

Below the critical Reynolds number, a laminar flow occurs near the sensor, and increases the response time of the measurement. The laminar flow boundary layer is thinner than the molecular temperature boundary layer (von Driest 1959):

Fig. 6.17. Temperature hut according to Gill for micrometeorological investigations (Photograph: R. M. YOUNG Company/ GWU-Umwelttechnik GmbH)

Table 6.12. Maximal differences for 100 Ω platinum resistance thermometers (DIN-EN 1996)

temperature	maximal difference			
	class A		class B	
°C	K	Ω	K	Ω
−100	± 0.35	± 0.14	± 0.8	± 0.32
0	± 0.15	± 0.06	± 0.3	± 0.12
100	± 0.35	± 0.13	± 0.8	± 0.30

$$\frac{\delta}{\delta_T} = \sqrt{Pr} = 0.85 \, (air) \tag{6.38}$$

Except for thermocouples, platinum wires have prevailed over all other temperature measurement sensors because of the stable temperature-resistance dependency. The resistance can be determined with the following equation

$$R(T) = R(0°C) \cdot \left(1 + \alpha T + \beta T^2\right), \tag{6.39}$$

where α is a temperature-dependent coefficient which ranges from 0.00385 K^{-1} to 0.00392 K^{-1} depending on the purity of the platinum. By adding iridium, the brittleness of the wire can be reduced so that lower temperature coefficients are typical. Under these circumstances β is about $-5.85 \cdot 10^{-7}$ K^{-2}. In the meteorological measuring range of −50°C–50°C nearly linear temperature dependence is obtained. Typically, platinum thermometers are produced with a resistance of $R\,(0°C) = 100$ Ω (also is 1000 Ω available). The quality of resistance thermometers in Germany and other countries is standardized (DIN-EN 1996). The sensors are separated into classes A and B (Table 6.12), where selected resistance thermometers with differences of only $\frac{1}{3}$ or $\frac{1}{10}$ of the DIN-class are also available. Regularly, thermometers are used with $\frac{1}{3}$DIN class B, which is better than class A.

The electrical measurement of the resistance is made with bridge circuits. The classical Wheatstone bridge with two and three wire circuits is seldom used because of non-linearity and a poor compensation for wire resistances. Recently, *e.g.* Thompson bridges with four-wire circuits are used.

The progress of resistance measurements using integrated electronic circuits, has widely replaced the thermistor (Rink 1961) as a measuring sensor, because even though it has a 10-fold higher temperature sensitivity, it has nonlinear characteristics. The thermistor must be age hardening with alternating high and low temperatures before it can be used, and during its use calibrations are often necessary. The temperature dependence is given by

$$R(T) = R(0°C) \cdot e^{\left(\frac{\alpha}{T} + \frac{\beta}{T^3}\right)} \tag{6.40}$$

with typical values of $\alpha \sim 4500$ K and $\beta \sim -1.5 \cdot 10^7$ K^3 (Brock and Richardson 2001). The most important areas of application are radiosondes and the measurements of body and dome temperatures in radiation sensors.

Humidity Measurement

The reliable device for measuring humidity, which can also be used for comparison measurements, is Assmann's aspirated psychrometer (see temperature measurement). This instrument has a second thermometer with a wet bulb for the measurement of the wet temperature. It uses the effect of evaporative cooling. The water vapor pressure can be determined from the temperature difference between the dry (t) and wet (t') temperatures using Sprung's psychrometric equation ($p_0 = 1000$ hPa, $t = 20°C$: psychrometer constant $\gamma = 0{,}666$ hPa K^{-1}):

$$e = E(t') - \gamma \cdot \frac{p}{p_0} \cdot (t - t') \tag{6.41}$$

For high accuracies, there are dew-point hygrometers. These expensive sensors can be used for comparison experiments (Sonntag 1994). In meteorological networks, capacitive sensors are widely used. These have replaced the hair hygrometer, which is still used for temperatures below 0°C when the psychrometer has insufficient accuracy.

For many micrometeorological measurements of turbulent fluxes, the absolute humidity is necessary. For relative humidity sensors such as capacitive and hair hygrometers, the absolute humidity must be calculated with the temperature. This introduces temperature sensitivity, and makes temperature-independent humidity measurements impossible.

For the measurement of turbulent humidity fluctuations optical measuring methods are chiefly used, which are based on Lambert-Beer's law

$$I = I_0 \cdot e^{-k \cdot d \cdot \frac{c}{c_0}},$$ (6.42)

where k is the absorption coefficient, d is the path length, and I_0 the radiation intensity at absorber concentration c_0. The measurement is made per unit volume. The measurement principle is schematically illustrated in Fig. 6.18. Devices with UV and IR radiation are used (Table 6.13). Devices in the UV-range are preferred for the measurement of low absolute humilities, and for water vapor pressures > 10 hPa the IR-range is preferred. Due to the low absorptivity of the IR-range relatively long measuring paths greater than 0.12 m are necessary. For a suitable wavelength, IR devices can be used to measure carbon dioxide concentration. The lifetime of UV devices is limited due to changes of the lamp after about 1000 h. For these sensors, the surfaces of the optical windows are treated with hygroscopic material such as magnesium fluoride, which must be taken into account for high humidities.

The calibration characteristics can be change during the application time, and the devices working in the UV range are affected more than those in the IR range. Two calibration methods are commonly used: One method uses the calibration within a gas flow with a constant concentration of water vapor or other trace gases, and the other method uses a climate chamber. Also, the in-situ calibration is possible by changing the path length for nearly constant background humidity (Foken *et al.* 1998). This method works because according to Eq. (6.42) the trace gas concentration as well as the path length is in the exponent.

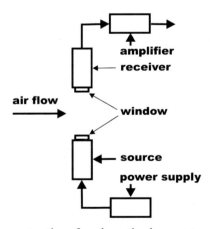

Fig. 6.18. Schematically construction of an absorption hygrometer

Table 6.13. Selected spectral lines for the water vapor absorption

range	wave length	radiation source	measuring length	absorber
UV	0.12156 μm	atomic hydrogen (Lyman-α)	3–10 mm	H_2O
	0.12358 μm	krypton	5–15 mm	H_2O,
	0.11647 μm			(O_2, O_3)
IR-A.B	different wave length	stable electric light bulb	0.125–1 m	H_2O, CO_2

Table 6.14. Recently mainly used commercial UV and IR hygrometers (Foken *et al.* 1995, modified)

range	sensor type	Producer	gases
UV	KH20-Krypton hygrometer	Campbell Sci. (USA)	H_2O
UV	Lyman-alpha hygrometer	MIERIJ METEO (NL)	H_2O
UV	Lyman-alpha hygrometer	Wittich and Visser (NL)	H_2O
UV	AIR-LA-1 Lyman-alpha hygrometer	AIR (USA) now: Vaisala (Finland)	H_2O
IR	LI 6262 (closed-path)	LI-COR Inc. (USA)	H_2O, CO_2
IR	LI 7500 (open-path)	LI-COR Inc. (USA)	H_2O, CO_2
IR	OP-2 (open-path)	ATC Bioscientific Ltd. (U.K.)	H_2O, CO_2
IR	Ophir IR-2000 (open-path, very large)	Ophir Corporation (USA)	H_2O
IR	Gas Analyzer E-009 (open-path, very large)	Kysei Maschin. Trading Co. Ltd. (Japan)	H_2O, CO_2

The application of very sensitive tunable diode lasers (Edwards *et al.* 1994) increased during the last few years because different gases such as methane, nitrogen oxides, and carbon isotopes can be measured.

The application of these techniques for turbulence measurements in the IR range are based on Elagina (1962) and in the UV range on Buck (1973), Kretschmer and Karpovitsch (1973), and Martini *et al.* (1973). Many commercial developments followed, which were summarized by Foken *et al.* (1995). Today only commercial sensors are used, and the most important ones are listed in Table 6.14.

Another form of the IR measuring technique are systems in which the air is withdrawn near the sonic anemometer and measured some meters away in an IR measuring cell using the so-called closed-path sensor (Leuning and Judd 1996; Moncrieff *et al.* 1997) instead of the usual open-path hygrometer. These systems have the benefit that with heated tubes density fluctuations do not occur, and the

WPL correction (see Chap. 4.1.2) can be dropped. However, these are very complicated dynamical systems with low-pass characteristics. The time delay between the wind and concentration measurements is especially important and must be taken into account for flux calculations. Such systems are widely used in the international carbon dioxide flux network (Aubinet *et al.* 2000).

6.2.4 Precipitation Measurements

Precipitation measurements are one of the standard meteorological measuring techniques. In Germany, the rain gauge according to Hellmann should have a 200 cm^2 collection area, but in other countries it is 500 cm^2. The daily emptying of a collection bucket is often replaced by automatic rain gauges. The measuring principles are very divers. The *tipping-bucket* gauge collects a certain amount of water in a bucket, and when full the bucket tips, empties, and gives an electrical impulse. The *droplet counter* counts the number of droplets of a uniform size. Recently, weight scale rain gauges are used; this type has the benefit that the danger of freezing is low.

Precipitation measurements play an important role in the calculations of water balance. Note that precipitation data in climatological tables are not corrected. The moisten error and the wind error (Richter 1995; Sevruk 1981) must be corrected. For solid precipitation, the wind error can be very large (Fig. 6.19). The correction equation for both errors is:

$$N_{korr} = N_{mess} + b\, N_{mess}^{\varepsilon} \qquad (6.43)$$

The coefficients are given in Table 6.15. Correction values are for different seasons and regions in Germany (Richter 1995).

For wind exposure on slopes with 40% inclination, up to 10% more precipitation can be expected due to the increasing horizontal contribution. A calculation of this effect was presented by Junghans (1967)

Table 6.15. Coefficients of the precipitation correction according Eq. (6.43) according to Richter (1995)

type of precipitation	ε	horizontal shading (b-value)			
		2°	5°	9,5°	16°
rain (summer)	0.38	0.345	0.310	0.280	0.245
rain(winter)	0.46	0.340	0.280	0.240	0.190
mixed precipitation	0.55	0.535	0.390	0.305	0.185
Snow	0.82	0.720	0.510	0.330	0.210

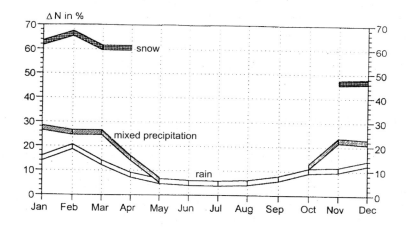

Fig. 6.19. Measuring error of precipitation measurements (Richter 1995)

$$\frac{N_H}{N} = 1 + 0,113 \, tg \, g \, \sin A_j \,, \tag{6.44}$$

where N_H is the precipitation on the slope, N is the measured precipitation on a horizontal surface, g is the inclination of the slope, and A_j is the azimuth of the jth wind direction .

6.2.5 Remote Sensing Methods

Increasingly, remote sensing techniques are being used in micrometeorology and applied meteorology. The techniques use sound waves (Sodar) or radio waves (RADAR), and by applying the Doppler Effect the wind velocity can be measured remotely. The use of lasers and microwaves makes moisture measurements possible. These methods supply measuring data only about 20–50 m above the ground. But for the micrometeorologist, this is an important region of the atmospheric boundary layer. However, in the lower PBL scintillometers can be used to measure the structure functions of the refraction index over horizontal path lengths and thus determine fluxes indirectly. The electromagnetic spectrum, which is used by the remote sensing methods, is summarized in Table 6.16.

Table 6.16. Ranges of the electromagnetic spectrum, which are used for remote sensing techniques in the atmospheric boundary layer

frequency/wave length	description	sounding technique
1-5 kHz	sound	Sodar
100 MHz – 3 GHz	ultra short waves (VHF), decimetre waves (UHF)	RASS, windprofiler
3–30 GHz	centimetre waves	microwave scintillometer
0.8 – 5 μm	infrared	scintillometer

Sodar-RASS

The development of profiling techniques using sound are based on Russian theoretical work on wave propagation in the atmosphere (*e.g.* Tatarski 1961), while the first Sodar devices (**S**onic **d**etecting **a**nd **r**anging) by Kallistratova (1959) and McAllister *et al.* (1969) were constructed and used. The measuring principle is based on the fact that a sound pulse emitted into the atmosphere will be backscattered by temperature inhomogeneities. The degree of backscatter is directly proportional to the temperature structure function parameter. The classical range for application of these "monostatic" Sodars was the detection of inversion layers for air pollution purposes. By calibration of theses devices, the determination of the sensible heat flux using Eq. (2.139) is possible.

Subsequently, the Doppler Effect was explored as a way to measure the three-dimensional wind field. These investigations lead to bistatic devices, which have low vertical operation range, and monostatic devices with one vertical and two sloping antenna. The accuracy of these Sodars is in the range of mechanical sensors (Table 6.17). However, these measurements are volume averages, and the horizontal wind components are measured in different volume, which may be far apart at high altitudes. For the standard deviation of the vertical wind it was found that the high values are underestimated and the low values are overestimated; this can be corrected.

Table 6.17. Measuring range and accuracy of Doppler-Sodar and RASS systems (Neff and Coulter 1986)

parameter	max. height in m	resolution in m	accuracy
horizontal wind velocity	1000	15–50	± 0.5 m s^{-1}
vertical wind velocity	1000	15–50	± 0.2 m s^{-1}
structure function parameter	1500	2–10	$\pm 30\%$
temperature (RASS)	600	15–50	± 0.5 K
boundary layer height	600	2–10	$\pm 10\%$
sensible heat flux	500	15–50	± 50 W m^{-2}

Fig. 6.20. Sodar-RASS system consistent on a "phased-array" Sodar (middle) and two radar antenna (Photograph: METEK GmbH)

Modern Sodars use a "phased-array" measuring system, where a large number of loudspeakers are triggered in such a way that the sound pulse can be emitted in any direction. Thus, ground echoes and obstacles can be excluded. Also, the emission of single frequency pulses can be replaced by a continuous multi-frequency emission. The measuring frequency must be chosen in such a way that the cross sensitivity to the humidity is minimal.

The combinations of a Sodar and a RADAR (Fig. 6.20) makes the measurement of the temperature field possible (RASS: **R**adio **A**coustic **S**ounding **S**ystem). In the RASS system, a Sodar emits vertically an acoustic wave, which is observed by the radar. Using the propagation speed of the sound wave, the sound temperature can be determined with Eqs. (4.17) and (6.28). Periodic inhomogeneities (turbulence elements) can only be detected if their size is approximately half of the wavelength of the sounding signal (Bragg condition).

While Sodar and Sodar-RASS systems offer reliable data at heights as low as 10–30 m, wind profilers have their first values at larger heights. The boundary-layer wind profiler works in the UHF range and is still interesting for micrometeorological applications; however, the tropospheric wind profiler works in the VHF rang and gives little data below 500 m.

Scintillometer

The so-called scintillometer (Hill 1997; Hill *et al.* 1980) measures the refraction structure function parameter, C_n^2, and offers a method for the determination of the sensible heat flux. The bases for this are Eqs. (2.137) and (2.139) as well as other parameterizations of the dependency of the sensible heat flux on the temperature structure function parameter, C_T^2. The measuring principle is schematically illustrated in Fig. 6.21. Temperature or humidity inhomogeneities (IR or micro-wave scintillometer) cause a scintillation of the measuring beam, which can be evaluated.

Essentially, scintillometers are separated into two classes (DeBruin 2002) the large aperture scintillometer (LAS) and the small aperture scintillometer (SAS). The latter instrument is commercially available as a displaced-beam small aperture laser scintillometer (DBSAS). The LAS works in the IR range, has a measuring path length of several kilometers, and can determine only the sensible heat flux. In contrast, the DBSAS works with two laser beams over a distance of about 100 m (Andreas 1989). These systems can determine also the path-length-averaged tur-bulence scale, which is associated with the energy dissipation according to Eqs. (2.105) and (5.65). By using the TKE equation Eq. (2.41), the direct determination of the friction velocity is possible when a stability dependence is taken into ac-count (Thiermann and Grassl 1992). Note that scintillometers are not able to de-termine the sign of the sensible heat flux. Additional measurements (temperature gradient) are necessary, if the sign is not be simply determined by the daily cycle because possibly unclear situations can occur in the afternoon. Scintillometers have a sensitivity that is in the middle of the measuring path rather than near the transmitter and receiver. This must be taken into account for footprint analyses of the measuring sector (Meijninger *et al.* 2002). Both scintillometer types have been successfully used in long-term measuring programs (Beyrich *et al.* 2002a; De-Bruin *et al.* 2002). Relatively new is the microwave scintillometer, which meas-ures the humidity structure function parameter C_q^2, see Eq. (2.136), and has the possibility determining the latent heat flux (Meijninger *et al.* 2002; 2006).

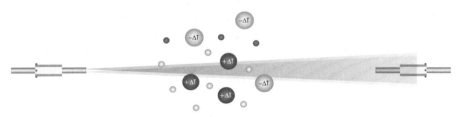

Fig. 6.21. Measuring principle of a scintillometer; temperature inhomogeneities in the laser beam (left transmitter, right receiver) cause a scintillation of the light

6.2.6 Other Measuring Techniques

From the micrometeorological point of view, parameters other than those discussed are of special interest. In this section, a short overview of important measurements is given. But in some cases, the study of additional literature is necessary (Brock and Richardson 2001; DeFelice 1998).

Measurements in the Soil

Soil temperature, soil moisture, and the soil heat flux are especially relevant micrometeorological parameters (see Chap. 1.4.2). The soil temperature is a standard measurement using water-protected thermometers at 5, 10, 20, 50, and 100 cm depth. Micrometeorological investigations often need an increase of the measurement density close to the surface, *e.g.* in 2 cm depth. The heterogeneity of the soil and the plant cover make the positioning of the sensors difficult.

The soil moisture as a control parameter for evaporation is of special importance. The most accurate version for measurement is still the gravimetric method, where the soil probe is made with a soil core sampler and the soil is weighted after drying at a temperature of 105 °C. The measured water content is the gravimetric (θ_g). Using a cylinder of known volume the volumetric water content (θ_v) can be determined. For a conversion of both parameters, the knowledge of the bulk density of the soil (dry density ρ_b) is necessary (ρ_w water density):

$$\theta_v = \theta_g \frac{\rho_b}{\rho_w} \qquad (6.45)$$

A special electrical method for dry soil is the use of gypsum blocks, which include electrodes for measuring the electric conductivity. However, these devices must be calibrated individually in the laboratory. Note that the gypsum blocks have a strong hysteresis between moistening and drying.

For not very dry soils, tensiometers are widely used for the determination of the soil moisture tension by capillary forces. The measuring technique uses a fine porous ceramic cup filled with water and placed in good contact with the moist soil. A balance of water pressure develops between the inner and outer sides of the cup. Using a glass tube extended into the cup, the pressure can be measured directly or the tube is closed with a septum and the pressure is measured with puncture devices. The matrix potential, ψ, in the depth, L (distance between the middle of the cup and the top of the tensiometer) can be calculated by

$$\psi = \psi_0 + \left(L - L_{bubble}\right) \qquad (6.46)$$

where L is reduced by the height of the air bubble, L_{bubble}, in the glass tube. The pressure measured in the bubble (negative) is expressed as a length (ψ_0). There is a soil-type specific and characteristic connection (water tension curve) between the matrix potential and the volumetric water content that is used for the calculation. The use of tensimeters is diminishing because of the effort needed for the measurements and the indirect calculation.

Widely used today, is the capacitive TDR-method (time domain reflection). The instrument consists of two or more electrodes separated 2–5 cm and put into the soil. The dielectric constant is measured by the travel time of an electric wave in the medium. The reflection properties are influenced in a characteristic way by the soil moisture. Because the electrical field develops over a large soil volume, an absolutely exact determination of the measuring height is not possible. With this method only the volumetric soil moisture is measured.

Soil heat flux can be measured using soil heat flux plates. These consist of two metallic plates separated by a layer of resin, which the same heat conductance assumed for the soil. The temperature difference measurement between both plats is made with thermocouples, where according to Eq. (1.12) the output signal is proportional to the soil heat flux. The plates require calibration.

There are numerous sources of errors for the soil heat flux plates. Especially important are the differences of the heat conductivities between the plates and the soil and at the edges of the plates, which are insufficiently included in the temperature measurement (van Loon *et al.* 1998). The first published and widely used correction method is the Philip correction (Philip 1961). The correction factor f between the measured soil heat flux, Q_G', and the soil heat flux through the surrounding soil, Q_G, is given by

$$f = \frac{Q_G'}{Q_G} = \frac{\varepsilon}{1 + (\varepsilon - 1)H},\qquad(6.47)$$

where ε is the ratio of the heat conductivities $\lambda_{plate}/\lambda_{soil}$. H is dependent on the geometry of the heat flux plate and for square plates is:

$$H = 1 - \frac{1.70\,T}{L}\qquad(6.48)$$

and for round plates is:

$$H = 1 - \frac{1.92\,T}{D}\qquad(6.49)$$

where T is the thickness of the plate, L is the length of the square plate, and D is the diameter of the round plate. The Philip correction is recommended by Fuchs (1986), but it is doubted by other authors (van Loon *et al.* 1998). Also different types of plates have significant differences (Sauer *et al.* 2002). By applying self-calibrating heat flux plates (Hukseflux HFP01SC), which determines a correction value by using a short and well-defined heating, the Philip correction can be more or less in applicable (Liebethal and Foken 2006). In contrast, these sensors (Hukseflux HFP01SC and TF01) are less suitable to determine the heat conductance and the volumetric heat capacity by in-situ measurements.

Generally, for micrometeorological measurements only the soil heat flux at the surface is necessary. Therefore, according to Eq. (1.14) the storage term above the heat flux plate must be determined by additional temperature measurements. It is also possible to eliminate the heat flux plates, if in a certain depth according to Eq. (1.12) the heat flux can be determined from the temperature gradient. It is recommended to determine the temperature gradient or to locate the heat flux plates at a depth of 10–20 cm to reduce errors. Both methods (using soil heat flux plates or temperature gradients for determining the storage) have approximately the same accuracy in an undisturbed soil profile (Liebethal *et al.* 2005).

The volumetric heat capacity, necessary for the determination of the storage term, can be determined with the method of de Vries (1963)

$$C_G = C_{G,m} x_m + C_{G,o} x_o + C_{G,w} \theta,$$ (6.50)

with the heat capacity of the mineral and organic compounds ($C_{G,m} = 1.9 \cdot 10^6$ J m^{-3} K^{-1}, $C_{G,o} = 2.479 \cdot 10^6$ J m^{-3} K^{-1}) and for water ($C_{G,w} = 4.12 \cdot 10^6$ J m^{-3} K^{-1}). x_m is the contribution of mineral components (assumed mineral 2650 kgm^{-3}) and can be determined by volume measurements of the soil. For depth up to 20 cm, x_o can often neglected. The volumetric moisture of the soil is θ and given in m^3 m^{-3} as is x_m.

For the calculation of soil heat flux from the temperature gradient, the coefficient of heat conductance, a_G, is necessary:

$$a_G = C_G v_T$$ (6.51)

where v_T is the thermal diffusion coefficient and C_G is given by Eq. (6.50).

The thermal diffusion coefficient can also be determined using the method by Horton et al. (1983), in which the temperatures sensors are installed at three depths (10, 15 und 20 cm) and the temperature difference between two time steps ($\Delta t = 1$min) is determined:

$$\frac{T_{15cm}^{n+1} - T_{15cm}^{n}}{v_T \, \Delta t} = \frac{T_{20cm}^{n} - 2T_{15cm}^{n} + T_{10cm}^{n}}{(\Delta z)^2}$$ (6.52)

A relatively simple yet reliable approach (Liebethal and Foken 2007) for the determination of the soil heat flux was presented by Braud et al. (1993). It based on a heat flux measurement at $z = 10$ cm depth and temperature measurements at 1 and 10 cm depth as well as an estimate of the soil moisture for the determination of the volumetric heat capacity:

$$Q_G(0,t) = Q_G(-z,t) +$$
$$C_G \, z \, \frac{T_1(t) - T_1(t - \Delta t) + 0.5 \left[\Delta T(t - \Delta t) - \Delta T(t) \right]}{\Delta t}$$ (6.53)

The time interval for such investigations is ten minutes.

Measurements at Plants

Even though measurements at plants are not the task of meteorologists, many plant parameters are used for modeling the energy and matter exchange (see Chap. 5). Probably the most important parameter is the leaf-area-index (LAI), which gives the contribution of the leaf surface per area element (see Table 5.11). It can be determined by spectral radiation measurements in the photosynthetic active range (PAR). Thus, the radiation below leafs will be compared with the radiation without any influence of the biomass. Note that the functional response comes into saturation for LAI > 5–6, and thus high LAI-values cannot be reliable determined with optical methods.

The leaf area index can be relatively simply determined with remote sensing methods. The satellite images must be corrected for atmospheric influences before application to ensure the comparability of different pictures. For evaluation of single pictures, this can be done without it (Song *et al.* 2001). For the determination of the *LAI* value the Normalized Difference Vegetation Index (*NDVI*) is applied, which is the difference of the spectral channels of the red light (0.63 – 0.69 μm) and the near infrared (0.76 – 0.90 μm):

$$NDVI = \frac{NIR - red}{NIR + red} \tag{6.54}$$

Often the vertical distribution of the leaf area density (*LAD*) is also necessary and sometimes even the available leaf mass is of interest, but for this determination a harvesting of leaves is necessary.

Direct Evaporation Measurement

The determination of the evaporation is a very important task. Before starting with micrometeorological methods especially in agrometeorology various devices were developed such as the evaporimeter according to Piche with absorbent paper as evaporation area or the evaporation measuring instrument according to Czeratzki with porous clay plates, all of which are no longer used.

For evaporation measurements in agrometeorology lysimeters are applied, which are a good direct measuring method if exactly used. The benefit is that contrary to micrometeorological measurements, only small areas are necessary.

Table 6.18. Coefficient *Kp* for the determination of the evaporation with the Class-A-Pan according to Eq. (6.55) for meadows or grain (in brackets for bare soil) in the vicinity of the device (Doorenbos and Pruitt 1977; Smajstrla *et al.* 2000)

mean wind velocity at the measuring date	extension of the area in the surrounding in m	minimum of the relative humidity at the measuring date	
		> 40%	< 40%
low	1	0.65 (0.80)	0.75 (0.85)
≤ 2 ms^{-1}	10	0.75 (0.70)	0.85 (0.80)
	100	0.80 (0.65)	0.85 (0.75)
	1000	0.85 (0.60)	0.85 (0.70)
moderate	1	0.60 (0.75)	0.65 (0.80)
2.1 to 4.4 ms^{-1}	10	0.70 (0.65)	0.75 (0.70)
	100	0.75 (0.60)	0.80 (0.65)
	1000	0.80 (0.55)	0.80 (0.60)
strong	1	0.50 (0.65)	0.60 (0.70)
≥ 4.5 ms^{-1}	10	0.60 (0.55)	0.65 (0.65)
	100	0.65 (0.50)	0.70 (0.60)
	1000	0.70 (0.45)	0.75 (0.55)

Especially in hydrological networks, evaporation pans are still used, mainly the Class-A-Pan (DeFelice 1998). It is a round pan with 1.14 m² water surface and 0.2 m water depth. The evaporation is measured by the water loss in the pan, and complicated corrections (Linacre 1994; Sentelhas and Folegatti 2003) based on wind velocity, air moisture, and water temperature must be made. A greatly simplified but reliable method for daily sums of the evaporation (Smajstrla *et al.* 2000) uses the height difference of the water level in the evaporation pan corrected with the precipitation

$$Q_E = Kp \left(h_{day\ before} - h_{measuring\ day} \right) \tag{6.55}$$

(in mm d^{-1}). Corrections factors, which are functions of wind velocity, minimum relative humidity, and the conditions in the surrounding environment of the pan are given in Table 6.18.

6.3 Quality Assurance

Measurements are now widely automatic, and based on modern measuring devices and electronic data storage technique. But it is a fallacy to believe this saves manpower, because uncontrolled measured data have only a low reliability or may be worthless. The expense of measurements has moved from manpower for observation to manpower for quality assurance (QA), which requires measures that are formally covered by visual observations. Quality assurance is a package of considerations, which are partly connected each other (DeFelice 1998; Shearman 1992).

During the planning of a measuring system, numerous questions must be taken into account. This starts with the data user giving clear instructions about the measurement resolutions in time and space, the requested representativeness, the accuracy, and the availability of the data, *etc.* This sounds easy, but it is often a big problem because users often have little knowledge of measuring methodology and technology, and unrealistic requirements can occur. An interdisciplinary and iterative planning is necessary before a specification of the parameters can be made. This includes the technical parameters of the sensors, the units of the whole data sampling, transmitting and storing of the data, and data transfer to the user.

After this, follows the choice of suitable sensors, measuring places, and data systems. Often large differences in price exist for apparently similar sensors. Therefore, detailed knowledge of the sensor characteristics and their influence on the whole project is necessary. Often in brochures, important information is missing and cannot be found elsewhere, because many low cost instruments are insufficiently investigated. Sensor lifetime, maintenance expenditure, and application under expected or extreme weather conditions are the deciding factors for high-end instruments. Sometimes special measuring sensors must be developed. The micrometeorological requirements for the measuring places are often so great that compromises are necessary, such as the exclusion of wind sectors or nighttime conditions with stable stratification. This must be done in agreement with the user requirements. For expensive measuring programs, it is required to test the measuring place with a pre-experiment.

The largest part of the costs in running an experiment is the necessity for calibration and maintenance. Therefore, the periods for calibration (every 6–12 month for most of the systems) and maintenance (from a few days to several weeks depending on the specific maintenance work) must be determined, as well as the type of maintenance without data loss. This needs partly also investigation into calibration systems. Consideration of work safety is an important issue. In this complex, necessary corrections must be included, not only based on the calibration but also on the weather conditions.

A very important point is the definition of quality control (QC). This includes a possible daily control (visual or partly automatic) so that data failure or other defects can be immediately found. Only controlled and marked data must stored in the data base or go to the user. Appropriate possibilities are extensively discussed in Chap. 6.3.1.

Because the data quality depends not only on the state of the sensor but also on the meteorological constraints, a complex quality management is necessary. Accordingly, the data should be flagged regarding the data control, if the data have the necessary quality for usage, or if it can only be used for orientation. Such a system for the eddy-covariance method is shown in Chap. 4.1.3.

Feedback from the data user on further qualification of the system should not be underestimated. It is definitely a duty of the user and the initiator of the measuring program to insure that the data fulfill the desired aims regarding the kind of the measurements and their quality. This can require some improvements or even reductions of the expenditure.

Quality assurance is a work package, which should not be underestimated with respect to the scope of the experiment. An increased attention for the running of the system helps in all cases.

6.3.1 Quality Control

Quality control has extremely important considerations. These are the control of the measuring data under different points of view, and the signature of the data quality depending on the sensor and the meteorological conditions. Finally the measuring value gets a quality stamp.

The quality control can be made in different steps. The first step is the exclusion of missing data and obviously false data especially based on an electronic plausibility. The second step is assurance that the measuring values are in the range of the measuring devices and in the possible meteorological range, which can change with the season. The resolution of the measuring signal and its dependence on the digitalization must be controlled for the further calculations.

The meteorological tests follow, where typically comparisons are made with other measuring data. Complicated meteorological measurements need test models, *e.g.* boundary layer models, which test the combinations of all measured parameters. As a result of the test, a decision is necessary about the use of a manual or an automatic data correction, and the further use of the data. The storage in the database follows with a sufficient flagging, and information about the controls made, and probably also a quality stamp.

After the initial tests, which can be automated, special tests on single parameters must be done (Fiebrich and Crawford 2001). The simplest cases are plausibility tests, *e.g.* for wind velocity and wind direction as shown in Table 6.19.

Stronger are the tests on radiation components, which are orientated on other meteorological parameters (Gilgen *et al.* 1994), where at least the components of the longwave radiation are visually not easy to access. For the upwelling longwave radiation, the test is to see if the radiation is within a certain difference of the radiation according to Stefan-Boltzmann law with the body temperature of the sensor:

$$\sigma(T_G - 5K)^4 \leq I \uparrow \leq \sigma(T_G + 5K)^4 \qquad (6.56)$$

For strong longwave upwelling radiation during the night or strong heating of a dry soil, the threshold value of 5 K must be increased. For the atmospheric downwelling longwave radiation, the test is related to a black body (dome covered with water) and a grey body (clear sky with a radiation temperature of $-55°C$):

$$0{,}7\,\sigma\,T_G^4 \leq I \uparrow \leq \sigma\,T_G^4 \qquad (6.57)$$

For shortwave radiation the dependence on astronomical parameters and the transmission of the atmosphere is tested. For the reflected shortwave radiation the test is made with the albedo (Table 1.1), where the albedo should only be calculated if the reflected radiation is >20–50 Wm^{-2}, otherwise large measuring errors occur.

For many micrometeorological measurements, it is necessary to test if a developed turbulent regime is present. For eddy-covariance measurements, this can be done with a test on flux-variance similarities (Foken and Wichura 1996), as already shown in Chap. 4.1.3.

For large measuring programs, it is possible to compare different measurements. This can be in the simplest case the pure comparison of all wind and temperature data, where thresholds or altitude-dependent functions must be given. Also useful is the control of the energy balance, but the residual should be tested before selecting the measuring places, the sensors, and the underlying surface. This should be taken into account by developing criteria (see Chap. 3.7). Models, ranging from analytical up to meso-scale can be used for tests if the measuring quantities are consistent with the model (Gandin 1988).

The quality control of meteorological measurements, especially of larger continuously measuring systems is still a developing question. Semi-automatic and fully automatic systems are only partly available. This is not only a question of software but also of research. Nevertheless, for available systems cases will be found where only experts are able to accurately separate meteorological effects from errors.

Table 6.19. Plausibility tests for wind velocity and direction (DeGaetano 1997). Measuring values should be deleted if the data fall above or below the given thresholds.

wind velocity	wind direction
< 0 ms^{-1}	
	$< 0°$ or $> 360°$
$<$ threshold velocity	for $1°$ to $360°$
$>$ threshold velocity	for $0°$ (calm)
> 60 ms^{-1}	
(height < 600 m a.s.l.)	

6.3.2 Intercomparison of Measuring Devices

Intercomparison experiments are an important requirement for the applications of sensors, because calibrations in wind tunnels and climate chambers have only a restricted meaningfulness in the turbulent atmosphere. Such experiments also have high requirements on the terrain, which must be free of obstacles and guaranties of an undisturbed installation of the sensors. The measuring heights should be chosen so that the near-surface gradients are small and do not have an influence on the measuring result. The sensors for comparison should be in excellent condition and should have had a basic calibration.

Because in meteorology no absolute instruments exist for the comparisons, standards must be applied. These are devices, which are proved internationally over a long time and have shown constant results in previous comparisons. The sensor calibrations specification should be available.

For radiation measurements at the radiation centers, good standards are available. For turbulence measurements, this is a problem because wind tunnel calibrations of anemometers cannot simply be transferred to the turbulent flow. Nevertheless proven devices should be used, which have taken part in comparison experiments for at least 3–5 years.

The evaluations of comparisons are described with statistical numbers. One is the bias

$$b = \overline{(x_1 - x_2)} = \frac{1}{N} \sum_{i=1}^{N} (x_{1i} - x_{2i}), \tag{6.58}$$

where subscript 1 is the device to be compared and subscript 2 the standard (etalon). The comparability gives the differences due to the device and the medium:

$$c = \left[\frac{1}{N} \sum_{i=1}^{N} (x_{1i} - x_{2i})^2 \right]^{\frac{1}{2}}$$

(6.59)

The precision shows the systematic differences:

$$s = \left(c^2 - b^2 \right)^{\frac{1}{2}}$$

(6.60)

If the measuring accuracy and the influences of the surrounding medium for both devices are nearly equal, a mean regression line can be calculated, which can also be used for correction purposes. Note, if there is no real independent device available and for correlation coefficient < 0.99, than two regression lines must be calculated and averaged.

The results of in-situ comparison experiments are not comparable with laboratory measurements because the stochastic character of atmospheric turbulence and many uncontrolled influencing factors always create scatter even for equal sensor types. Therefore in Figs. 6.22 and 6.23 examples of a comparison experiment under ideal conditions during the experiment EBEX-2000 (Mauder *et al.* 2007b) are shown. In all comparisons, the standard deviation of the vertical wind component has the best results. In contrast, comparisons of fluxes are significantly poorer, and the largest scatter occurs for the friction velocity. Reasons for this are the different spectra of the vertical and horizontal wind velocity, especially with respect to the frequency of the maximum of energy. Typical differences for sonic anemometers are shown in Table 6.20.

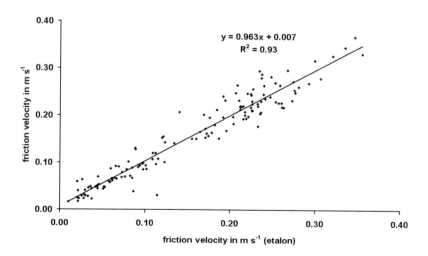

Fig. 6.22. Comparison of the standard deviation of the vertical wind of two sonic anemometers of the type CSAT3 during EBEX-2000 according to Mauder (2002)

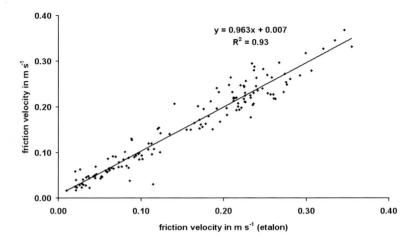

Fig. 6.23. Comparison of the friction velocity of two sonic anemometers of the type CSAT3 during EBEX-2000 according to Mauder (2002)

Table 6.20. Typical differences of different parameters during comparison experiments of sonic anemometers, in brackets for the same sensor type (Foken 1990, modified)

measuring parameter	difference in %	correlation coefficient
σ_w^2	0–10 (0–3)	0.90–0.99 (0.95–0.99)
σ_u^2	0–15 (0–5)	0.90–0.95 (0.95–0.98)
σ_T^2	5–20 (0–5)	0.90–0.95 (0.95–0.98)
$-\overline{w'u'}$	5–25 (2–5)	0.80–0.90 (0.90–0.95)
$\overline{w'T'}$	5–20 (0–5)	0.90–0.95 (0.95–0.98)
$\overline{w'q'}$	5–20 (0–5)	0.85–0.95 (0.95–0.98)

7 Microclimatology

Microclimatology investigates mean states and permanent repeated phenomena on the micrometeorological scale. These are small-scale circulation systems such as mountain and valley winds, land-sea wind circulations, and katabatic winds. These phenomena are the subjects of many textbooks (Bailey *et al.* 1997; Hupfer and Kuttler 2005; Oke 1987; Stull 1988). Therefore, the following Chapter does not provide a comprehensive overview, but rather discusses some typical microclimatological phenomena. These phenomena are present under special weather situations, and influence the small-scale climate in typical ways. Many of these local effects are described only in regional publications. The impressive wind system of the foene is not described because it exists on the larger meteorological scale.

7.1 Climatological Scales

The scales of climate are not as strict and uniform as those of meteorology, as proposed by Orlanski (1975). The term *microclimate* is often applied to space scales up to 100 m as in micrometeorology. It follows that *mesoclimate* has an extension of up to 100 km which is about one class of scales lower in comparison to Orlanski's (1975) classes in meteorology (Hupfer 1991). In the small-scale range, terms such as urban climate, topo or area climate (Knoch 1949), ecoclimate, local climate, and others are usual. A comparison of different classifications is given in Table 7.1, where the classifications by Kraus (1983), formerly used in the Western part of Germany, and Hupfer (1989) are most likely comparable to the classifications by Orlanski (1975). The term *microscale* is often used for the much smaller scales. In what was formerly East Germany, the microscale classification applied to scales below 1 m (Böer 1959).

Table 7.1. Different classifications of climatological scales (Hupfer 1991)

km	Orlanski (1975)	Böer (1959)	Kraus (1983)	Hupfer (1989)
10^4	makro-β	large climate	macro range	global climate
10^3	meso-α	range	synopt. range	zonal climate
10^2	meso-β		meso-	landscape climate
10^1	meso-γ	local climate	range	
10^0	mikro-α	range	micro range	plot climate
10^{-1}	mikro-β		topo range	small scale climate
10^{-2}	mikro-γ			
10^{-3}		micro climate range		border layer climate
10^{-4}				

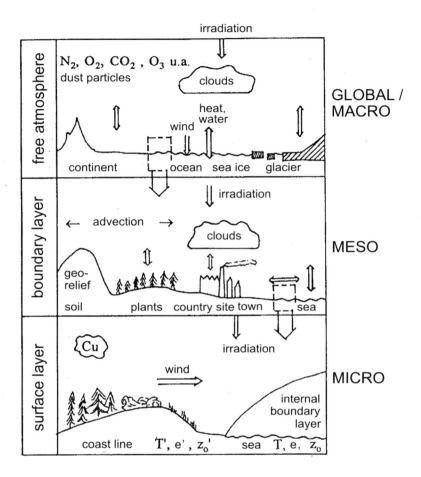

Fig. 7.1. The climate system in the macro, meso and micro scale (Hupfer 1996)

Figure 7.1 is an illustration of the various climate scales and the phenomena associated with them. Microclimate is mainly associated with processes in the surface layer, *e.g.* energy and matter exchange, radiation processes close to the ground surface, effects of the underlying surface, *etc.* However, cumulus convection is not considered a part of the microclimate.

7.2 Small-Scale Changes of Climate Elements

Microclimatological effects often result in strikingly different phenomena if the climate elements have relevant differences for certain weather situations or at certain time periods. Elements such as global and diffuse radiation in the absence of clouds show little climatological differences; however, differences can occur in mountainous regions. Exceptions may occur if long-duration local circulation systems are connected with clouds (see Chap. 7.3). For example, in the summer cumulus clouds will develop only over the land during a land-sea breeze. In the winter, the opposite effect occurs, *i.e.* clouds develop over the warmer sea.

Large-scale pressure fields cause wind speed and direction; however, the topography or obstacles near the measuring station also influence the winds, and these can lead to small-scale climate differences. Similar effects occur for precipitation where differences often cannot be clearly separated from measuring errors.

Air temperature and moisture in a uniform air mass and at the same altitude show very little differences, except for nights with high longwave up-welling radiation and cooling. Thus, the temperature minima near the ground may show remarkable differences (see Chap. 7.4). Similar results are found for the net radiation, which may be very variable on small scales due to the dependence on the albedo and the surface humidity.

Relevant climate elements in microclimatology are listed in Table 7.2. It is shown under which circumstances hardly any microclimatologically-caused differences occur.

Table 7.2. Microclimatological relevant climate elements

climate element	range and reason of microclimatological differences	hardly microclimatological differences
global and diffuse radiation	hardly available	unlimited horizon, no climatological caused clouds
net radiation	partly significant due to differences in albedo and surface temperature	unlimited horizon and uniform underlying surface
wind velocity and wind direction	partly significant in complex terrain and in the case of obstacles	large fetch over uniform surfaces and no obstacles
temperature (general) and air humidity	often small	open location
minimum of the temperature	partly significant, especially in valleys and hollows (also in very small scales)	open location
precipitation	partly significant but mostly in the range of the measuring error	open location

7.3 Microclimate Relevant Circulations

7.3.1 Land-Sea Wind Circulation

Land-sea wind circulations arise form weak background winds and large temperature differences between land and sea. The heating over the land generates low-
pressure areas with rising air. For compensation, cold air from the sea forms a sea
breeze. Above the sea, is a high-pressure area with descending air, *i.e.* subsidence.
The sea breeze front penetrates inland distances ranging from a few decameters up
to kilometers, and causes remarkable temperature differences over small areas, illustrated in Fig. 7.2. After the beginning of the land-sea wind circulation, the temperature at the beach may be unpleasantly cool, while a little distance inland
pleasant temperatures are found. During the night, these relations are opposite but
less developed.

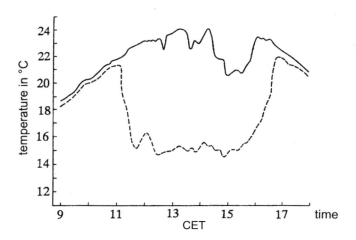

Fig. 7.2. Temperature cycle at the beach of Zingst (Baltic Sea, broken line) and 200 m inland
with offshore wind of 4-7 ms^{-1} over land on May 17, 1966 according to Nitzschke (1970) from
Hupfer (1996)

7.3.2 Mountain-Valley Circulation

Cooling during the night caused by radiation is the reason that cold air from
mountaintops and slopes descends under gravity into valleys and fills cold-air reservoirs. If drainage winds pass through narrow valleys, a considerable katabatic

wind can be occur. The strongest katabatic wind are found in Antarctica (King and Turner 1997). On smaller scales, katabatic flows have a significant role (see. Chap. 7.4).

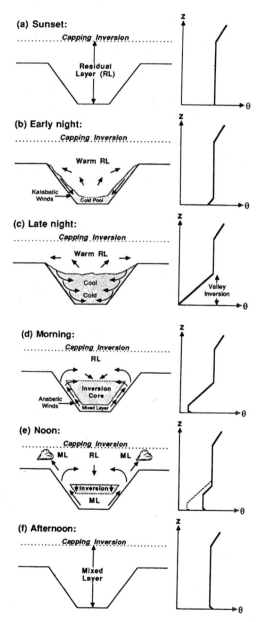

Fig. 7.3. Schematic view of the mountain-valley circulation (Stull 1988)

During the day, the valley and the slopes are heated faster than the mountain-tops (low wind velocities, less air exchange), and warm air moves upward along the slopes. These flows slowly lift the inversion layer until the mixed layer is fully developed. This process is illustrated in Fig. 7.3 following the classical paper by Defant (1949).

The strength of the katabatic wind can be determined from the temperature difference between the cold air flowing downlsope and the surrounding area. Applying a local equilibrium between bouyancy/surpression and stress the velocity of the katabatic flow is (Stull 2000)

$$u_k = \left(\frac{g \cdot (T - T_k)}{T} \cdot \frac{h}{C_D} \cdot \sin \alpha \right)^{1/2}, \qquad (7.1)$$

h is the depth of the cold air layer, C_D is the drag coefficient, α is the angle of inclination of the slop, g is the acceleration due to gravity, T is the surrounding air temperature, and T_K is the temperature of the cold air layer. Also, Eq. (7.1) can be added as a buoyancy term into the Navier-Stokes equation Eq. (2.1).

7.4 Local Cold-Air Flows

Cold-air flows are the most spread of the microclimtological phenomena. Source areas include, for example, open hilltops, forested slopes, and other inclined surface areas. The cold-air flow can be imagined as a flow of compact air parcels that can be interrupted or damped by obstacles. With a trained eye, the relevant cold-air flows can be localized according to the form of the landscape and the vegetation. This assessment can be made using a classification. Steps of the classification are detailed in Tables 7.3 and 7.4. The verifications can be made with well-aimed measurements (see Chap. 7.5). At very high risk, are cold air nights with strong longwave radiation due to cloudless skies. This is especially the case following synoptically-caused cold air advection (e.g. Three Saints' Days, a European weather phenomenon occurring in May). Two useful preventive measures are sprinkling with water so that more energy is necessary for the freezing of the water droplets, and fogging with smoke so that the cooling by long-wave radiation is above the plant canopy.

Cold-air flows can have a strong effect on urban climate. Because the buildings and streets within a city have high heat capacities, urban air cools more slowly at night than the surrounding rural areas forming a *heat island*. If cold air is able to flow into a city, the heat island effect can be reduced. Green areas in cities are often very small and separated by buildings, and are therefore unable to produce enough cold air for the surrounding areas. Urban parks and grasslands should be larger than 200x200 m² and wet (Spronken-Smith and Oke 1999; Spronken-Smith *et al.* 2000).

Besides the importance of cold-air flows in urban climate, the properties of the heat island are relevant. To investigate both, special measurements are necessary, as described in Chap. 7.5, because classical climate maps cannot describe all the processes. For special bioclimatological investigations, Table 7.5 gives some suggestions for assessment criteria by the superposition of heat, pressure, and ventilation (Gerth 1986).

An impressive example of a small-scale cold-air flow is shown in Fig. 7.4. For a slope with an inclination angle of 2°, the measured temperatures at 5 cm height are shown along a 100 m transect from a radiation-cooling forest, to a knoll in an open meadow, and to a groove. The cooling effect of the longwave radiation was about 6 K and with the additional cold-air flow of 8 K ground frost was observed. However, the air temperature at 2 m height in the open meadow was never below 7°C. Such local cold-air flows can be found by a visual assessment.

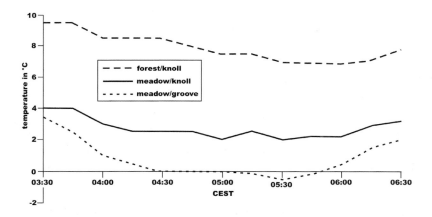

Fig. 7.4. Temperature plot on a night at a slope with 2° inclination between the point knoll and groove (150 m away) in the Ecological-Botanical Garden of the University of Bayreuth on May 14, 1998

Table 7.3. Classification of the frost risk (cloudiness <2/8, wind velocity < 3 ms^{-1}) according to Schumann (personal communication)

risk class	indication	comparison to normal conditions
1	favored	+1 to +5 K
2	normal condition	−1 to +1 K
3	weak to moderate frost risk	−2 to −4 K
4	high to very high frost risk	−5 to −8 K

Table 7.4. Risk class according to the relief of the terrain according to Schumann (personal communication)

relief form	cold air inflow, production	cold air drainage	risk class*
closed basin	existing	–	3
ground of the valley, low slope	existing	weak	3
ground of the valley, moderate slope	existing	moderate	3
plates, Δh > 10m	–	–	2
slope, low inclination (1–3°)	–	moderate	2
slope, high inclination (>15°)	–	very good	1
hills h > 50 m, inclination > 10°	–	very good	1

* marshy soil +1

Table 7.5. Superposition of criteria for a bioclimatological assessment (Gerth 1986)

	high	moderate	low
bioclimatological pressure:			
heat pressure	> 25 days	all other	≤ 25 days
wind velocity	< 2 m s^{-1}	combinations	≥ 2 m s^{-1}
frequency of inversions	> 30% a^{-1}		≤ 30% a^{-1}
ventilation:			
wind velocity	≥ 2 m s^{-1}	all other combinations	< 2 m s^{-1}

Cold-air generation in valleys augmented by cold-air flows can lead to radiation fog if the air is cooled below its dew point. After the generation of the fog, the strongest cooling by radiation occurs at the top of the fog. Because of this, the coldest temperatures are no longer found in the valley but along the slopes above the fog. The reduced longwave radiation from the ground causes a lifting of the fog, and in the low radiation season this is the reason for daylong low stratus within closed valleys or basins.

A special form of the fog is sea smoke, a cumulus-like cloud of some meters in thickness. This is generated by cold air flowing over warmer water. Because this convective effect can happen only for water-air temperature differences greater than 10 K (Tiesel and Foken 1987), sea fog occurs most often in late spring, when small lakes are already warm, or in autumn and early winter over large lakes or the oceans.

7.5 Microclimatological Measurements

Microclimatological measurements have a high relevance for consulting activities (see Chap. 7.4). They are necessary for urban and health-spa climate assessments. Also, many applications are found in agriculture, for example assessing frost risk in orchards and vineyards. Engineers with different degrees of experience often make these assessments. Assessments made using single situations are not sufficient for a generalization. It is important to find those processes that are relevant to the observations. Therefore, the special conditions of the near-surface layer discussed in Chap. 3 must be taken into account. Predominantly, only relevant state parameters are mapped out, *e.g.* cold-air flows or the heating of city centers. There are only a few approaches where atmospheric stability and turbulent fluxes are considered. These are important parameters that must be compared for instance in assessing deposition potential.

Microclimatological measurements cannot be done independently from climatologically measurements. This means that climatologically measurements must be made first. This requires the selection of a relevant basic climate station with measurements over many years and made by meteorological services. Furthermore, measurements in the area of investigation over 1 to 2 years are necessary to transfer the climatological data to this area. Such station can be small automatic weather stations, which should be placed at relevant points according to the classification of the landscape. The real microclimatological measurements are made with moving instrument packages such as cars, vans *etc.* or intensive field campaigns. Therefore, it is necessary that a weather situation must be chosen during which the phenomena being investigated are clearly seen. For cold-air risk investigations, clear nights after cold air advection are necessary. During these mobile measurements, the vehicles stop at the relevant measuring stations to correct the temporal trends. Mobile measurements can be supported by thermal pictures from aircraft. These are not alternatives to mobile measurements because of the missing wind information, which is important for the dynamics of the processes. Otherwise, there is a risk that single measurements can be over interpreted, which are absolutely different for other wind conditions.

Vertical soundings of the atmospheric boundary layer with balloons or indirect measuring technique (Sodar) may be useful. The aim is to study the local development of inversion layers and therefore the dilution of emitted air pollutions and depth of the exchange volume.

In heavily built up areas, radiation measurements are also necessary, *e.g.* the measurements of the reflected shortwave radiation or the longwave emitted radiation close to buildings. Because of climate change, these effects have an increasing relevance during hot summer days. In the last few years, the contribution of turbulent fluxes have been not only been better understood, but also measured in many urban experiments (Rotach *et al.* 2005).

8 Applied Meteorology

In the previous chapters, the basics of applied meteorology as defined in Chap. 1.1 were presented. In this chapter, some examples of applications are presented which are imported for practical work. Certainly, all countries have their own standards and other rules for expert reports, methods, and software programs; thus, in this chapter we can give only some hints that are not greatly different form the German regulations, which are also available in English and given in Appendix A7.

8.1 Distribution of Air Pollution

The simplest models for the distribution of air pollution are Gaussian models. A more precise calculation needs a more sophisticated boundary layer model as described in the literature (Arya 1999; Blackadar 1997). Of micrometeorological relevance, is the determination of the stratification, and so measurements in the surface layer are necessary. The following air pollution situations which depend on wind velocity and stratification are typical and can be observed, (Blackadar 1997; Kraus 2004):

- For high wind velocities the turbulent eddies are small and the plume of pollutants in the atmosphere expands quickly. Note that for wind velocities above 6 ms^{-1}, nearly always neutral stratification can be assumed.
- If the size of the turbulent eddies approximately equals the width of the plume, then serpentine forms of sideways and up-and-down movements of the plume occur. This is typical for low wind and sunny conditions and is referred to as *looping*.
- For smaller eddies, instead of looping the plume shape is conical, and is referred top as *coning*.
- An almost completely horizontal distribution of the plume exists for only low movements of air and on clear nights. This condition is referred to as *fanning*.
- In the case of an inversion layer and neutral or slightly unstable stratification near the surface, as in the morning hours, the plume can come in contact with the ground surface. This condition is referred to as *fumigation*.

The estimation of the stratification is made using stability classes. These classes are based on meteorological observations or on the stability-dependent fluctuation of the wind direction. (Note that for averaging of wind directions, the jump between 360° and 0° must be taken into account!) If the wind direction fluctuation parameters are not observed, then a parameterization with the mixed layer z_i, the roughness parameter z_0, and integrated stability function $\psi(z/L)$, dependent on the Obukhov-length L, are possible (Blackadar 1997):

$$\sigma_\varphi \approx \frac{\sigma_v}{\overline{u}} = \frac{\kappa\left(12 - 0.5\, z_i/L\right)^{\frac{1}{3}}}{\ln\left(z/z_0\right) - \psi\left(z/L\right)} \qquad (8.1)$$

With this and Table 8.1, the stability classes according to Pasquill (1961) can be determined. Note that in complex terrain the standard deviation of the wind direction is greater than that over flat terrain and a more unstable stratification is predicted. In the case of available direct measurements of the stratification, these should have priority.

An estimation of the stratification is also possible with the wind speed, the radiation, and the cloudiness (Table 8.2). The comparison of different stability classes is of special interest because the Pasquill classes are not used in all countries. In Germany, for example, the classes according to Klug-Manier (Klug 1969; Manier 1975) are used. A comparison including the Obukhov length is given in Table 8.3. With the methods to measure the energy exchange given in Chap. 4 or the models given in Chap. 5, the necessary stratification for dispersion models can be determined.

The dispersion of air pollutions can be described with probability density functions (Arya 1999; Blackadar 1997):

$$\int_{-\infty}^{\infty} F(x)\, dx = 1 \qquad (8.2)$$

For a three dimensional distribution follows:

$$\int_{-\infty}^{\infty}\int_{-\infty}^{\infty}\int_{-\infty}^{\infty} F(x)G(y)H(z)\, dx\, dy\, dz \qquad (8.3)$$

For a point source with constant emission rate $Q\, dt$ and constant horizontal wind velocity the distribution density function is:

$$F(x) = \frac{1}{u\, dt} \qquad (8.4)$$

For the transverse horizontal and vertical distributions, the Gaussian distribution functions are used:

Table 8.1. Definition of the Pasquill classes with the standard deviation of the wind direction (Blackadar 1997)

Pasquill class	description	σ_φ
A	extreme unstable	25
B	unstable	20
C	light unstable	15
D	neutral	10
E	light stable	5
F	stable	2.5

Table 8.2. Determination of the Pasquill stability class from meteorological parameters (Blackadar 1997)

surface wind m s^{-1}	irradiation at day			cloudiness at night	
	strong	moderate	low	thin clouds or $\geq$ 4/8	$\leq$ 3/8
< 2	A	A–B	B		
2	A–B	B	C	E	F
4	B	B–C	C	D	E
6	C	C–D	D	D	D
> 6	C	D	D	D	D

Table 8.3. Comparison of different stability classes

	Klug/Manier	Pasquill	Obukhov length L	z/L for z=10 m
very unstable	V	A	− 30	− 0.33
unstable	IV	B	− 100	− 0.1
neutral to light unstable	III/2	C	− 300	− 0.033
neutral to light stable	III/1	D (neutral)	5000	0.002
stable	II	E (light stable)	250	0.04
very stable	I	F (stable)	60	0.17

$$G(y) = \frac{1}{\sqrt{2\pi}\,\sigma_v} \exp\left(-\frac{y^2}{2\sigma_v}\right) \tag{8.5}$$

$$H(z) = \frac{1}{\sqrt{2\pi}\,\sigma_w} \exp\left(-\frac{z^2}{2\sigma_w}\right) \tag{8.6}$$

The concentration distribution can be also calculated with Fick's diffusion law:

$$\frac{\partial \chi}{\partial t} + \bar{u}\frac{\partial \chi}{\partial x} = \frac{\partial}{\partial x}\left(K_x \frac{\partial \chi}{\partial x}\right) + \frac{\partial}{\partial y}\left(K_y \frac{\partial \chi}{\partial y}\right) + \frac{\partial}{\partial z}\left(K_z \frac{\partial \chi}{\partial z}\right) \tag{8.7}$$

The parameterization of the diffusion coefficients is made with error functions (see also Eq. (3.2)):

$$\sigma_{u_x}^2 = 2K_x t \qquad \sigma_{u_y}^2 = 2K_y t \qquad \sigma_{u_z}^2 = 2K_z t \tag{8.8}$$

The concentration distribution for constant source strength and horizontal wind speed in x-direction is given by:

$$\chi(x,y,z) = \frac{Q}{2\pi \bar{u}\, \sigma_{u_y}\, \sigma_{u_z}} \exp\left(-\frac{y^2}{2\sigma_{u_y}} - \frac{z^2}{2\sigma_{u_z}}\right) \tag{8.9}$$

In the case of no meteorological data, the standard deviations of the wind components can be parameterized with micrometeorological approaches as described in Chap. 2.4.1.

Presently, a number of dispersion models are available. One widely distributed model is that developed by Gryning *et al.* (1983), which is the basis for the footprint model by Schmid (1997). Differences of various models are the parameterization of the wind components and the applications of three-dimensional wind fields in flat and complex terrain. Recent and more sophisticated models use numerical boundary layer approaches with parameterizations of the profiles of the turbulent wind components rather than the Gaussian distribution functions.

8.2 Meteorological Conditions of Wind Energy Use

In the beginning of wind energy use, meteorology was especially important in the investigations of the operational requirements and efficiency. The reason for this is that the capacity of wind power stations depends on the cube of the wind speed, u. Thus, slightly higher wind speeds make a significant higher wind power capacity (F: area covered over by the rotor, ρ: air density):

$$P = \frac{\rho}{2} F u^3 \tag{8.10}$$

A first comprehensive study to the meteorological aspects of wind energy use was presented by the World Meteorological Organization (WMO 1981). In that study, the remarkable influence of the surface roughness and internal boundary layers were mainly discussed (see Chaps. 3.1 and 3.2).

Significant benefits for the meteorological support of wind energy were made with the European Wind Atlas (Troen and Peterson 1989). The basis for expert reports on the capacity of wind power stations is the Weibull frequency distribution, where the skewness of the wind velocity distribution is well presented for $k < 3.6$. Using this distribution function, the power density can be determined:

$$P = A^3 F_p(k) \tag{8.11}$$

The Weibull parameters A and k and the distribution $F_P(k)$ are tabulated for wind direction classes for all relevant meteorological stations. The basic idea of the wind atlas is to correct long-term climate measurements of meteorological reference stations regarding local influences, and to determine Weibull distributions that are free of these influences for these reference stations. For an expert review of a new location of a wind power station, the nearest reference station is used and the Weibull distribution will be corrected by the local influences of the location. From this, the possible wind power capacity of the new station can be estimated.

Some of the correction methods used are of general interest (compare also Chap. 3.3). One method is the determination of the surface roughness, which is dependent on the distribution of obstacles. For a uniform distribution of obstacles it follows

$$z_0 = 0.5 \frac{h\,S}{A_H},$$
(8.12)

where h is obstacle height, S is the area of the cross section of the obstacle in wind direction, A_H is the mean horizontal area where $A_H \gg S$). Over the ocean, the Charnock-equation Eq. (3.5) is used, and over land only four roughness classes are applied (sand/water, agricultural area with an open image, agricultural area with a complex image, and built-up areas/suburban areas). For the averaging of the roughness parameter, a special area averaging method was developed (Table 3.2). The method used for the determination of the wind shadow behind an obstacle was already discussed in Chap. 3.3. Furthermore, the wind atlas includes a current model, which also can calculate the increase of the wind speed over hills.

In Europe and other regions, the European Wind Atlas together with the application software is presently applied in expert reviews. It is designed for flat lands, and has only limited use for mountain regions. In mountain regions, the very complex turbulence and flow structures cause a high mechanical pressure on the wind power station, and the operation is not always optimal relative to the turbulence spectra (Hierteis *et al.* 2000).

The determination of the wind energy potential over a large area is made with meso-scale models (Mengelkamp 1999). The increasing use of wind energy production in hilly regions, large wind parks, and *off-shore* requires further meteorological research and a specific recommendation because these questions are insufficiently covered in the wind atlas.

8.3 Sound Propagation in the Atmosphere

The propagation of sound in the atmosphere is an important issue because of increasing noise pollution for example by vehicle traffic on roads and aircraft. The noise is very disturbing during the night; however, these periods are not included in laws and are also insufficiently investigated from the scientific point of view.

The sound velocity can be determined in the dry atmosphere with the Laplace-equation

$$c = \sqrt{\gamma_d p/\rho},$$
$$\gamma_d = c_p/c_v,$$
(8.13)

where p is the air pressure. From the application of the ideal gas equation, which requires that the sound pressure $p_s \ll p$ and the frequency of the free molecule movement $f_s \ll 10^5$ kHz, and $\mu_d = 28.97$ kg kmol^{-1}, follows:

$$c_d = \sqrt{\gamma_d \, R_L \, T / \mu_d} \tag{8.14}$$

For moist air:

$$c_m = \sqrt{\gamma_d \frac{R_L \, T}{\mu_d} \left(1 + 0.28 \frac{e}{p}\right)} \tag{8.15}$$

The sound velocity of moist air can also be determined as an additional term in the sound velocity for dry air:

$$c_m = c_d \sqrt{1 + 0.28 \frac{e}{p}} \tag{8.16}$$

$$c_m = c_d \left(1 + 0.14 \frac{e}{p}\right) \qquad \text{for } t < 30°C, \ e/p < 4\ \%$$

Due to the strong vertical gradients of wind and temperature near the ground surface, there are gradients of the sound velocity, which can cause deflections of sound either to or from the surface

$$\frac{dc}{dz} = 0.6 \frac{dT}{dz} + \cos\varphi \frac{du}{dz}, \tag{8.17}$$

where φ: angle of the sound direction to the vertical direction.

Consequently, sound deflection to the ground is connected with an increase of the sound velocity with the height. This happens when the temperature increase with the height (inversion), and the deflection is stronger downwind of the sound source than upwind. The opposite case occurs for unstable stratification, where the sound deflection is upward and increases more upwind of the sound source than downwind. Downwind of the sound source, shadow zones are found near the ground surface. The situations are illustrated in Fig. 8.1.

The international standards (ISO 1996) combine the meteorological influences on sound velocity into a mean measure, which represents neutral and slightly unstable stratification. But the sound propagation in the stable case is complex, especially when the sound is deflected to the ground and the atmosphere is divided into many thin layers. Sound can follow these layers without significant damping. Increasing traffic-generated sound is connected with increasing noisiness especially at night. This connection is not adequately reflected in laws and assessment procedures. Micrometeorologist can help solve such problems, but the ability to describe stable stratification conditions is limited (see Chap. 3.6).

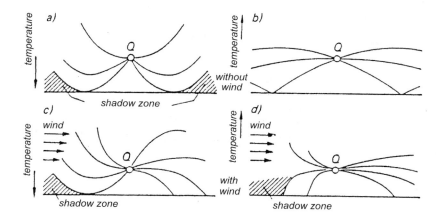

Fig. 8.1. Sound propagation close to the surface (VDI 1988)

8.4 Human Biometeorology

Human biometeorology describes the specific influence of weather and climate on people (Hentschel 1982; Tromp 1963). This is separated into effect complexes: The *thermo effect* complex includes temperature, humidity, wind velocity, short-wave, and long-wave radiation. The *actinic effect* complex includes the visible and ultraviolet radiation with a direct biological effectiveness (no heating). The *air hygienic* effect complex includes solid, liquid, and gaseous natural and anthropogenic air pollution. Furthermore noise, wind, and smells affect humans.

Fanger (1972) developed a Predicted Mean Vote (*PMV*) to combine the effects of these complexes on humans as an empirical heat balance equation. This is a very complex empirical function, which includes the internal heat production, the heat isolation by clothing, the air temperature, the water vapor pressure, the radiation temperature of the surface, and the wind velocity:

$$PMV = \left[0.028 + 0.303 \exp\left(-0.036 \frac{M}{A_{Du}} \right) \right] \cdot$$

$$\cdot \left[\frac{H}{A_{Du}} - Q_{Ed} - Q_{Esw} - Q_{Ea} - Q_{Ha} - Q_{H} - I \uparrow \right] \qquad (8.18)$$

For the determination of the *PMV*, the internal heat production per unit area is H/A_{Du}, in Wm^{-2}, where H is the sum of the total energy turnover (metabolic value), and M is the mechanical energy. From these values, the energy loss by the diffusion of water vapor through the skin Q_{Ed}, the evaporation of sweat Q_{Esw}, the latent Q_{Ea} and the sensible heat loss Q_{Ha} by respiration, the convective heat loss Q_H, and the longwave radiation loss $I\uparrow$ are subtracted. The equations are bulk equations similar to Eq. (4.24) to (4.26) and the Stefan-Boltzmann equation Eq. (1.4), but the coefficients used are not comparable. The normalization of the equation is made in such a way that *PMV* values of ± 0.5 are related to comfort conditions. The range from 0 to –3.5 is related to increasing cold stress, and from 0 to 3.5 to increasing heat stress.

In Germany, the model for the *PMV* calculation is called *Klimamichel* (a word meaning the plane honest German) model (Höppe 1984, 1992; Jendritzky *et al.* 1990) which is shown schematically in Fig. 8.2. The calculations are based on a uniform man or woman. Therefore single values have no great significance. But frequency distributions give information about the heat stress in different regions. A transformation of the *PMV* index into a simple value for the heat stress on a human is the *physiological equivalent temperature* (*PET*, Höppe 1999) given in degrees centigrade. A comparison of both parameters is given in

Table 8.4. These values also include related feeling classes and stress factors. With the *PET* index, urban climatologically and climatologically classifications of larger areas can be described (Matzarakis *et al.* 1999).

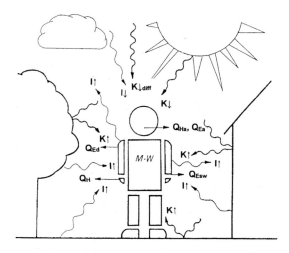

Fig. 8.2. Illustration of the *PMV* determination with the *Klimamichel* model (VDI 1998)

Table 8.4. Comparison table of the Predicted Mean Vote (*PMV*) and the physiological equivalent temperature (*PET*) and related feelings and stress effects (Matzarakis and Mayer 1997)

PMV	PET in °C	thermal feeling	physiological pressure class
		very cold	extreme cold stress
−3.5	4		
		cold	strong cold stress
−2.5	8		
		cool	moderate cold stress
−1.5	13		
		slightly cool	light cold stress
−0.5	18		
		comfortable	no heat pressure
0.5	23		
		slightly warm	light heat pressure
1.5	29		
		warm	moderate heat pressure
2.5	35		
		hot	strong heat pressure
3.5	41		
		very hot	extreme heat pressure

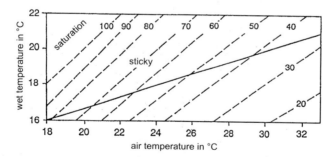

Fig. 8.3. Determination of the sticky range according to Hentschel (1982) in the modification by Hupfer (1996)

For the feeling of the stress due to the weather, which can also be different because of microclimatological effects, the cooling effect of the wind velocity (wind chill) and the sticky feeling are very meaningful parameters. The wind chill temperature is a hypothetic temperature for which under calm conditions the skin passes the same amount of heat for a given temperature and wind velocity. Recently, the wind chill temperature has been repeatedly modified (Osczevski and Bluestein 2005; Osczevski 2000; Tikusis and Osczevski 2002; 2003), and is used from the American Meteorological Service in the following form:

$$t_{Wind-Chill} = 13.12 + 0.6215\,t - 11.37\,u^{0,16} + 0.3965\,t\,u^{0,16} \qquad (8.19)$$

$$t\ in\ ^{\circ}C,\quad u\ in\ km\,h^{-1}$$

The stickiness depends on the air temperature and the air moisture. In Fig. 8.3 the sticky range is dependent on the wet temperature measured with an aspirated psychrometer.

8.5 Perspectives of the Applied Meteorology

It should be noted that the methods used in applied meteorology are often phenomenological or work with simple parameterizations. This may be in contrast to detailed theoretical explanations found in this book. This can be explained by the engineering work in applied meteorology, which uses simple analytical relations. In air pollution and wind energy investigations, in addition to simple standards numerical models are more often applied. The international tendency is to replace the description of state variables with the description of processes. Modern flux measuring methods and measuring devices as described in Chaps. 4 and 6 are increasingly being applied. The limiting application conditions of the methods in areas with obstacles and in complex terrain have been overcome with a quality assessment of the data and a combination of in-situ and wind tunnel measurements.

This book should be a basis for increasing the theoretical and experimental levels of applied meteorology. In this period of climate change, it is important to make qualitative statements, but it is even more important to make quantitative statements. Therefore basic knowledge is necessary.

Appendix

A1 Further Monographs

Following monographs are listed, which are recommended for a further study. These are limited on micrometeorological and measuring technique textbooks.

Further Micrometeorological Literature

Arya, SP (1999) Air pollution meteorology and dispersion. Oxford University Press, New York, Oxford, 310 pp.

Arya, SP (2001) Introduction to micrometeorology. Academic Press, San Diego, 415 pp.

Bailey, WG, Oke, TR, Rouse, WR (Editors) (1997) The surface climate of Canada. Mc Gill-Queen's University Press, Montreal, Kingston, 369 pp.

Bendix, J (2004) Geländeklimatologie. Borntraeger, Berlin, Stuttgart, 282 pp.

Blackadar, AK (1997) Turbulence and diffusion in the atmosphere. Springer, Berlin, Heidelberg, 185 pp.

Campbell, GS, Norman, JM (1998) Introduction to environmental biophysics. Springer, New York, 286 pp.

Garratt, JR (1992) The atmospheric boundary layer. Cambridge University Press, Cambridge, 316 pp.

Geiger, R, Aron, RH, Todhunter, P (1995) The climate near the ground. Friedr. Vieweg & Sohn Verlagsges. mbH, Braunschweig, Wiesbaden, 528 pp.

Helbig, A, Baumüller, J, Kerschgens, J (Editors) (1999) Stadtklima und Luftreinhaltung. Springer, Berlin, Heidelberg, 467 pp.

Jones, HG (1992) Plants and microclimate. Cambridge Univ. Press, Cambridge, 428 pp.

Kaimal, JC, Finnigan, JJ (1994) Atmospheric boundary layer flows: Their structure and measurement. Oxford University Press, New York, NY, 289 pp.

Kantha, LH, Clayson, CA (2000) Small scale processes in geophysical fluid flows. Academic Press, San Diego, 883 pp.

Lee, X, Massman, WJ, Law, B (Editors) (2004) Handbook of micrometeorology: A guide for surface flux measurement and analysis. Kluwer, Dordrecht, 250 pp.

Monteith, JL, Unsworth, MH (1990) Principles of environmental physics. Edward Arnold, London, 291 pp.

Oke, TR (1987) Boundary layer climates. Methuen, New York, 435 pp.

Stull, RB (1988) An Introduction to boundary layer meteorology. Kluwer Acad. Publ., Dordrecht, Boston, London, 666 pp.

Further Measuring Technique Literature

Bentley, JP (2005) Principles of measurement systems. Pearson Prentice Hall, Harlow, 528 pp.

Brock, FV, Richardson, SJ (2001) Meteorological measurement systems. Oxford University Press, New York, 290 pp.

DeFelice, TP (1998) An introduction to meteorological Instrumentation and measurement. Prentice Hall, Upper Saddle River, 229 pp.

Dobson, F, Hasse, L, Davis, R (Editors) (1980) Air-sea interaction, Instruments and methods. Plenum Press, New York, 679 pp.

Kaimal, JC, Finnigan, JJ (1994) Atmospheric boundary layer flows: Their structure and measurement. Oxford University Press, New York, NY, 289 pp.

A2 Use of SI-Units

The following table includes important SI-units used in the book. The basic units are **bold** highlighted.

name	SI-unit	unit	calculation
length	**meter**	**m**	
time	**second**	**s**	
velocity		$m\ s^{-1}$	$1\ km\ h^{-1} = (1/3.6)\ m\ s^{-1}$
acceleration		$m\ s^{-2}$	
mass	**kilogram**	**kg**	
density		$kg\ m^{-3}$	
impulse		$kg\ m\ s^{-1}$	$1\ kg\ m\ s^{-1} = 1\ N\ s$
force	Newton	N	$1\ N = 1\ kg\ m\ s^{-2}$
pressure, friction	Pascal	Pa	$1\ Pa = 1\ N\ m^{-2}$
			$1\ Pa = 1\ kg\ m^{-1}\ s^{-2}$
air pressure	hektopascal	hPa	$1\ hPa = 100\ Pa$
work, energy	Joule	J	$1\ J = 1\ N\ m = 1\ W\ s$
			$1\ J = 1\ kg\ m^2\ s^{-2}$
power	Watt	W	$1\ W = 1\ J\ s^{-1} = 1\ N\ m\ s^{-1}$
			$1\ W = 1\ kg\ m^2\ s^{-3}$
energy flux density		$W\ m^{-2}$	$1\ W\ m^{-2} = 1\ kg\ s^{-3}$
temperature	**Kelvin**	**K**	
Celsius-temperature		°C	$0°C = 273.15\ K$
temperature difference		K	

A3 Constants and Important Parameters

Even though the accuracy of meteorological measurements is at most only 3–5 significant digits, the following physical constants are given with errors expressed in ppm, because sometimes in the literature different values are reported.

The values listed are taken from the 1986 calculations of Cohen and Taylor (1986) and the international temperature scale ITS-90 as given by Sonntag (1990):

constant	symbol	value	error
standard values			
standard air pressure	p_0	1013.25 hPa	
standard temperature	T_0	273.15 K = 0°C	
temperature of the triple point of water		273.16 K	
acceleration due to gravity at lat.. 45°	g	9.80665 m s^{-2}	
general constants			
velocity of the light in vacuum	c	299 792 458 m s^{-1}	exact
Planck's constant	h	6.626 0755(40) 10^{-34} J s	0.60
physical-chemical constants			
Avogadro number	N_A	6.022 1367(36) 10^{23} mol^{-1}	0.59
atom mass ^{12}C/12	m_u	1.660 5402(10) 10^{-27} kg	0.59
universal gas constant	R	8.314 510(70) J mol^{-1} K^{-1}	8.4
Boltzmann constant R/N_A	k	1.380 658(12) J K^{-1}	8.4
molar volume (ideal gas)	RT_0/p_0	22.414 10(19) l mol^{-1}	8.4
Stefan-Boltzmann constant	σ_{SB}	5.670 51(19) 10^{-8} W m^{-2} K^{-4}	34
Wien's constant	$\lambda_{max} T$	2.897 756(24) 10^{-3} m K	8.4
thermo-dynamical constants			
molar mass of dry air	M_L	0.028 9645(5) kg mol^{-1}	17
molar mass of water vapor	M_W	0.018 01528(50) kg mol^{-1}	27
ratio M_W/M_L	γ	0.62198(2)	33
gas constant of dry air	R_L	287.058 6(55) J kg^{-1} K^{-1}	19
gas constant of water vapor	R_W	461.525 (13) J kg^{-1} K^{-1}	29

The following quantities are valid for 1013.25 hPa and 15°C, if no other remark is given (Stull 1988):

quantity	symbol	value	re-mark
air			
specific heat of dry air at constant pressure	$(c_p)_L$	1004.67 J kg^{-1} K^{-1}	1)
specific heat of moist air at constant pressure	c_p	$= c_{pL}(1+0.84\,q)$, q in kg kg^{-1}	
specific heat of dry air at constant volume	$(c_v)_L$	718 J kg^{-1} K^{-1}	1)
ratio of the specific heats	$(c_p/c_v)_L$	$= 7/5 = 1.4$	1)
ratio of the gas constant and the specific heat for dry air	$(R_L/c_p)_L$	$= 2/7 = 0.286$	1)
density	ρ_L	1.225 kg m^{-3}	
	ρ_{L0}	1.2923 kg m^{-3}, 0°C	1)
kinematic molecular viscosity	υ	1.461 10^{-5} m^2 s^{-1}	
molecular thermal conductivity	υ_T	2.06 10^{-5} m^2 s^{-1}	
dynamic molecular viscosity	$\mu = \rho_L/\upsilon$	1.789 10^{-5} kg m^{-1} s	
molecular thermal diffusivity	$a_T =$ $\upsilon_T \rho_L \cdot c_{PL}$	2.53 10^{-2} W m^{-1} K^{-1}	
psychrometric constant (water)		6.53 10^{-4} (1+0.000944 t') p K^{-1}	2)
psychrometric constant (ice)		5.75 10^{-4} p K^{-1}	2)
water and water vapor			
specific heat of water vapor at constant pressure	$(c_p)_W$	1846 J kg^{-1} K^{-1}	
specific heat of water vapor at constant volume	$(c_v)_W$	1389 J kg^{-1} K^{-1}	
ratio of the specific heats	$(c_p/c_v)_W$	$= 4/3 = 1.333$	
ratio of the gas constant and the specific heat	$(R_L/c_p)_W$	$= 1/4 = 0.25$	
density of water	ρ_W	1025 kg m^{-3}	
latent heat of vaporization	λ	(2.501 −0.00237 t) 10^6 J kg^{-1}	
other quantities			
Coriolis parameter	f	1.458 10^{-4} sin φ s^{-1}	1)
solar constant	S	energy units: − 1368 W m^{-2} kinematic units: − 1.125 K m s^{-1}	3)
constant gravity acceleration	g_0	9.80 m s^{-2}	

1) identical data in Landolt-Börnstein (Fischer 1988) according to WMO recommendations
2) Sonntag (1990)
3) Glickman (2000) and Houghton (2001)

Temperature dependent quantities (Fischer 1988):

temperature	specific heat capacity in 10^3 J kg^{-1} K^{-1}		latent heat in 10^6 J kg^{-1}		
	ice	water	vaporization	fusion	sublimation
−20	1.959	4.35	2.5494	0.2889	2.8387
−10	2.031	4.27	2.5247	0.3119	2.8366
0	2.106	4.2178	2.50084	0.3337	2.8345
5		4.2023	2.4891		
10		4.1923	2.4774		
15		4.1680	2.4656		
20		4.1818	2.4535		
25		4.1797	2.4418		
30		4.1785	2.4300		
35		4.1780	2.4183		
40		4.1785	2.4062		

A4 Further Equations

Calculation of Astronomical Quantities

In some applications, it is often necessary to determine the solar inclination angle as a function of time. The following approximations for some of these calculations must be applied in several steps:

To determine the declination of the sun, δ, the latitude of the sun, φ_s, must first be calculated (Holtslag and van Ulden 1983)

$$\varphi_S = 4.871 + 0.0175\,DOY + 0.033\,\sin(0.0175\,DOY), \qquad (A1)$$

where DOY is the day of the year where the 1st of January has the number 1.

The declination angle is given by:

$$\delta = \arcsin(0.398\sin\varphi_s) \qquad (A2)$$

To determine the position of the sun, the "hour angle" h

$$h = \frac{2\pi t_H}{\Delta t_d} \qquad (A3)$$

must be calculated which gives the angular difference between δ and the zenith of the sun (Liou 1992), where t_H is the time distance to culmination of the sun in seconds and $\Delta t_d = 86400$ s is the duration of a full rotation of the Earth. It is necessary to apply the equation of time (EQT), which gives the difference between the true and the averaged local time. From tables of the time equation (Neckel and Montenbruck 1999) for 15° E and 50 °N and the year 2000 Göckede (2000) calculated an approximation equation:

$$EQT = \sum_{n=0}^{12} x_n \, DOY^n \qquad \text{(A4)}$$

The coefficients are listed in the following table:

n	x_n	n	x_n
12	1.67050145E-27	5	-3.35088054E-09
11	-3.94339477E-24	4	2.29121835E-07
10	4.05537695E-21	3	-9.77373856E-06
9	-2.37686237E-18	2	1.23393733E-04
8	8.72332216E-16	1	6.69458473E-03
7	-2.07900028E-13	0	4.87707231E-02
6	3.25188650E-11		

The time distance to culmination of the sun, t_H, for Central European Time is given by

$$t_H = \left\{ t - \left[12 + EQT + \frac{(15 - \lambda)4}{60} \right] \right\} 3600, \qquad \text{(A5)}$$

where t: time in hours, and λ is longitude.

With the latitude ψ in radians of a location, the angle of inclination of the sun can be determined for any time:

$$\sin \varphi = \sin \delta \sin \psi + \cos \delta \cos \psi \cos h \qquad \text{(A6)}$$

To determine the incoming extraterrestrial radiation at the upper border of the atmosphere from the solar constant the variability of the distance between the sun and the Earth must be taken into account

$$K \downarrow_{extraterr.} = S \left(\frac{r_0}{r} \right)^2 \sin \varphi, \qquad \text{(A7)}$$

where r_0 is the mean distance of the Earth from the Sun (149 597 870.66 km) and r is the actual distance. The ratio of both can be determined according to Hartmann (1994) as a Fourier series

$$\left(\frac{r_0}{r} \right)^2 = \sum_{n=0}^{2} a_n \cos (n \theta_d) + b_n \sin (n \theta_d) \qquad \text{(A8)}$$

with

$$\theta_d = \frac{2\pi\, DOY}{265} \tag{A9}$$

where in the case of a leap year the denominator is 266. The coefficients for Eq. (A8) are given in the following table:

n	a_n	b_n
0	1.000110	
1	0.034221	0.001280
2	0.000719	0.000077

Universal Functions

Even though the universal function formulated by Businger *et al.* (1971) and later modified by Högström (1988) are widely used, knowledge of other universal functions may be quite useful for different research activities. The following table is based on works of Dyer (1974), Yaglom (1977), Foken (1990), and Andreas (2002). The values the von-Kármán constant, κ, used in the formulations are given. The notation, 0.40*, indicates that the original function was re-calculated by Högström (1988) using $\kappa = 0.40$.

reference	κ	universal function for momentum exchange
Swinbank (1964)	–	$\frac{z}{L}\left(1 - e^{z/L}\right)^{-1}$ $z/L < 0$
Swinbank (1968)	0.40	$0.613\left(-z/L\right)^{-0.2}$ $-0.1 \geq z/L \geq -2$
Tschalikov (1968)	0.40	$1 + 7.74\, z/L$ $z/L > 0.04$
Zilitinkevich and. Tschalikov (1968)	0.434	$1 + 1.45\, z/L$ $-0.15 < z/L < 0$
		$0.41\left(-z/L\right)^{-1/3}$ $-1.2 < z/L < -0.15$
		$1 + 9.9\, z/L$ $0 < z/L$
	0.40*	$1 + 1.38\, z/L$ $-0.15 < z/L < 0$
		$0.42\left(-z/L\right)^{-1/3}$ $-1.2 < z/L < -0.15$
		$1 + 9.4\, z/L$ $0 < z/L$
Webb (1970)	–	$1 + 4.5\, z/L$ $z/L < -0.03$
Dyer and Hicks (1970)	0.41	$\left(1 - 16\, z/L\right)^{-1/4}$ $-1 < z/L < 0$

reference	κ	universal function for momentum exchange
Businger *et al.* (1971)	0.35	$\left(1-15\frac{z}{L}\right)^{-\frac{1}{4}}\qquad -2<\frac{z}{L}<0$ $1+4.7\frac{z}{L}\qquad 0<\frac{z}{L}<1$
	0.40*	$\left(1-19,3\frac{z}{L}\right)^{-\frac{1}{4}}\qquad -2<\frac{z}{L}<0$ $1+6\frac{z}{L}\qquad 0<\frac{z}{L}<1$
Dyer (1974)	0.41	$\left(1-16\frac{z}{L}\right)^{-\frac{1}{4}}\qquad -1<\frac{z}{L}<0$ $1+5\frac{z}{L}\qquad 0<\frac{z}{L}$
Dyer (1974)	0.40*	$\left(1-15.2\frac{z}{L}\right)^{-\frac{1}{4}}\qquad -1<\frac{z}{L}<0$ $1+4.8\frac{z}{L}\qquad 0<\frac{z}{L}$
Skeib (1980), see also: Foken and. Skeib (1983) and Foken (1990)	0.40 0.40*	$1\qquad -0.0625<\frac{z}{L}<0.125$ $\left(\dfrac{\frac{z}{L}}{-0.0625}\right)^{-\frac{1}{4}}\qquad -2<\frac{z}{L}<-0.0625$ $\dfrac{\frac{z}{L}}{0.125}\qquad 0.125<\frac{z}{L}<2$
Gavrilov and Petrov (1981)	0.40	$\left(1-8\frac{z}{L}\right)^{-\frac{1}{3}}\qquad \frac{z}{L}<0$ $1+5\frac{z}{L}\qquad 0<\frac{z}{L}$
Dyer and. Bradley (1982)	0.40 0.40*	$\left(1-28\frac{z}{L}\right)^{-\frac{1}{4}}\qquad \frac{z}{L}<0$
Beljaars and Holtslag (1991)	0.40	$1+\frac{z}{L}+\frac{2}{3}\frac{z}{L}\left(6-0.35\frac{z}{L}\right)\cdot e^{-0.35\frac{z}{L}}\quad 0<\frac{z}{L}$
King *et al.* (1996)	0.40	$1+5.7\frac{z}{L}\le12\qquad 0<\frac{z}{L}$
Handorf *et al.* (1999)	0.40	$1+5\frac{z}{L}\qquad 0<\frac{z}{L}<0.6$ $4\qquad 0.6<\frac{z}{L}$

reference	κ	universal function for the exchange of sensible heat, $\alpha_0=1$
Swinbank (1968)	0.40	$0.227\left(-\frac{z}{L}\right)^{-0.44}\qquad -0.1\ge\frac{z}{L}\ge-2$
Tschalikow (1968)	0.40	$1+5.17\frac{z}{L}\qquad \frac{z}{L}>0.04$

reference	κ	universal function for the exchange of sensible heat, $\alpha_0 = 1$
Zilitinkevich and Tschalikov (1968)	0.434	$1 + 1.45\,z/L$ $\qquad -0.15 < z/L < 0$ $0.41\left(-z/L\right)^{-\frac{1}{3}}$ $\qquad -1.2 < z/L < -0.15$ $1 + 9.9\,z/L$ $\qquad 0 < z/L$
Zilitinkevich and Tschalikov (1968)	0.40*	$0.95 + 1.31\,z/L$ $\qquad -0.15 < z/L < 0$ $0.40\left(-z/L\right)^{-\frac{1}{3}}$ $\qquad -1.2 < z/L < -0.15$ $0.95 + 8.9\,z/L$ $\qquad 0 < z/L$
Webb (1970)	–	$1 + 4.5\,z/L$ $\qquad z/L < -0.03$
Dyer and Hicks (1970)	0,41	$\left(1 - 16\,z/L\right)^{-\frac{1}{2}}$ $\qquad -1 < z/L < 0$
Businger *et al.* (1971)	0,35	$0.74\left(1 - 9\,z/L\right)^{-\frac{1}{2}}$ $\qquad -2 < z/L < 0$ $0.74 + 4.7\,z/L$ $\qquad 0 < z/L < 1$
	0,40*	$0.95\left(1 - 11.6\,z/L\right)^{-\frac{1}{2}}$ $\qquad -2 < z/L < 0$ $0.95 + 7.8\,z/L$ $\qquad 0 < z/L < 1$
Dyer (1974)	0,41	$\left(1 - 16\,z/L\right)^{-\frac{1}{2}}$ $\qquad -1 < z/L < 0$ $1 + 5\,z/L$ $\qquad 0 < z/L$
	0,40*	$0.95\left(1 - 15.2\,z/L\right)^{-\frac{1}{2}}$ $\qquad -1 < z/L < 0$ $0.95 + 4.5\,z/L$ $\qquad 0 < z/L$
Skeib (1980), see also: Foken and Skeib (1983) and Foken (1990)	0,40	1 $\qquad -0.0625 < z/L < 0.125$ $\left(\dfrac{z/L}{-0.0625}\right)^{-\frac{1}{2}}$ $\qquad -2 < z/L < -0.0625$ $\left(\dfrac{z/L}{0.125}\right)^{2}$ $\qquad 0.125 < z/L < 2$

reference	κ	universal function for the exchange of sensible heat, $\alpha_0 = 1$	
Skeib (1980), see also: Foken and Skeib (1983) and Foken (1990)	0.40*	0.95	$-0.0625 < {}^z/_L < 0.12$
		$0.95 \left(\dfrac{{}^z/_L}{-0.0625} \right)^{-1/2}$	$-2 < {}^z/_L < -0.0625$
		$0.95 \left(\dfrac{{}^z/_L}{0.125} \right)^{2}$	$0.125 < {}^z/_L < 2$
Gavrilov and Petrov (1981)	0.40	$0.65 \left[\left(1 - 35\,{}^z/_L\right)^{-1/2} + \dfrac{0.25}{1 + 8\left({}^z/_L\right)^2} \right]$	${}^z/_L < 0$
		$0.9 + 6\,{}^z/_L \qquad 0 < {}^z/_L$	
Dyer and Bradley (1982)	0.40 0.40*	$\left(1 - 14\,{}^z/_L\right)^{-1/2} \qquad {}^z/_L < 0$	
Beljaars and. Holtslag (1991)	0.40	$1 + {}^z/_L\left(1 + \tfrac{2}{3}\,{}^z/_L\right)^{1/2} + \tfrac{2}{3}\,{}^z/_L\left(6 - 0.35\,{}^z/_L\right)e^{-0.35\,{}^z/_L}$	
		$0 < {}^z/_L$	
King et al. (1996)	0.40	$0.95 + 4.99\,{}^z/_L \le 12 \qquad 0 < {}^z/_L$	
Handorf et al. (1999)	0.40	$1 + 5\,{}^z/_L \qquad 0 < {}^z/_L < 0.6$	
		$4 \qquad\qquad 0.6 < {}^z/_L$	

reference	κ	Universal function for the energy dissipation	
Wyngaard and Coté (1971)	0.35	$\left[1 + 0.5\left|{}^z/_L\right|^{2/3}\right]^{3/2}$	${}^z/_L < 0$
		$\left[1 + 2.5\left({}^z/_L\right)^{3/5}\right]^{3/2}$	${}^z/_L > 0$
Thiermann and Graßl (1992)		$\left(1 - 3\,{}^z/_L\right)^{-1} - {}^z/_L$	${}^z/_L < 0$
		$\left[1 + 4\,{}^z/_L + 16\left({}^z/_L\right)^2\right]^{-1/2}$	${}^z/_L > 0$
Kaimal and Finnigan (1994)		$\left[1 + 0.5\left|{}^z/_L\right|^{2/3}\right]^{3/2}$	${}^z/_L < 0$
		$1 + 5\,{}^z/_L$	${}^z/_L > 0$

reference	κ	Universal function for the energy dissipation
Frenzen and Vogel (2001)		$0.85\left[\left(1-16\,z/_L\right)^{-2/3} - z/_L\right]$ $\quad z/_L < 0$ $0.85 + 4.26\,z/_L + 2.58\cdot\left(z/_L\right)^2$ $\quad z/_L > 0$
Hartogensis and DeBruin (2005)		$0.8 + 2.5\cdot z/_L$ $\qquad\qquad z/_L > 0$

reference	κ	universal function for the temperature structure function parameter
Wyngaard et al. (1971b)	0.35	$4.9\left(1-7\,z/_L\right)^{-2/3}$ $\qquad z/_L < 0$ $4.9\left[1+2.75\left(z/_L\right)\right]$ $\qquad z/_L > 0$
Foken and Kretschmer (1990)	0.4	$\left(0.95/_\kappa\right)^2\left(1-11.6\,z/_L\right)^{-1/2}$ $\quad -2 < z/_L < 0$ $0.95/_{\kappa^2}\left(0.95+7.8\,z/_L\right)$ $\qquad 0 < z/_L < 1$
Thiermann and Graßl (1992)		$6.34\left[1-7\,z/_L+75\left(z/_L\right)^2\right]^{-1/3}$ $\quad z/_L < 0$ $6.34\left[1-7\,z/_L+20\left(z/_L\right)^2\right]^{1/3}$ $\quad z/_L > 0$
Kaimal and Finnigan (1994)		$5\left(1+6.4\left|z/_L\right|\right)^{-2/3}$ $\qquad z/_L < 0$ $4\left(1+3\,z/_L\right)$ $\qquad\qquad z/_L > 0$
Hartogensis and DeBruin (2005)		$4.7\left[1+1.6\left(z/_L\right)^{2/3}\right]$ $\qquad\qquad z/_L > 0$

Integral Turbulence Characteristics in the Surface Layer

reference	$\sigma_u/_{u_*}$	$\sigma_v/_{u_*}$	stratification
Lumley and Panofsky (1964), Panofsky and Dutton (1984)	2.45	1.9	neutral, unstable
McBean (1971)	2.2	1.9	unstable
Beljaars et al. (1983)	2.0	1.75	unstable

reference	σ_u/u_*	σ_v/u_*	stratification
Sorbjan (1986)	2.3		stable
Sorbjan (1987)	2.6		
Foken *et al.* (1991)	2.7 $4.15(z/L)^{1/8}$		$-0.032 < z/L < 0$ $z/L < -0.032$
Thomas and Foken (2002)	$0.44\ln\left(\dfrac{1\,m\,f}{u_*}\right)+3.1$		$-0.2 < z/L < 0.4$

reference	σ_w/u_*	σ_T/T_*	stratification
Lumley and Panofsky (1964), Panofsky and Dutton (1984)	1.45		neutral, unstable
McBean (1971)	1.4	1.6	unstable
Panofsky *et al.* (1977)	$1.3(1-2\,z/L)^{1/3}$		unstable
Caughey u. Readings (1975)		$(z/L)^{-1/3}$	unstable
Hicks (1981)	$1.25(1-2\,z/L)^{1/3}$	$0.95(z/L)^{-1/3}$	unstable
Caughey u. Readings (1975)		$(z/L)^{-1/3}$	unstable
Beljaars *et al.* (1983)		$0.95(z/L)^{-1/3}$	unstable
Sorbjan (1986)	1.6	2.4	stable
Sorbjan (1987)	1.5	3.5	stable
Foken *et al.* (1991)	1.3 $2.0(z/L)^{1/8}$		$-0.032 < z/L < 0$ $z/L < -0.032$

reference	σ_w/u_*	σ_T/T_*	stratification
Foken et al. (1991; 1997a)		$1.4\left(z/_L\right)^{-\frac{1}{4}}$	$0.02 < z/_L < 1$
		$0.5\left(z/_L\right)^{-\frac{1}{2}}$	$-0.062 < z/_L < 0.02$
		$\left(z/_L\right)^{-\frac{1}{4}}$	$-1 < z/_L < -0.062$
		$\left(z/_L\right)^{-\frac{1}{3}}$	$z/_L < -1$
Thomas and. Foken (2002)	$0.21\ln\left(\dfrac{1\,m\cdot f}{u_*}\right)+6.3$		$-0.2 < z/_L < 0.4$

A5 Overall View of Experiments

Experiments for the Investigation of the Surface Layer

The following table lists the important micrometeorological experiments which gave special consideration of the surface layer (Foken 1990; Foken 2006a; Garratt and Hicks 1990; McBean *et al.* 1979); in this table ITCE means *International Turbulence Comparison Experiment*.

experiment	location, time	reference
O'Neill	O'Neill, USA 1953	Lettau (1957)
Kerang	Kerang, Australien 1962	Swinbank and Dyer (1968)
Hay	Hay, Australien 1964	
Hanford	Hanford, USA 1965	Businger *et al.* (1969)
Wangara	Hay, Australia 1967	Hess *et al.* (1981)

experiment	location, time	reference
KANSAS 1968	Kansas, USA. 1968	Izumi (1971)
ITCE-1968	Vancouver, Canada 1968	Miyake et al. (1971)
ITCE-1970	Tsimlyansk, Russia 1970	Tsvang et al. (1973)
Koorin	Koorin, Australia 1974	Garratt (1980)
ITCE-1976	Conargo, Australia 1976	Dyer (1982)
ITCE-1981	Tsimlyansk, Russia 1981	Tsvang et al. (1985)
Lövsta	Lövsta, Sweden 1986	Högström (1990)

Experiments Over Heterogeneous Landscapes

In the last 30 years, many micrometeorological experiments were conducted over heterogeneous landscapes and also included boundary layer processes and air chemical measurements (Mengelkamp et al. 2006, added).

experiment	location, time	reference
HAPEX-MOBILHY	France 1986	André et al. (1990)
FIFE	Kansas, 1987–1989	Sellers et al. (1988)
KUREX-88	Kursk, Russia 1988	Tsvang et al. (1991)
HAPEX-SAHEL	Niger, 1990–1992	Goutorbe et al. (1994)
SANA	Eisdorf, Melpitz, Germany, 1991	Seiler (1996)
EFEDA	Spain 1990–1991	Bolle et al. (1993)
BOREAS	Canada, 1993–1996	Sellers et al. (1997)
SHEBA	Arctic 1998	Uttal et al. (2002)
LITFASS-98	Lindenberg, Germany, 1998	Beyrich et al. (2002b)
CASES-99	Kansas, USA 1999	Poulos et al. (2002)
EBEX-2000	near Fresno CA, USA, 2000	Oncley et al. (2007)
LITFASS-2003	Lindenberg, Germany, 2003	Beyrich and Mengelkamp (2006)

Other Experiments Referred to in the Text

The following table gives some information about further experiments mentioned in this book.

experiment	location, time	reference
Greenland	Greenland, summer 1991	Ohmura (1992)
FINTUREX	Neumayer station, Antarktica, Jan.–Febr. 1994	Foken (1996), Handorf *et al.* (1999)
LINEX-96/2	Lindenberg, Germany, June 1996	Foken *et al.* (1997a)
LINEX-97/1	Lindenberg, Germany, June 1997	Foken (1998b)
WALDATEM-2003	Waldstein, Germany, May–July 2003	Thomas and Foken (2007a)

A6 Meteorological Measuring Stations

In Chap. 6.2, different types of meteorological measuring stations were defined (Table 6.1). The measuring parameters of these stations (VDI 2006a) are given in the following table, where X indicates necessary parameters and o indicates desirable additional parameters. The measuring parameters include air temperature (t_a), air moisture (f_a), wind velocity (u), precipitation (R_N), global radiation (G), net radiation (Q_s). surface temperature (t_{IR}), photosynthetic active radiation (PAR), soil temperature (t_b), soil heat flux (Q_G), air pressure (p), state of the weather (ww), sensible heat flux (Q_H), latent heat flux (Q_E), deposition (Q_c), shear stress (τ).

The most important parameters are:

type of the station	t_a	f_a	u	dd	R_N	G	Q_S	p	ww
agrometeorological	X	X	X	X	X	X	o	o	
micrometeorological	X	X	X	X	X	o		o	o
micrometeorological with turbulence measurements	X	X	X	X	X	o	X	o	
air pollution	o	o	X	X		X	o	o	o
immission measuring	X	X	X	X	X	X			
disposal site	X	X	X	X	X	o	X		
noise measuring	X		X	X					
traffic measuring		X	X	X					o

type of the station	t_a	f_a	u	dd	R_N	G	Q_S	p	ww
hydrological	o	o	o		X		o		
forest climate	X	X	X	X	X	X	o		
nowcasting	X	X	X	X	X	o		o	X
hobby	X	X	o	o				o	

Additional parameters measured at selected stations are:

type of station	t_{IR}	PAR	t_b	f_b	Q_G	Q_H	Q_E	Q	τ
agrometeorological		o	X	X	o				
micrometeorological									
micrometeorological with turbulence measurements					o	X	X	o	X
air pollution									o
immission measuring									
disposal site									
noise measuring									
traffic measuring	o		o						
hydrological									
forest climate		o	o	o					
nowcasting									
hobby									

A7 Micrometeorological Standards used in Germany

In Germany, Austria, and Switzerland meteorological measuring techniques and some applied meteorological methods are standardized. Some of these standards were incorporated in the international ISO standards. Because these standards are available in English, only the most important standards are given. The relevant standards appear in volume 1B of the "VDI/DIN Handbook on Keeping the Air Clean" (Queitsch 2002; VDI 2006b):

VDI/DIN	sheet	content
3781	2, 4	Dispersion of pollutants in the atmosphere; stack heights
3782	1, 3, 5, 7	Environmental meteorology - Atmospheric dispersion models

VDI/DIN	sheet	content
3783	1, 2, 4, 5, 6, 8, 9, 10, 12	Dispersion of pollutants in the atmosphere; dispersion of emissions by accidental releases
3784	1, 2	Environmental meteorology; cooling towers
3786	1–17	Environmental meteorology – Meteorological measurements
3787	1, 2, 5, 9,10	Environmental meteorology – Climate and air pollution
3788	1	Environmental meteorology – Dispersion of odorants
3789	2, 3	Environmental meteorology – Interactions between atmosphere and surfaces
3790	1, 2, 3	Environmental meteorology – Emissions of gases
3945	1, 3	Environmental meteorology – Atmospheric dispersion models

Because of a lack in the literature on meteorological measuring systems, the VDI/DIN 3786 (Sheets 1–17) and the relevant ISO standards may be of interest to a wide readership:

VDI/DIN	title
DIN ISO 16622	Meteorology – Sonic anemometers/thermometers – Acceptance test methods for mean wind measurements (ISO 16622:2002)
DIN ISO 17713-1	Meteorology – Wind measurements – Part 1: Wind tunnel test methods for rotating anemometer performance (ISO 17713-1:2007)
DIN ISO 17714	Meteorology – Air temperature measurements – Test methods for comparing the performance of thermometer shields/screens and defining important characteristics (ISO/DIS 17714:2004)
VDI 3786 Sheet 1	Environmental meteorology – Meteorological measurements – Fundamentals
VDI 3786 Sheet 2	Environmental meteorology – Meteorological measurements concerning questions of air pollution – Wind
VDI 3786 Sheet 3	Meteorological measurements concerning questions of air pollution; air temperature
VDI 3786 Sheet 4	Meteorological measurements concerning questions of air pollution; air humidity
VDI 3786 Sheet 5	Meteorological measurements concerning questions of air pollution; global radiation, direct solar radiation and net total radiation
VDI 3786 Sheet 6	Meteorological measurements of air pollution; turbidity of ground-level atmosphere standard visibility
VDI 3786 Sheet 7	Meteorological measurements concerning questions of air pollution; precipitation
VDI 3786 Sheet 8	Meteorological measurements; concerning questions of air pollution; aerological measurements

VDI/DIN	title
VDI 3786 Sheet 9	Environmental Meteorology – Meteorological measurements – Visual weather observations
VDI/DIN	title
VDI 3786 Sheet 10	Environmental meteorology; measurement of the atmospheric turbidity due to aerosol particles with sunphotometers
VDI 3786 Sheet 11	Environmental meteorology; determination of the vertical wind profile by Doppler SODAR systems
VDI 3786 Sheet 12	Environmental meteorology – Meteorological measurements – Turbulence measurements with sonic anemometers
VDI 3786 Sheet 13	Environmental meteorology – Meteorological measurements – Measuring station
VDI 3786 Sheet 14	Environmental meteorology – Ground-based remote sensing of the wind vector – Doppler wind LIDAR
VDI 3786 Sheet 15	Environmental meteorology – Ground-based remote sensing of visual range – Visual-range lidar
VDI 3786 Sheet 16	Environmental meteorology – Measurement of atmospheric pressure
VDI 3786 Sheet 17	Environmental meteorology – Ground-based remote sensing of the wind vector – Wind profiler radar

A8 Available Eddy-Covariance Software

The following software can be used for eddy-covariance measurements (VDI 2008) and some of them were compared in Mauder (2008; 2007b):

name	reference
EdiRe	The University of Edinburgh
	Institute of Atmospheric and Environmental Science
	Crew Building, The King's Buildings
	West Mains Road
	EDINBURGH EH9 3JN
	http://www.geos.ed.ac.uk/abs/research/micromet/EdiRe/
ECPack	Department of Meteorology and Air Quality
	Wageningen University and Research
	Duivendaal 2
	NL 6701 AP Wageningen
	http://www.met.wau.nl/index.html?http://www.met.wau.nl/projects /jep/

name	reference
EddyMess	Dr. Olaf Kolle
	Max-Planck-Institute for Biogeochemistry
	Hans-Knöll-Str. 10
	D-07749 Jena, Germany
TK2	University of Bayreuth
	Dept. of Micrometeorology
	D-95440 Bayreuth, Germany
	http://www.bayceer.uni-bayreuth.de/mm/de/software/software/software_dl.php

A9 Glossary

Advection: Transport of properties of the air (momentum, temperature, water vapor, *etc.*) by the wind. As a rule, horizontal transport is understood. Because of advection, air properties change in both horizontal coordinates, and the conditions are no longer homogeneous. Vertical advection is the vertical movement air due to mass continuity rather than buoyancy, *i.e.* convection.

Atmospheric window: Frequency range of electromagnetic waves, which pass through the atmosphere with little absorption. Within this frequency range, remote sensing methods of the surface properties can be applied. The most important atmospheric windows are in the visible range from 0.3 to 0.9 μm, in the IR-range from 8 to 13 μm, and in the microwave range for wave lengths greater than 1 mm.

Bolometer: A device for measuring radiant energy by using thermally-sensitive electric sensors (thermocouple, thermistor, platinum wire).

Calm: State of the atmosphere with no discernible air motion. The calm-threshold is the threshold speed of cup anemometers, which is about 0.3 ms^{-1}. A wind direction can not be determined under these circumstances.

CET: Abbreviation of Central European Time. It is the mean local time at 15° longitude, and differs from UTC by 1 h.

Clausius-Clapeyron equation: Clapeyron 1834 established and Clausius 1850 gave the reasons for the equations for the temperature dependence of equilibrium water-vapor pressure at saturation. Because of the equation's strong exponential dependence on temperature, the atmosphere can take significantly more water at high temperatures than at lower temperatures, and therefore can store more latent heat.

Climate element: Meteorological and other parameters that characterize (singly or in combinations) different climate types. These are state variables and fluxes.

Coherence: In generally, coherence is a constant phase relation between two waves. Coherent structures in atmospheric turbulence research are velocity, temperature, and other structures, which are significantly larger and longer-lived than the smallest local eddies (*e.g.* squall lines, convective cells).

Coriolis force: Fictitious force in a rotating coordinate system, and named after the mathematician Coriolis (1792–1843). It is a force normal to the velocity vector causing a deflection to the right in the Northern hemisphere and to the left in the Southern hemisphere.

Coriolis parameter: Twice the value of the angular velocity of the Earth for a certain location of the latitude φ: $f=2\,\Omega \sin \varphi$. At the equator $f = 0$, on the Northern hemisphere positive and on the Southern hemisphere negative.

Dissipation: Conversion of kinetic energy by work against the viscous stresses. Under turbulent conditions, it is the conversion of the kinetic energy of the smallest eddies into heat.

Element, meteorological: *see* climate element

Entrainment: Exchange process at the top of the atmospheric boundary layer, and is due to the actions of eddies that are smaller than those in the mixed layer.

Fetch: Windward distance from a measuring point to a change of the surface properties or an obstacle; extent of a measuring field for micrometeorological research.

Froude number: Dimensionless ratio of the inertia force to the gravity force $Fr=V^2\,L^{-1}\,g^{-1}$ where V is a characteristic velocity and L is a characteristic length. For flow over hills, L is the characteristic distance between hills or obstacles. In these cases, the external Froude number with the Brunt-Väisälä frequency N is used, see Eq. (3.36): $Fr= V\,N^{-1}\,L^{-1}$.

Gas constant: Proportionality factor of the equation of state for ideal gases, and is expressed in mol. In meteorology, the gas constant is expressed in mass units, and a special gas constant for dry air is used. In moist air, the temperature in the equation of state must be replaced by the virtual temperature (see below).

Hysteresis: The change between two states that depends on the way the change occurs, for example, the characteristics of a moisture sensor are different for wetting and drying.

Inversion: An air layer where the temperature increases with the altitude instead of the usual decrease. Inversions are of two types; surface inversion due to longwave radiation from the ground, and elevated or free inversions *e.g.* at the top of the atmospheric boundary layer.

Kelvin-Helmholtz instability: A dynamic instability caused by strong wind shear resulting in breaking waves or billow clouds (Sc, Ac lent). Typically, these occur at inversions or above hills. They also can occur over obstacles and forests.

Leaf area density: The vertical probability density function of the leaf area.

Leaf area index: Ratio of the leaf area (upper side) within a vertical cylinder to the bottom area of the cylinder.

Low-level jet: Vertical band of strong winds in the lower part of the atmospheric boundary layer. For stable stratification, the low-level jet develops at the upper border of the nocturnal surface inversion. Typical heights are 100–300 m, and sometimes lower.

Matrix potential: A measure of the absorption and capillary forces of the solid soil matrix on the soil water. Its absolute value is called tension.

Mixed layer: A layer of strong vertical mixing due to convection resulting in vertically-uniform values of potential temperature and wind speed but decreasing values of moisture. It is often capped by an inversion layer (see above).

MLT: Abbreviation for Mean Local Time. Time related to the meridian of the location and for all locations of the same longitude. The mean local time is the solar time measured from the lower culmination of the sun. It is calculated by addition of 4 min to the universal time (UT) for each degree of longitude in eastward direction (Brockhaus 2003).

Parameterization: Representation of complicated relations in models by more simple combinations of parameters, which are often only valid under certain circumstances.

RLT: Abbreviation for Real Local Time. It is the time related to the meridian of all points on the same longitude. The real local time is the solar time measured from the daily lower culmination of the sun and changes with the time equation, which is the difference between the real and the mean solar time. The time equation is positive if the real solar time earlier culminates than the mean solar time (sun day). It changes between –14 min 24 s (approx. middle of February) and +16 min 21 s (approx. beginning of November). For an approximation relation see A4 (Brockhaus 2003).

Rossby similarity: In the free atmosphere, the Rossby number is the ratio of the inertial force to the Coriolis force. In the atmospheric boundary layer, the roughness Rossby number is the ratio of the friction velocity to the Coriolis parameter, $Ro = u_*/(f \, z_0)$. The friction Rossby number is an assessment of the ageostrophic component of the wind.

Stability of the stratification: The static stability separates turbulent and laminar flow conditions depending on the gradient of the potential temperature (see below). If the potential temperature decreases with height, then the stratification is unstable, but if it increases with height, then the stratification is stable. Due to the effects of vertical wind sheer, the statically-stable range is turbulent up to the critical Richardson number.

Temperature, potential: The temperature of a dry air parcel that is moved adiabatically to a pressure of 1000 hPa, see Eq. (2.60).

Temperature, virtual: The temperature of a dry air parcel if it had the same density as a moist air parcel. The virtual temperature is slightly higher than the temperature of moist air, see Eq. (2.69).

Transmission: Permeability of the atmosphere for radiation. The radiation can be reduced *e.g.* by gases, aerosols, particles, and water droplets.

UTC: Abbreviation for Universal Time Coordinated, a time scale based on the international atomic time by setting the zero point to the zero meridian (Greenwich Meridian) , with the mean solar day as a basic unit. It is the basis for political and scientific time (Brockhaus 2003).

Wind, geostrophic: Wind above the atmospheric boundary layer where pressure gradient force and Coriolis force (see above) are in equilibrium.

References

Akima, H (1970) A new method of interpolation and smooth curve fitting based on local procedures. J Assc Comp Mach 17: 589–602

Albertson, JD, Parlange, MB (1999) Natural integration of scalar fluxes from complex terrain. Adv Water Res 23: 239–252

Albrecht, F (1927) Thermometer zur Messung der wahren Temperatur. Meteorol Z 24: 420–424

Albrecht, F (1940) Untersuchungen über den Wärmehaushalt der Erdoberfläche in verschiedenen Klimagebieten. Reichsamt Wetterdienst, Wiss Abh Bd. VIII, Nr. 2: 1–82

Allen, RG, Pereira, LS, Raes, D, Smith, M (2004) Crop evaporation. FAO Irrigation Drainage Pap 56: XXVI + 300 pp.

Amiro, BD (1990) Comparison of turbulence statistics within three boreal forest canopies. Boundary-Layer Meteorol 51: 99–121

Amiro, BD, Johnson, FL (1991) Some turbulence features within a boreal forest canopy during winter. 20th Conference on Agricultural and Forest Meteorology, Salt Lake City, 10–13 September 1991, Am. Meteorol. Soc., 135–138.

Amiro, BD (1998) Footprint climatologies for evapotranspiration in a boreal catchment. Agric Forest Meteorol 90: 195–201

André, J-C, Bougeault, P, Goutorbe, J-P (1990) Regional estimates of heat and evaporation fluxes over non-homogeneous terrain, Examples from the HAPEX-MOBILHY programme. Boundary-Layer Meteorol 50: 77–108

Andreas, EL (1989) Two-wavelength method of measuring path-averaged turbulent surface heat fluxes. J Atm Oceanic Techn 6: 280–292

Andreas, EL, Cash, BA (1999) Convective heat transfer over wintertime leads and polynyas. J Geophys Res 104: 15.721–725.734

Andreas, EL (2002) Parametrizing scalar transfer over snow and ice: A review. J Hydrometeorol 3: 417–432

Andreas, EL, Claffey, KJ, Fairall, CW, Grachev, AA, Guest, PS, Jordan, RE, Persson, POG (2004) Measurements of the von Kármán constant in the atmospheric surface layer – further discussions. 16th Conference on Boundary Layers and Turbulence, Portland ME, Am. Meteorol. Soc., 1–7, paper 7.2.

Andreev, EG, Lavorko, VS, Pivovarov, AA, Chundshua, GG (1969) O vertikalnom profile temperatury vblizi granicy rasdela more – atmosphera (About the vertical temparature profile close to to ocean – atmosphere interface). Okeanalogija 9: 348–352

Antonia, RA, Luxton, RE (1971) The response of turbulent boundary layer to a step change in surface roughness, Part 1. J Fluid Mech 48: 721–761

Antonia, RA, Luxton, RE (1972) The response of turbulent boundary layer to a step change in surface roughness, Part 2. J Fluid Mech 53: 737–757

Arya, SP (1999) Air pollution meteorology and dispersion. Oxford University Press, New York, Oxford, 310 pp.

Arya, SP (2001) Introduction to Micrometeorology. Academic Press, San Diego, 415 pp.

Assmann, R (1887) Das Aspirationspsychrometer, ein neuer Apparat zur Ermittlung der wahren Temperatur und Feuchtigkeit der Luft. Das Wetter 4: 245–286

Assmann, R (1888) Das Aspirationspsychrometer, ein neuer Apparat zur Ermittlung der wahren Temperatur und Feuchtigkeit der Luft. Das Wetter 5: 1–22

Aubinet, M, Grelle, A, Ibrom, A, Rannik, Ü, Moncrieff, J, Foken, T, Kowalski, AS, Martin, PH, Berbigier, P, Bernhofer, C, Clement, R, Elbers, J, Granier, A, Grünwald, T, Morgenstern, K, Pilegaard, K, Rebmann, C, Snijders, W, Valentini, R, Vesala, T (2000) Estimates of the annual net carbon and water exchange of forests: The EUROFLUX methodology. Adv Ecol Res 30: 113–175

Aubinet, M, Clement, R, Elbers, J, Foken, T, Grelle, A, Ibrom, A, Moncrieff, H, Pilegaard, K, Rannik, U, Rebmann, C (2003a) Methodology for data acquisition, storage and treatment. In:

R Valentini (Editor), Fluxes of carbon, water and energy of European forests. Ecological Studies, Vol. 163. Springer, Berlin, Heidelberg, pp. 9–35.

Aubinet, M, Heinesch, B, Yernaux, M (2003b) Horizontal and vertical CO_2 advection in a sloping forest. Boundary-Layer Meteorol 108: 397–417

Aubinet, M, Berbigier, P, Bernhofer, C, Cescatti, A, Feigenwinter, C, Granier, A, Grünwald, T, Havrankova, K, Heinesch, B, Longdoz, B, Marcolla, B, Montagnani, L, Sedlak, P (2005) Comparing CO_2 storage and advection conditions at night at different CarboEuroflux sites. Boundary-Layer Meteorol 116: 63–94

Avissar, R, Pielke, RA (1989) A parametrization of heterogeneous land surface for atmospheric numerical models and its impact on regional meteorology. Monthly Weather Review 117: 2113–2136

Bailey, WG, Oke, TR, Rouse, WR (Editors) (1997) The surface climate of Canada. Mc Gill-Queen's University Press, Montreal, Kingston, 369 pp.

Baldocchi, D, Hicks, BB, Camara, P (1987) A canopy stomatal resistance model for gaseous deposition to vegetated surfaces. Atmos Environm 21: 91–101

Baldocchi, D (1988) A multi-layer model for estimating sulfor dioxid deposition to a deciduous oke forest canopy. Atmos Environm 22: 869–884

Baldocchi, D (1998) Flux footprints within and over forest canopies. Boundary-Layer Meteorol 85: 273–292

Baldocchi, D, Falge, E, Gu, H., L, Olson, R, Hollinger, D, Running, S, Anthoni, P, Bernhofer, C, Davis, K, Evans, R, Fuentes, J, Goldstein, A, Katul, G, Law, B, Lee, XH, Malhi, Y, Meyers, T, Munger, W, Oechel, W, PawU, KT, Pilegaard, K, Schmid, HP, Valentini, R, Verma, S, Vesala, T (2001) FLUXNET: A new tool to study the temporal and spatial variability of ecosystem-scale carbon dioxide, water vapor, and energy flux densities. Bull Amer Meteorol Soc 82: 2415–2434

Barkov, E (1914) Vorläufiger Bericht über die meteorologischen Beobachtungen der Deutschen Antarktisexpedition 1911–1912. Meteorol Z 49: 120–126

Barr, AG, King, KM, Gillespie, TJ, Hartog, GD, Neumann, HH (1994) A comparison of Bowen ratio and eddy correlation sensible and latent heat flux measurements above deciduous forest. Boundary-Layer Meteorol 71: 21–41

Barrett, EW, Suomi, VE (1949) Preliminary report on temperature measurement by sonic means. J Meteorol 6: 273–276

Bartels, J (1935) Zur Morphologie geophysikalischer Zeitfunktionen. Sitzungsberichte Preuß Akad Wiss 30: 504–522

Baumgartner, A (1956) Untersuchungen über den Wärme- und Wasserhaushalt eines jungen Waldes. Ber Dt Wetterdienstes 28: 53 pp.

Behrens, J, Rakowsky, N, Hiller, W, Handorf, D, Läuter, M, Päpke, J, Dethloff, K (2005) amatos: parallel adaptive mesh generator for atmospheric and oceanic simulation. Ocean Modelling 10: 171–183

Beljaars, ACM, Schotanus, P, Nieuwstadt, FTM (1983) Surface layer similarity under nonuniform fetch conditions. J Climate Appl Meteorol. 22: 1800–1810

Beljaars, ACM, Holtslag, AAM (1991) Flux parametrization over land surfaces for atmospheric models. J Appl Meteorol 30: 327–341

Beljaars, ACM (1995) The parametrization of surface fluxes in large scale models under free convection. Quart J Roy Meteorol Soc 121: 255–270

Beljaars, ACM, Viterbo, P (1998) Role of the boundary layer in a numerical weather prediction model. In: AAM Holtslag, PG Duynkerke (Editors), Clear and cloudy boundary layers. Royal Netherlands Academy of Arts and Sciences, Amsterdam, pp. 287–304.

Beniston, M (1998) From turbulence to climate. Springer, Berlin, Heidelberg, 328 pp.

Bentley, JP (2005) Principles of measurement systems. Pearson Prentice Hall, Harlow 528 pp.

Bergström, H, Högström, U (1989) Turbulent exchange above a pine forest. II. Organized structures. Boundary-Layer Meteorol 49: 231–263

Bernhardt, K-H (1995) Zur Interpretation der Monin-Obuchovschen Länge. Meteorol Z 4: 81–82

Bernhardt, K (1970) Der ageostrophische Massenfluß in der Bodenreibungsschicht bei beschleunigungsfreier Strömung. Z Meteorol 21: 259–279

Bernhardt, K (1972) Lecture'Dynamik der Atmosphäre', Humboldt-Universität zu Berlin.

Bernhardt, K (1975) Some characteristics of the dynamic air-surface interaction in Central Europe. Z Meteorol 25: 63–68

Bernhardt, K (1980) Zur Frage der Gültigkeit der Reynoldsschen Postulate. Z Meteorol 30: 261–268

Bernhardt, K, Piazena, H (1988) Zum Einfluß turbulenzbedingter Dichteschwankungen auf die Bestimmung turbulenter Austauschströme in der Bodenschicht. Z Meteorol 38: 234–245

Bernhofer, C (1992) Estimating forest evapotranspiration at a non-ideal site. Agric Forest Meteorol 60: 17–32

Best, MJ, Beljaars, A, Polcher, J, Viterbo, P (2004) A proposed structure for coupling tiled surfaces with the planetary boundary layer. J Hydrometeorol 5: 1271–1278

Beyrich, F, DeBruin, HAR, Meijninger, WML, Schipper, JW, Lohse, H (2002a) Results from one-year continuous operation of a large aperture scintillometer over a heterogeneous land surface. Boundary-Layer Meteorol 105: 85–97

Beyrich, F, Herzog, H-J, Neisser, J (2002b) The LITFASS project of DWD and the LITFASS-98 Experiment: The project strategy and the experimental setup. Theor Appl Climat 73: 3–18

Beyrich, F, Richter, SH, Weisensee, U, Kohsiek, W, Lohse, H, DeBruin, HAR, Foken, T, Göckede, M, Berger, FH, Vogt, R, Batchvarova, E (2002c) Experimental determination of turbulent fluxes over the heterogeneous LITFASS area: Selected results from the LITFASS-98 experiment. Theor Appl Climat 73: 19–34

Beyrich, F, Kouznetsov, RD, Leps, J-P, Lüdi, A, Meijninger, WML, Weisensee, U (2005) Structure parameters for temperature and humidity from simultaneous eddy-covariance and scintillometer measurements. Meteorol Z 14: 641–649

Beyrich, F, Mengelkamp, H-T (2006) Evaporation over a heterogeneous land surface: EVA_GRIPS and the LITFASS-2003 experiment – an overview. Boundary-Layer Meteorol 121: 5–32

Bilger, RW (1980) Turbulent flows with nonpremixed reactants. In: PA Libby, FA Willimas (Editors), Turbulent Reacting Flows. Springer, Berlin, Heidelberg, pp. 65–113.

Bjutner, EK (1974) Teoreticeskij rascet soprotivlenija morskoj poverchnosti (Theoretical calculation of the resistance at the surface of the ocean). In: AS Dubov (Editor), Processy perenosa vblizi poverchnosti razdela okean – atmosfera (Exchange processes near the ocean – atmosphere interface). Gidrometeoizdat, Leningrad, pp. 66–114

Blackadar, AK (1976) Modeling the nocturnal boundary layer. 4th Symp Atm Turbul, Diff and Air Poll, Raylaigh, NC, October 19–22, 1976, Am. Meteorol. Soc., 46–49

Blackadar, AK (1997) Turbulence and Diffusion in the Atmosphere. Springer, Berlin, Heidelberg, 185 pp.

Blöschl, G, Sivapalan, M (1995) Scale issues in hydrological modelling – a review. Hydrol Processes 9: 251–290

Blümel, K (1998) Estimation of sensible heat flux from surface temperature wave and one-time-of-day air temperature observations. Boundary-Layer Meteorol 86: 193–232

Blyth, EM (1995) Comments on 'The influence of surface texture on the effective roughness length' by H. P. Schmid and D. Bünzli (1995, 121, 1–21). Quart J Roy Meteorol Soc 121: 1169–1171

Böer, W (1959) Zum Begriff des Lokalklimas. Z Meteorol 13: 5–11

Bolle, H-J, André, J-C, Arrie, JL, Barth, HK, Bessemoulin, P, A., B, DeBruin, HAR, Cruces, J, Dugdale, G, Engman, ET, Evans, DL, Fantechi, R, Fiedler, F, Van de Griend, A, Imeson, AC, Jochum, A, Kabat, P, Kratsch, P, Lagouarde, J-P, Langer, I, Llamas, R, Lopes-Baeza, E, Melia Muralles, J, Muniosguren, LS, Nerry, F, Noilhan, J, Oliver, HR, Roth, R, Saatchi, SS, Sanchez Diaz, J, De Santa Olalla, M, Shutleworth, WJ, Sogaard, H, Stricker, H, Thornes, J, Vauclin, M, Wickland, D (1993) EFEDA: European field experiment in a desertification-threatened area. Annales Geophysicae 11: 173–189

Boussinesq, J (1877) Essai sur la théorie des eaux courantes. Mem Savants Etrange 23: 46 pp.

Bovscheverov, VM, Voronov, VP (1960) Akustitscheskii fljuger (Acoustic rotor). Izv AN SSSR, ser Geofiz 6: 882–885

Bowen, AJ, Teunissen, HW (1986) Correction factors for the directional response of Gill propeller anemometer. Boundary-Layer Meteorol 37: 407–413

Bowen, IS (1926) The ratio of heat losses by conduction and by evaporation from any water surface. Physical Review 27: 779–787

Bowling, DR, Delany, AC, Turnispseed, AA, Baldocchi, DD, Monson, RK (1999) Modification of the relaxed eddy accumulation technique to maximize measured scalar mixing ratio differences in updrafts and downdrafts. J Geophys Res 104: 9121–9133

Bowling, DR, Tans, PP, Monson, RK (2001) Partitioning Net Ecosystem Carbon Exchange with Isotopic Fluxes of CO_2. Global Change Biology 7: 127–145

Braden, H (1995) The model AMBETI – A detailed description of a soil-plant-atmosphere model. Ber Dt Wetterdienstes 195: 117 pp

Bradley, EF (1968a) A micrometeorological study of velocity profiles and surface drag in the region modified by change in surface roughness. Quart J Roy Meteorol Soc 94: 361–379

Bradley, EF (1968b) A shearing stress meter for micrometeorological studies. Quart J Roy Meteorol Soc 94: 380–387

Braud, J, Noilhan, P, Bessemoulin, P, Mascart, P, Haverkamp, R, Vauclin, M (1993) Bare ground surface heat and water exchanges under dry conditions. Boundary-Layer Meteorol 66: 173–200

Brock, FV, Richardson, SJ (2001) Meteorological measurement systems. Oxford University Press, New York, 290 pp.

Brockhaus (2003) Der Brockhaus Naturwissenschaft und Technik. Bibliographisches Institut & F.A. Brockhaus AG, Spektrum Akademischer Verlag GmbH, Mannheim, Heidelberg, 2259 pp.

Brocks, K, Krügermeyer, L (1970) Die hydrodynamische Rauhigkeit der Meeresoberfläche. Ber Inst Radiometeorol Marit Meteorol 14: 55 pp.

Brook, RR (1978) The influence of water vapor fluctuations on turbulent fluxes. Boundary-Layer Meteorol 15: 481–487

Brutsaert, WH (1982) Evaporation into the atmosphere: Theory, history and application. D. Reidel, Dordrecht, 299 pp.

Buck, AL (1973) Development of an improved Lyman-alpha hygrometer. Atm Technol 2: 213–240

Businger, JA, Miyake, M, Inoue, E, Mitsuta, Y, Hanafusa, T (1969) Sonic anemometer comparison and measurements in the atmospheric surface layer. J Meteorol Soc Japan 47: 1–12

Businger, JA, Wyngaard, JC, Izumi, Y, Bradley, EF (1971) Flux-profile relationships in the atmospheric surface layer. J Atmos Sci 28: 181–189

Businger, JA, Yaglom, AM (1971) Introduction to Obukhov's paper "Turbulence in an atmosphere with a non-uniform temperature". Boundary-Layer Meteorol 2: 3–6

Businger, JA (1982) Equations and concepts. In: FTM Nieuwstadt, H Van Dop (Editors), Atmospheric turbulence and air pollution modelling: A course held in The Hague, 21–25 September 1981. D. Reidel Publ. Co., Dordrecht, pp. 1–36.

Businger, JA (1986) Evaluation of the accuracy with which dry deposition can be measured with current micrometeorological techniques. J Appl Meteorol 25: 1100–1124

Businger, JA (1988) A note on the Businger-Dyer profiles. Boundary-Layer Meteorol 42: 145–151

Businger, JA, Oncley, SP (1990) Flux measurement with conditional sampling. J Atm Oceanic Techn 7: 349–352

Campbell, GS, Norman, JM (1998) Introduction to environmental biophysics. Springer, New York, 286 pp.

Caughey, SL, Readings, CJ (1975) Turbulent fluctuations in convective conditions. Quart J Roy Meteorol Soc 101: 537–542

Chamberlain, AC (1961) Aspects of travel and deposition of aerosol and vapour clouds. Rep. HP/R 1261 (RP/14), A.E.R.E., Harwell, Berkshire, 38 pp.

Charnock, H (1955) Wind stress on water surface. Quart J Roy Meteorol Soc 81: 639–642

Chundshua, GG, Andreev, EG (1980) O mechanizme formirovanija inversii temperatury v privodnoim sloe atmosfery nad morem (About a mechanism of the developement of a temperature inversion in the near surface layer above the ocean). Dokl AN SSSR 255: 829–832

Cionco, RM (1978) Analysis of canopy index values for various canopy densities. Boundary-Layer Meteorol 15: 81–93

Claussen, M (1991) Estimation of areally-averaged surface fluxes. Boundary-Layer Meteorol 54: 387–410

Claussen, M, Walmsley, JL (1994) Modification of blending procedure in a proposed new PBL resistance law. Boundary-Layer Meteorol 68: 201–205

Cohen, ER, Taylor, BN (1986) The 1986 adjustment of the fundamental physical constants. International Council of Scientific Unions (ICSU), Committee on Data for Science and Technology (CODATA). CODATA-Bull No. 63: 36 pp.

Collineau, S, Brunet, Y (1993a) Detection of turbulent coherent motions in a forest canopy. Part II: Time-scales and conditional averages. Boundary-Layer Meteorol 66: 49–73

Collineau, S, Brunet, Y (1993b) Detection of turbulent coherent motions in a forest canopy. Part I: Wavelet analysis. Boundary-Layer Meteorol 65: 357–379

Corrsin, S (1951) On the spectrum of isotropic temperature fluctuations in an isotropic turbulence. J Appl Phys 22: 469–473

Csanady, GT (2001) Air-sea interaction, Laws and mechanisms. Cambridge University Press, Cambridge, New York, 239 pp.

Culf, AD, Foken, T, Gash, JHC (2004) The energy balance closure problem. In: P Kabat et al. (Editors), Vegetation, water, humans and the climate. A new perspective on an interactive system. Springer, Berlin, Heidelberg, pp. 159–166.

Davenport, AG, Grimmond, CSB, Oke, TR, Wieringa, J (2000) Estimating the roughness of cities and shelterred country. 12th Conference on Applied Climatology, Ashville, NC, American Meteorological Society, 96–99.

Davidan, IN, Lopatuhin, LI, Rogkov, VA (1985) Volny v okeane (Waves in the ocean). Gidrometeoizdat, Leningrad, 256 pp.

de Vries, DA (1963) Thermal Properties of Soils. In: WR van Wijk (Editor), Physics of the Plant Environment. North-Holand Publ. Co., Amsterdam, pp. 210–235.

Deardorff, JW (1972) Numerical investigation of neutral und unstable planetary boundary layer. J Atmos Sci 29: 91–115

DeBruin, HAR, Holtslag, AAM (1982) A simple parametrization of the surface fluxes of sensible and latent heat during daytime compared with the Penman-Monteith concept. J Climate Appl Meteorol. 21: 1610–1621

DeBruin, HAR, Moore, JC (1985) Zero-plane displacement and roughness length for tall vegetation, derived from a simple mass conservation hypothesis. Boundary-Layer Meteorol 38: 39–49

DeBruin, HAR, Bink, NJ, Kroon, LJ (1991) Fluxes in the surface layer under advective conditions. In: TJ Schmugge, JC André (Editors), Workshop on Land Surface Evaporation, Measurement and Parametrization. Springer, New York, pp. 157–169.

DeBruin, HAR (2002) Introduction: Renaissance of scintillometry. Boundary-Layer Meteorol 105: 1–4

DeBruin, HAR, Meijninger, WML, Smedman, A-S, Magnusson, M (2002) Displaced-beam small aperture scintillometer test. part I: The WINTEX data-set. Boundary-Layer Meteorol 105: 129–148

Defant, F (1949) Zur Theorie der Hangwinde, nebst Bemerkungen zur Theorie der Berg- und Talwinde. Archiv Meteorol Geophys Bioklim, Ser A 1: 421–450

DeFelice, TP (1998) An introduction to meteorological instrumentation and measurement. Prentice Hall, Upper Saddle River, 229 pp.

DeGaetano, AT (1997) A quality-control routine for hourly wind observations. J Atm Oceanic Techn 14: 308–317

Denmead, DT, Bradley, EF (1985) Flux-gradient relationships in a forest canopy. In: BA Hutchison, BB Hicks (Editors), The forest-atmosphere interaction. D. Reidel Publ. Comp., Dordrecht, Boston, London, pp. 421–442.

Desjardins, RL (1977) Description and evaluation of a sensible heat flux detector. Boundary-Layer Meteorol 11: 147–154

Desjardins, RL, MacPherson, JI, Schuepp, PH, Karanja, F (1989) An evaluation of aircraft flux measurements of CO_2, water vapor and sensible heat. Boundary-Layer Meteorol 47: 55–69

DIN-EN (1996) Industrielle Platin-Widerstandsthermometer und Platin-Meßwiderstände (Industrial platinum resistance thermometer sensors). DIN-EN 60751

Dlugi, R (1993) Interaction of NO_x and VOC's within vegetation. In: PW Borrell (Editor), Proceedings EUROTRAC-Symposium 92. SPB Acad. Publ., The Hague, pp. 682–688.

Dobson, F, Hasse, L, Davis, R (Editors) (1980) Air-sea interaction, Instruments and methods. Plenum Press, New York, 679 pp.

Doetsch, G (1985) Anleitung zum praktischen Gebrauch der Laplace-Transformation und der Z-Transformation. Oldenbourg, München, Wien, 256 pp.

Dommermuth, H, Trampf, W (1990) Die Verdunstung in der Bundesrepublik Deutschland, Zeitraum 1951-1980, Teil 1. Deutscher Wetterdienst, Offenbach, 10 pp.

Doorenbos, J, Pruitt, WO (1977) Guidelines for predicting crop water requirements. FAO Irrigation Drainage Pap 24, 2nd ed.: 156 pp.

Doran, JC, Verholek, MG (1978) A note on vertical extrapolation formulas for Weibull velocity distribution parameters. J Climate Appl Meteorol. 17: 410–412

Drinkov, R (1972) A solution to the paired Gill-anemometer response function. J Climate Appl Meteorol. 11: 76–80

Dugas, WA, Fritschen, LJ, Gay, LW, Held, AA, Matthias, AD, Reicosky, DC, Steduto, P, Steiner, JL (1991) Bowen ratio, eddy correlation, and portable chamber measurements of sensible and latent heat flux over irrigated spring wheat. Agric Forest Meteorol 56: 12–20

Dutton, JA (2002) The Ceaseless Wind: An Introduction to the Theory of Atmospheric Motion. Dover Publications, Mineola, NY, 640 pp.

DVWK (1996) Ermittlung der Verdunstung von Land- und Wasserflächen. DVWK-Merkblätter zur Wasserwirtschaft 238: 134 pp.

Dyer, AJ, Hicks, BB, King, KM (1967) The Fluxatron – A revised approach to the measurement of eddy fluxes in the lower atmosphere. Journal Applied Meteorology 6: 408–413

Dyer, AJ, Hicks, BB (1970) Flux-gradient relationships in the constant flux layer. Quart J Roy Meteorol Soc 96: 715–721

Dyer, AJ (1974) A review of flux-profile-relationships. Boundary-Layer Meteorol 7: 363–372

Dyer, AJ (1981) Flow distortion by supporting structures. Boundary-Layer Meteorol 20: 363–372

Dyer, AJ, Bradley, EF (1982) An alternative analysis of flux-gradient relationships at the 1976 ITCE. Boundary-Layer Meteorol 22: 3–19

Dyer, AJ, Garratt, JR, Francey, RJ, McIlroy, IC, Bacon, NE, Hyson, P, Bradley, EF, Denmead, DT, Tsvang, LR, Volkov, JA, Kaprov, BM, Elagina, LG, Sahashi, K, Monji, N, Hanafusa, T, Tsukamoto, O, Frenzen, P, Hicks, BB, Wesely, M, Miyake, M, Shaw, WJ (1982) An international turbulence comparison experiment (ITCE 1976). Boundary-Layer Meteorol 24: 181–209

Edwards, GC, Neumann, HH, den Hartog, G, Thurtell, GW, Kidd, G (1994) Eddy correlation measurements of methane fluxes using a tunable diode laser at the Kinosheo Lake tower site during the Northern Wetlands Study (NOWES). J Geophys Res 99 D1: 1511–1518

Elagina, LG (1962) Optitscheskij pribor dlja izmerenija turbulentnych pulsacii vlaschnosti (Optical sensor for the measurement of turbulent humidity fluctuations). Izv AN SSSR, ser Geofiz 12: 1100–1107

Elliott, WP (1958) The growth of the atmospheric internal boundary layer. Trans Am Geophys Union 39: 1048–1054

ESDU (1972) Characteristics of wind speed in the lowest layers of the atmosphere near the ground: strong winds. 72026, Engl. Sci. Data Unit Ltd. Regent St., London.

Etling, D (2002) Theoretische Meteorologie. Springer, Berlin, Heidelberg, 354 pp.

Eugster, W, Senn, W (1995) A cospectral correction for measurement of turbulent NO_2 flux. Boundary-Layer Meteorol 74: 321–340

Falge, E, Baldocchi, D, Olson, R, Anthoni, P, Aubinet, M, Bernhofer, C, Burba, G, Ceulemans, R, Clement, R, Dolman, H, Granier, A, Gross, P, Grunwald, T, Hollinger, D, Jensen, NO, Katul, G, Keronen, P, Kowalski, A, Lai, CT, Law, BE, Meyers, T, Moncrieff, H, Moors, E, Munger, JW, Pilegaard, K, Rannik, U, Rebmann, C, Suyker, A, Tenhunen, J, Tu, K, Verma, S, Vesala, T, Wilson, K, Wofsy, S (2001) Gap filling strategies for long term energy flux data sets. Agric Forest Meteorol 107: 71–77

Falge, EM, Ryel, RJ, Alsheimer, M, Tenhunen, JD (1997) Effects on stand structure and physiology on forest gas exchange: A simulation study for Norway spruce. Trees 11: 436–448

Fanger, PO (1972) Thermal comfort: Analysis and applications in environmental engineering. McGraw Hill, New York, 244 pp.

Farge, M (1992) Wavelet transforms and their application to turbulence. Annu. Rev. Fluid Mech. 24: 395–347

Fiebrich, CA, Crawford, KL (2001) The impact of unique meteorological phenomena detected by the Oklahoma Mesonet and ARS Micronet on automatic quality control. Bull Amer Meteorol Soc 82: 2173–2187

Fiedler, F, Panofsky, HA (1972) The geostrophic drag coefficient and the 'effective' roughness length. Quart J Roy Meteorol Soc 98: 213–220

Finlayson-Pitts, BJ, Pitts, JN (2000) Chemistry of the upper and lower atmosphere. Academic Press, San Diego, 963 pp.

Finn, D, Lamb, B, Leclerc, MY, Horst, TW (1996) Experimental evaluation of analytical and Lagrangian surface-layer flux footprint models. Boundary-Layer Meteorol 80: 283–308

Finn, D, Lamb, B, Leclerc, MY, Lovejoy, S, Pecknold, S, Schertzer, D (2001) Multifractal analysis of line-source plume concentration fluctuations in surface-layer flows. J Appl Meteorol 40: 229–245

Finnigan, J (2000) Turbulence in plant canopies. Ann Rev Fluid Mech 32: 519–571

Finnigan, JJ, Clement, R, Malhi, Y, Leuning, R, Cleugh, HA (2003) A re-evaluation of long-term flux measurement techniques, Part I: Averaging and coordinate rotation. Boundary-Layer Meteorol 107: 1–48

Fischer, G (Editor), (1988) Landolt-Börnstein, Numerical data and functional relationships in science and technology, Group V: Geophysics and space research, Volume 4: Meteorology, Subvolume b: Physical and chemical properties of the air. Springer, Berlin, Heidelberg, 570 pp.

Flemming, G (1991) Einführung in die Angewandte Meteorologie. Akademie-Verlag, Berlin, 168 pp.

Foken, T (1978) The molecular temperature boundary layer of the atmosphere over various surfaces. Archiv Meteorol Geophys Bioklim, Ser A 27: 59–67

Foken, T, Kitajgorodskij, SA, Kuznecov, OA (1978) On the dynamics of the molecular temperature boundary layer above the sea. Boundary-Layer Meteorol 15: 289–300

Foken, T, Kuznecov, OA (1978) Die wichtigsten Ergebnisse der gemeinsamen Expedition "KASPEX-76" des Institutes für Ozeanologie Moskau und der Karl-Marx-Universität Leipzig. Beitr. Meeresforsch. 41: 41–47

Foken, T (1979) Temperaturmessung mit dünnen Platindrähten. Z Meteorol 29: 299–307

Foken, T, Skeib, G (1980) Genauigkeit und Auswertung von Profilmessungen zur Energieaustauschbestimmung. Z Meteorol 30: 346–360

Foken, T, Kaiser, H, Rettig, W (1983) Propelleranemometer: Überblick und spezielle Entwicklungen am Meteorologischen Hauptobservatorium Potsdam. Veröff Meteorol Dienstes DDR 24: 48 pp.

Foken, T, Skeib, G (1983) Profile measurements in the atmospheric near-surface layer and the use of suitable universal functions for the determination of the turbulent energy exchange. Boundary-Layer Meteorol 25: 55–62

Foken, T (1984) The parametrisation of the energy exchange across the air-sea interface. Dynamics Atm Oceans 8: 297–305

Foken, T (1986) An operational model of the energy exchange across the air-sea interface. Z Meteorol 36: 354–359

Foken, T (1990) Turbulenter Energieaustausch zwischen Atmosphäre und Unterlage – Methoden, meßtechnische Realisierung sowie ihre Grenzen und Anwendungsmöglichkeiten. Ber Dt Wetterdienstes 180: 287 pp.

Foken, T, Kretschmer, D (1990) Stability dependence of the temperature structure parameter. Boundary-Layer Meteorol 53: 185–189

Foken, T, Skeib, G, Richter, SH (1991) Dependence of the integral turbulence characteristics on the stability of stratification and their use for Doppler-Sodar measurements. Z Meteorol 41: 311–315

Foken, T, Gerstmann, W, Richter, SH, Wichura, B, Baum, W, Ross, J, Sulev, M, Mölder, M, Tsvang, LR, Zubkovskii, SL, Kukharets, VP, Aliguseinov, AK, Perepelkin, VG, Zelený, J (1993) Study of the energy exchange processes over different types of surfaces during TARTEX-90. Dt Wetterdienst, Forsch. Entwicklung, Arbeitsergebnisse 4: 34 pp.

Foken, T, Dlugi, R, Kramm, G (1995) On the determination of dry deposition and emission of gaseous compounds at the biosphere-atmosphere interface. Meteorol Z 4: 91–118

Foken, T, Oncley, SP (1995) Results of the workshop 'Instrumental and methodical problems of land surface flux measurements'. Bull Amer Meteorol Soc 76: 1191–1193

Foken, T (1996) Turbulenzexperiment zur Untersuchung stabiler Schichtungen. Ber Polarforschung 188: 74–78

Foken, T, Wichura, B (1996) Tools for quality assessment of surface-based flux measurements. Agric Forest Meteorol 78: 83–105

Foken, T, Jegede, OO, Weisensee, U, Richter, SH, Handorf, D, Görsdorf, U, Vogel, G, Schubert, U, Kirzel, H-J, Thiermann, V (1997a) Results of the LINEX-96/2 Experiment. Dt Wetterdienst, Forsch. Entwicklung, Arbeitsergebnisse 48: 75 pp.

Foken, T, Richter, SH, Müller, H (1997b) Zur Genauigkeit der Bowen-Ratio-Methode. Wetter und Leben 49: 57–77

Foken, T (1998a) Die scheinbar ungeschlossene Energiebilanz am Erdboden – eine Herausforderung an die Experimentelle Meteorologie. Sitzungsberichte der Leibniz-Sozietät 24: 131–150

Foken, T (1998b) Ergebnisse des LINEX-97/1 Experimentes. Dt Wetterdienst, Forsch. Entwicklung, Arbeitsergebnisse 53: 38 pp.

Foken, T (1998c) Genauigkeit meteorologischer Messungen zur Bestimmung des Energie- und Stoffaustausches über hohen Pflanzenbeständen. Ann Meteorol 37: 513–514

Foken, T, Buck, AL, Nye, RA, Horn, RD (1998) A Lyman-alpha hygrometer with variable path length. J Atm Oceanic Techn 15: 211–214

Foken, T, Kukharets, VP, Perepelkin, VG, Tsvang, LR, Richter, SH, Weisensee, U (1999) The influence of the variation of the surface temperature on the closure of the surface energy balance. 13th Symposium on Boundary Layer and Turbulence, Dallas, TX., 10–15 January 1999, Am. Meteorol. Soc., 308–309

Foken, T, Mangold, A, Rebmann, C, Wichura, B (2000) Characterization of a complex measuring site for flux measurements. 14th Symposium on Boundary Layer and Turbulence, Aspen, CO, 07–11 August 2000, Am. Meteorol. Soc., Boston, 388–389

Foken, T, Wichura, B, Klemm, O, Gerchau, J, Winterhalter, M, Weidinger, T (2001) Micrometeorological conditions during the total solar eclipse of August 11, 1999. Meteorol Z 10: 171–178

Foken, T (2002) Some aspects of the viscous sublayer. Meteorol Z 11: 267–272

Foken, T, Göckede, M, Mauder, M, Mahrt, L, Amiro, BD, Munger, JW (2004) Post-field data quality control. In: X Lee, WJ Massman, B Law (Editors), Handbook of Micrometeorology: A Guide for Surface Flux Measurement and Analysis. Kluwer, Dordrecht, pp. 181–208.

Foken, T, Leclerc, MY (2004) Methods and limitations in validation of footprint models. Agric Forest Meteorol 127: 223–234

Foken, T (2006a) 50 years of the Monin-Obukhov similarity theory. Boundary-Layer Meteorol 119: 431–447

Foken, T (2006b) Angewandte Meteorologie, Mikrometeorologische Methoden. Springer, Berlin, Heidelberg, 2nd edition, 326 pp.

Foken, T, Mauder, M, Liebethal, C, Wimmer, F, Beyrich, F, Raasch, S, DeBruin, HAR, Meijninger, WML, Bange, J (2006a) Attempt to close the energy balance for the LITFASS-2003 experiment. 27th Symp Agricul Forest Meteorol, San Diego, 22-27 May 2006, American Meteorological Society, paper 1.11.

Foken, T, Wimmer, F, Mauder, M, Thomas, C, Liebethal, C (2006b) Some aspects of the energy balance closure problem. Atmos Chem Phys 6: 4395–4402

Foken, T (2008) The energy balance closure problem – An overview. Ecolog Appl 18:in print

Frankenberger, E (1951) Untersuchungen über den Vertikalaustausch in den unteren Dekametern der Atmosphäre. Ann Meteorol 4: 358–374

Frenzen, P, Vogel, CA (2001) Further studies of atmospheric turbulence in layers near the surface: Scaling the TKE budget above the roughness sublayer. Boundary-Layer Meteorol 99: 173–458

Friedrich, K, Mölders, N, Tetzlaff, G (2000) On the influence of surface heterogeneity on the Bowen-ratio: A theoretical case study. Theor Appl Climat 65: 181–196

Frisch, U (1995) Turbulence. Cambridge Univ. Press, Cambridge, 296 pp.

Fuchs, M, Tanner, CB (1970) Error analysis of Bowen ratios measured by differential psychrometry. Agricultural Meteorology 7: 329–334

Fuchs, M (1986) Heat flux. In: A Klute (Editor), Methods of Soil Analysis, Part 1: Physical and Mineralogical Methods. Agr. Monogr. ASA and SSSA, Madison, WI, pp. 957–968

Fuehrer, PL, Friehe, CA (2002) Flux correction revised. Boundary-Layer Meteorol 102: 415–457

Gandin, LS (1988) Complex quality control of meteorological observations. Monthly Weather Review 116: 1137–1156

Gao, W, Shaw, RH, PawU, KT (1989) Observation of organized structure in turbulent flow within and above a forest canopy. Boundary-Layer Meteorol 47: 349–377

Garratt, JR (1980) Surface influence upon vertical profiles in the atmospheric near surface layer. Quart J Roy Meteorol Soc 106: 803–819

Garratt, JR (1990) The internal boundary layer – A review. Boundary-Layer Meteorol 50: 171–203

Garratt, JR, Hicks, BB (1990) Micrometeorological and PBL experiments in Australia. Boundary-Layer Meteorol 50: 11–32

Garratt, JR (1992) The atmospheric boundary layer. Cambridge University Press, Cambridge, 316 pp.

Gash, JHC (1986) A note on estimating the effect of a limited fetch on micrometeorological evaporation measurements. Boundary-Layer Meteorol 35: 409–414

Gavrilov, AS, Petrov, JS (1981) Ocenka totschnosti opredelenija turbulentnych potokov po standartnym gidrometeorologitscheskim izmerenijam nad morem. Meteorol Gidrol: 52–59

Geernaert, GL (Editor), (1999) Air-Sea Exchange: Physics, Chemistry and Dynamics. Kluwer Acad. Publ., Dordrecht, 578 pp.

Geiger, R (1927) Das Klima der bodennahen Luftschicht. Friedr. Vieweg & Sohn, Braunschweig, 246 pp.

Geiger, R, Aron, RH, Todhunter, P (1995) The climate near the ground. Friedr. Vieweg & Sohn Verlagsges. mbH, Braunschweig, Wiesbaden, 528 pp.

Gerth, W-P (1986) Klimatische Wechselwirkungen in der Raumplanung bei Nutzungsänderungen. Ber Dt Wetterdienstes 171: 69 pp.

Gilgen, H, Whitlock, CH, Koch, F, Müller, G, Ohmura, A, Steiger, D, Wheeler, R (1994) Technical plan for BSRN data management. World Radiation Monitoring Centre (WRMC), Technical Report 1: 56 pp

Glickman, TS (Editor), (2000) Glossary of Meteorology. Am. Meteorol. Soc., Boston, MA, 855 pp.

Göckede, M (2000) Das Windprofil in den untersten 100 m der Atmosphäre unter besonderer Berücksichtigung der Stabilität der Schichtung. Dipl.-Arb., Universität Bayreuth, Bayreuth, 133 pp.

Göckede, M, Foken, T (2001) Ein weiterentwickeltes Holtslag-van-Ulden-Schema zur Stabilitätsparametrisierung in der Bodenschicht. Österreichische Beiträge zu Meteorologie und Geophysik 27: (Extended Abstract and pdf-file on CD) 210

Göckede, M, Rebmann, C, Foken, T (2004) A combination of quality assessment tools for eddy covariance measurements with footprint modelling for the characterisation of complex sites. Agric Forest Meteorol 127: 175–188

Göckede, M, Markkanen, T, Mauder, M, Arnold, K, Leps, J-P, Foken, T (2005) Validation footprint models using natural tracer Agric Forest Meteorol 135: 314–325

Göckede, M, Markkanen, T, Hasager, CB, Foken, T (2006) Update of a footprint-based approach for the characterisation of complex measuring sites. Boundary-Layer Meteorol 118: 635–655

Göckede, M, Thomas, C, Markkanen, T, Mauder, M, Ruppert, J, Foken, T (2007) Sensitivity of Lagrangian Stochastic footprints to turbulence statistics. Tellus 59B: 577–586

Goulden, ML, Munger, JW, Fan, F-M, Daube, BC, Wofsy, SC (1996) Measurements of carbon sequestration by long-term eddy covariance: method and critical evaluation of accuracy. Global Change Biol. 2: 159–168

Goutorbe, JP, Lebel, T, Tinga, A, Bessemoulin, P, Brouwer, J, Dolman, H, Engman, ET, Gash, JGC, Hoepffner, M, Kabat, P, Kerr, YH, Monteny, B, Prince, SD, Said, F, Sellers, P, Wallace, J (1994) HAPEX-SAHEL: A large scale study of land atmosphere interactions in the semi-arid tropics. Annales Geophysicae 12: 53–64

Graefe, J (2004) Roughness layer corrections with emphasis on SVAT model applications. Agric Forest Meteorol 124: 237–251

Graf, U (2004) Applied Laplace transforms and Z-transforms for scientists and engineers. Birkhäuser, Basel, 500 pp.

Grimmond, CSB, King, TS, Roth, M, Oke, TR (1998) Aerodynamic roughness of urban areas derived from wind observations. Boundary-Layer Meteorol 89: 1–24

Grimmond, CSB, Oke, TR (1999) Aerodynamic properties of urban areas derived from analysis of surface form. J Appl Meteorol 38: 1262–1292

Grimmond, CSB, Oke, TR (2000) Corrigendum. J Appl Meteorol 39: 2494

Gryning, S-E, van Ulden, AP, Larsen, S (1983) Dispersions from a ground level source investigated by a K model. Quart J Roy Meteorol Soc 109: 355–364

Gu, L, Falge, EM, Boden, T, Baldocchi, DD, Black, T, Saleska, SR, Sumi, T, Verma, SB, Vesala, T, Wofsy, SC, Xu, L (2005) Objective threshold determination for nighttime eddy flux filtering. Agric Forest Meteorol 128: 179–197

Guderian, R (Editor), (2000) Handbuch der Umweltveränderungen und Ökotoxikologie, Atmosphäre, Bd. 1A und 1B. Springer, Berlin, Heidelberg, 424, 516 pp.

Gurjanov, AE, Zubkovskij, SL, Fedorov, MM (1984) Mnogokanalnaja avtomatizirovannaja sistema obrabotki signalov na baze EVM (Automatic multi-channel system for signal analysis with electronic data processing). Geod Geophys Veröff, R II 26: 17–20

Gurtalova, T, Vidovitsch, J, Matejka, F, Mudry, P (1988) Dinamitscheskaja sherochovatost i biometritscheskie charakteristiki kukuruzy v ontogeneze (Dinamical roughness and biometric characteristics of corn during the Ontogenesis). Contr Geophys Inst Slovak Akad Sci 8: 46–53

Halldin, S, Lindroth, A (1992) Errors in net radiometry, comparison and evaluation of six radiometer designs. J Atm Oceanic Techn 9: 762–783

Hanafusa, T, Fujitana, T, Kobori, Y, Mitsuta, Y (1982) A new type sonic anemometer-thermometer for field operation. Papers in Meteorology and Geophysics 33: 1–19

Handorf, D, Foken, T (1997) Analysis of turbulent structure over an Antarctic ice shelf by means of wavelet transformation. 12th Symosium on Boundary Layer and Turbulence, Vancouver BC, Canada, 28 July – 1 August 1997, Am. Meteorol. Soc., 245–246

Handorf, D, Foken, T, Kottmeier, C (1999) The stable atmospheric boundary layer over an Antarctic ice sheet. Boundary-Layer Meteorol 91: 165–186

Hann, JF, Süring, R (1939) Lehrbuch der Meteorologie. Verlag von Willibald Keller, Leipzig, 480 pp.

Hartmann, DL (1994) Global physical climatology. Academic Press, San Diego, New York, 408 pp.

Hartogensis, OK, DeBruin, HAR (2005) Monin-Obukhov similarity functions of the structure parameter of temperature and turbulent kinetic energy dissipation in the stable boundary layer. Boundary-Layer Meteorol 116: 253–276

Hasager, CB, Jensen, NO (1999) Surface-flux aggregation in heterogeneous terrain. Quart J Roy Meteorol Soc 125: 2075–2102

Hasager, CB, Nielsen, NW, Jensen, NO, Boegh, E, Christensen, JH, Dellwik, E, Soegaard, H (2003) Effective roughness calculated from satellite-derived land cover maps and hedge-information used in a weather forecasting model. Boundary-Layer Meteorol 109: 227–254

Haude, W (1955) Bestimmung der Verdunstung auf möglichst einfache Weise. Mitteilungen des Deutschen Wetterdienstes 11: 24 pp.

Haugen, DA (1978) Effects of sampling rates and averaging periods on meteorological measurements. Fourth Symp Meteorol Observ Instr, Am Meteorol Soc: 15–18

Haugen, DH (Editor), (1973) Workshop on micrometeorology. Am. Meteorol. Soc., Boston, 392 pp.

Heinz, G, Handorf, D, Foken, T (1999) Direct visualization of the energy transfer from coherent structures to turbulence via wavelet analysis. 13th Symposium on Boundary Layer and Turbulence, Dallas, TX, 10–15 January 1999, Am. Meteorol. Soc., 664–665

Helbig, A, Baumüller, J, Kerschgens, MJ (Editors) (1999) Stadtklima und Luftreinhaltung. Springer, Berlin, Heidelberg, 467 pp.

Hentschel, G (1982) Das Bioklima des Menschen. Verlag Volk und Gesundheit, Berlin, 192 pp.

Herzog, H-J, Vogel, G, Schubert, U (2002) LLM – a nonhydrostatic model applied to high-resolving simulation of turbulent fluxes over heterogeneous terrain. Theor Appl Climat 73: 67–86

Hess, GD, Hicks, BB, Yamada, T (1981) The impact of the Wangara experiment. Boundary-Layer Meteorol 20: 135–174

Hesselberg, T (1926) Die Gesetze der ausgeglichenen atmosphärischen Bewegungen. Beitr Phys Atm 12: 141–160

Heusinkveld, BG, Jacobs, AFG, Holtslag, AAM, Berkowicz, SM (2004) Surface energy balance closure in an arid region: role of soil heat flux. Agric Forest Meteorol 122: 21–37

Hicks, BB (1981) An examination of the turbulence statistics in the surface boundary layer. Boundary-Layer Meteorol 21: 389–402

Hicks, BB, Baldocchi, DD, Meyers, TP, Hosker jr., RP, Matt, DR (1987) A preliminary multiple resistance routine for deriving dry deposition velocities from measured quantities. Water, Air and Soil Pollution 36: 311–330

Hicks, BB, Matt, DR (1988) Combining biology, chemistry and meteorology in modelling and measuring dry deposition. J Atm Chem 6: 117–131

Hierteis, M, Svoboda, J, Foken, T (2000) Einfluss der Topographie auf das Windfeld und die Leistung von Windkraftanlagen. DEWEK 2000, Wilhelmshaven, 07.-08.06.2000, Deutsches Windenergie-Institut, 272–276

Hill, R (1997) Algorithms for obtaining atmospheric surface-layer from scintillation measurements. J Atm Oceanic Techn 14: 456–467

Hill, RJ, Clifford, SF, Lawrence, RS (1980) Refractive index and absorption fluctuations in the infrared caused by temperature, humidity and pressure fluctuations. J Opt Soc Am 70: 1192–1205

Hillel, D (1980) Applications of soil physics. Academic Press, New York, 385 pp.

Högström, U (1974) A field study of the turbulent fluxes of heat water vapour and momentum at a 'typical' agricultural site. Quart J Roy Meteorol Soc 100: 624–639

Högström, U (1985) Von Kármán constant in atmospheric boundary flow: Reevaluated. J Atmos Sci 42: 263–270

Högström, U (1988) Non-dimensional wind and temperature profiles in the atmospheric surface layer: A re-evaluation. Boundary-Layer Meteorol 42: 55–78

Högström, U (1990) Analysis of turbulence structure in the surface layer with a modified similarity formulation for near neutral conditions,. J Atmos Sci 47: 1949–1972

Högström, U (1996) Review of some basic characteristics of the atmospheric surface layer. Boundary-Layer Meteorol 78: 215–246

Högström, U, Hunt, JCR, Smedman, A-S (2002) Theory and measurements for turbulence spectra and variances in the atmospheric neutral surface layer. Boundary-Layer Meteorol 103: 101–124

Högström, U, Smedman, A (2004) Accuracy of sonic anemometers: Laminar wind-tunnel calibrations compared to atmospheric in situ calibrations against a reference instrument. Boundary-Layer Meteorol 111: 33–54

Højstrup, J (1981) A simple model for the adjustment of velocity spectra in unstable conditions downstream of an abrupt change in roughness and heat flux. Boundary-Layer Meteorol 21: 341–356

Højstrup, J (1993) A statistical data screening procedure. Meas Sci Techn 4: 153–157

Holmes, P, Lumley, JL, Berkooz, G (1996) Turbulence, coherent structures, dynamical systems and symmetry. Cambridge University Press, Cambridge, 420 pp.

Holtslag, AAM, van Ulden, AP (1983) A simple scheme for daytime estimates of the surface fluxes from routine weather data. J Climate Appl Meteorol. 22: 517–529

Holtslag, AAM, Nieuwstadt, FTM (1986) Scaling the atmospheric boundary layer. Boundary-Layer Meteorol 36: 201–209

Höppe, P (1984) Die Energiebilanz des Menschen. Wiss. Mitt. Meteorol. Inst. Uni. München 49: 173 pp.

Höppe, P (1992) Ein neues Verfahren zur Bestimmung der mittleren Strahlungstemperatur im Freien. Wetter und Leben 44: 147–151

Höppe, P (1999) The physiological equivalent temperature – a universal index for biometeorological assessment of the thermal environment. Int J Biometeorol 43: 71–75

Horst, TW, Weil, JC (1992) Footprint estimation for scalar flux measurements in the atmospheric surface layer. Boundary-Layer Meteorol 59: 279–296

Horst, TW, Weil, JC (1994) How far is far enough?: The fetch requirements for micrometeorological measurement of surface fluxes. J Atm Oceanic Techn 11: 1018–1025

Horst, TW, Weil, JC (1995) Corrigenda: How far is far enough?: The fetch requirements for micrometeorological measurement of surface fluxes. J Atm Oceanic Techn 12: 447

Horst, TW (1999) The footprint for estimation of atmosphere-surface exchange fluxes by profile techniques. Boundary-Layer Meteorol 90: 171–188

Horton, R, Wieringa, PJ, Nielsen, DR (1983) Evaluation of methods for determining the apparent thermal diffusivity of soil near the surface. Soil Sci Soc Am J 47: 25–32

Houghton, DD (1985) Handbook of applied meteorology. Wiley, New York, XV, 1461 pp.

Houghton, JT, Ding, Y, Griggs, DJ, Noguer, M, van der Linden, PJ, Dai, X, Maskell, K, Johnson, CA (Editors) (2001) Climate change 2001, The scientific basic. Cambridge University Press, Cambridge, 881 pp.

Hsu, SA, Meindl, EA, Gilhousen, DB (1994) Determination of power-law wind-profile exponent under near- neutral stability conditions at sea. J Appl Meteorol 33: 757–765

Huang, CH (1979) A theory of dispersion in turbulent shear flow. Atmos Environm 13: 453–463

Hui, DF, Wan, SQ, Su, B, Katul, G, Monson, R, Luo, YQ (2004) Gap-filling missing data in eddy covariance measurements using multiple imputation (MI) for annual estimations. Agric Forest Meteorol 121: 93–111

Hupfer, P, Foken, T, Bachstein, U (1976) Fine structure of the internal boundary layer in the near shore zone of the sea. Boundary-Layer Meteorol 10: 503–505

Hupfer, P (1989) Klima im mesoräumigen Bereich. Abh Meteorol Dienstes DDR 141: 181–192

Hupfer, P (Editor), (1991) Das Klimasystem der Erde. Akademie-Verlag, Berlin, 464 pp.

Hupfer, P (1996) Unsere Umwelt: Das Klima. B. G. Teubner, Stuttgart, Leipzig, 335 pp.

Hupfer, P, Kuttler, W (Editors) (2005) Witterung und Klima, begründet von Ernst Heyer. B. G. Teubner, Stuttgart, Leipzig, 554 pp.

Hurtalová, T, Matejka, F, Vidovic, J (1983) Untersuchung der Oberflächenrauhigkeit und ihrer Variation auf Grund von Strukturveränderungen des Maisbestandes während der Ontogenese. Z Meteorol 33: 368–372

Hyson, P, Garratt, JR, Francey, RJ (1977) Algebraic und elektronic corrections of measured uw covariance in the lower atmosphere. Boundary-Layer Meteorol 16: 43–47

Ibrom, A, Dellwik, E, Flyvbjerg, H, Jensen, NO, Pilegaard, K (2007) Strong low-pass filtering effects on water vapour flux measurements with closed-path eddy correlation systems. Agric Forest Meteorol 147: 140–156

Inclan, MG, Forkel, R, Dlugi, R, Stull, RB (1996) Application of transilient turbulent theory to study interactions between the atmospheric boundary layer and forest canopies. Boundary-Layer Meteorol 79: 315–344

Irvin, JS (1978) A theoretical variation of the wind profile power-law exponent as a function of surface roughness and stability. Atmos Environm 13: 191–194

ISO (1996) Acoustics - Attenuation of sound during propagation outdoors – Part 2: General method of calculation, ISO 9613-2 Beuth-Verlag, Berlin, 26 pp.

Itier, B (1980) Une méthode simplifièe pour la mesure du flux de chaleur sensible. J Rech Atm 14: 17–34

Izumi, Y (1971) Kansas 1968 field program data report. Air Force Cambridge Research Papers, No. 379, Air Force Cambridge Research Laboratory, Bedford, MA, 79 pp.

Jacobs, AFG, McNaughton, KG (1994) The excess temperature of a rigid fast-response thermometer and its effects on measured heat fluxes. J Atm Oceanic Techn 11: 680–686

Jacobs, AFG, Heusinkveld, BG, Nieveen, JP (1998) Temperature behavior of a natural shallow water body during a summer periode. Theor Appl Climat 59: 121–127

Jacobson, MZ (2005) Fundamentals of atmospheric modelling. Cambridge University Press, Cambridge, 813 pp.

Jegede, OO, Foken, T (1999) A study of the internal boundary layer due to a roughness change in neutral conditions observed during the LINEX field campaigns. Theor. & Appl. Climatol. 62: 31–41

Jendritzky, G, Metz, G, Schirmer, H, Schmidt-Kessen, W (1990) Methodik zur raumbezogenen Bewertung der thermischen Komponente im Bioklima des Menschen. Beitr Akad Raumforschung Landschaftsplanung 114: 80 pp.

Joffre, SM (1984) Power laws and the empirical representation of velocity and directional sheer. J Climate Appl Meteorol. 23: 1196–1203

Johansson, C, Smedman, A, Högström, U, Brasseur, JG, Khanna, S (2001) Critical test of Monin-Obukhov similarity during convective conditions. J Atmos Sci 58: 1549–1566

Jones, HG (1992) Plants and microclimate. Cambridge Univ. Press, Cambridge, 428 pp.

Junghans (1967) Der Einfluß es Windes auf das Niederschlagsdargebot von Hängen. Archiv Forstw 16: 579–585

Kader, BA, Yaglom, AM (1972) Heat and mass transfer laws for fully turbulent wall flows. Int J Heat Mass Transfer 15: 2329–2350

Kader, BA, Perepelkin, VG (1984) Profil skorosti vetra i temperatury v prizemnom sloje atmosfery v uslovijach nejtralnoj i neustojtschivoj stratifikacii (The wind and temperature profile in the near surface layer for neutral and unstable stratification). Izv AN SSSR, Fiz Atm Okeana 20: 151–161

Kaimal, JC, Businger, JA (1963) A continuous wave sonic anemometer-thermometer. J Climate Appl Meteorol. 2: 156–164

Kaimal, JC, Wyngaard, JC, Izumi, Y, Coté, OR (1972) Spectral characteristics of surface layer turbulence. Quart J Roy Meteorol Soc 98: 563–589

Kaimal, JC, Wyngaard, JC (1990) The Kansas and Minnesota experiments. Boundary-Layer Meteorol 50: 31–47

Kaimal, JC, Gaynor, JE (1991) Another look to sonic thermometry. Boundary-Layer Meteorol 56: 401–410

Kaimal, JC, Finnigan, JJ (1994) Atmospheric boundary layer flows: Their structure and measurement. Oxford University Press, New York, NY, 289 pp.

Kallistratova, MA (1959) Eksperimentalnoje issledovanie rassejenija zvuka v turbulentnoj atmosfere (An experimental investigation in the scattering of sound in the turbulent atmosphere). Dokl AN SSSR 125: 69–72

Kanda, M, Inagaki, A, Letzel, MO, Raasch, S, Watanabe, T (2004) LES study of the energy imbalance problem with eddy covariance fluxes. Boundary-Layer Meteorol 110: 381–404

Kanemasu, ET, Verma, SB, Smith, EA, Fritschen, LY, Wesely, M, Fild, RT, Kustas, WP, Weaver, H, Steawart, YB, Geney, R, Panin, GN, Moncrieff, JB (1992) Surface flux measurements in FIFE: An overview. J Geophys Res 97: 18.547–18.555

Kantha, LH, Clayson, CA (2000) Small scale processes in geophysical fluid flows. Academic Press, San Diego, 883 pp.

Kasten, F (1985) Maintenance, calibration and comparison. WMO, Instruments Observ Methods WMO/TD 51: 65–84

Kiehl, J, Trenberth, KE (1997) Earth annual global mean energy budget. Bull Amer Meteorol Soc 78: 197–208

King, JC, Anderson, PS, Smith, MC, Mobbs, SD (1996) The surface energy and mass balance at Halley, Antarctica during winter. J Geophys Res 101(D14): 19119–19128

King, JC, Turner, J (1997) Antarctic Meteorology and Climatology. Cambridge University Press, Cambridge, 409 pp.

Kitajgorodskij, SA, Volkov, JA (1965) O rascete turbulentnych potokov tepla i vlagi v privodnom sloe atmosfery (The calculation of the turbulent fluxes of temperature and humidity in the atmosphere near the water surface) Izv AN SSSR, Fiz Atm Okeana 1: 1317–1336

Kitajgorodskij, SA (1976) Die Anwendung der Ähnlichkeitstheorie für die Bearbeitung der Turbulenz in der bodennahen Schicht der Atmosphäre. Z Meteorol 26: 185–204

Klaassen, W, van Breugel, PB, Moors, EJ, Nieveen, JP (2002) Increased heat fluxes near a forest edge. Theor Appl Climat 72: 231–243

Kleinschmidt, E (Editor), (1935) Handbuch der meteorologischen Instrumente und ihrer Auswertung. Springer, Berlin, 733 pp.

Kljun, N, Rotach, MW, Schmid, HP (2002) A three-dimensional backward Lagrangian footprint model for a wide range of boundary layer stratification. Boundary-Layer Meteorol 103: 205–226

Klug, W (1969) Ein Verfahren zur Bestimmung der Ausbreitungsbedingungen aus synoptischen Beobachtungen. Staub – Reinhalt Luft 29: 143–147

Knoch, K (1949) Die Geländeklimatologie, ein wichtiger Zweig der angewandten Klimatologie. Ber Dtsch Landesk 7: 115–123

Kohsiek, W (1982) Measuring C_T^2, C_Q^2, and C_{TQ} in the unstable surface layer, and relations to the vertical fluxes of heat and moisture. Boundary-Layer Meteorol 24: 89–107

Kohsiek, W, Liebethal, C, Foken, T, Vogt, R, Oncley, SP, Bernhofer, C, DeBruin, HAR (2007) The Energy Balance Experiment EBEX-2000. Part III: Behaviour and quality of radiation measurements. Boundary-Layer Meteorol 123: 55–75

Koitzsch, R, Dzingel, M, Foken, T, Mücket, G (1988) Probleme der experimentellen Erfassung des Energieaustausches über Winterweizen. Z Meteorol 38: 150–155

Kolmogorov, AN (1941a) Rassejanie energii pri lokolno-isotropoi turbulentnosti (Dissipation of energy in locally isotropic turbulence). Dokl AN SSSR 32: 22–24

Kolmogorov, AN (1941b) Lokalnaja struktura turbulentnosti v neschtschimaemoi schidkosti pri otschen bolschich tschislach Reynoldsa (The local structure of turbulence in incompressible viscous fluid for very large Reynolds numbers). Dokl AN SSSR 30: 299–303

Kondo, J, Sato, T (1982) The determination of the von Kármán constant. J Meteorol Soc Japan 60: 461–471

Kormann, R, Meixner, FX (2001) An analytical footprint model for non-neutral stratification. Boundary-Layer Meteorol 99: 207–224

Kramm, G, Dlugi, R, Lenschow, DH (1995) A re-evaluation of the Webb correction using density weighted averages. J Hydrol 166: 293–311

Kramm, G, Beier, M, Foken, T, Müller, H, Schröder, P, Seiler, W (1996a) A SVAT-skime for NO, NO_2, and O_3 - Model description and test results. Meteorol Atmos Phys 61: 89–106

Kramm, G, Foken, T, Molders, N, Muller, H, Paw U, KT (1996b) The sublayer-Stanton numbers of heat and matter for different types of natural surfaces. Contr Atmosph Phys 69: 417–430

Kramm, G, Foken, T (1998) Ucertainty analysis on the evaporation at the sea surface. Second Study Conference on BALTEX, Juliusruh, 25–29 May 1998, BALTEX Secretariat, 113–114

Kramm, G, Meixner, FX (2000) On the dispersion of trace species in the atmospheric boundary layer: a re-formulation of the governing equations for the turbulent flow of the compressible atmosphere. Tellus 51A: 500–522

Kramm, G, Dlugi, R, Mölders, N (2002) Sublayer-Stanton numbers of heat and matter for aerodynamically smooth surfaces: basic considerations and evaluations. Meteorol Atmos Phys 79: 173–194

Kraus, H (1983) Meso- und mikro-skalige Klimasysteme. Ann Meteorol 20: 4–7

Kraus, H (2004) Die Atmosphäre der Erde. Springer, Berlin, Heidelberg, 422 pp.

Kretschmer, SI (1954) Metodika izmerenija mikropulsacii skorosti vetra i temperatura v atmosfere (A method to measure the fluctuations of the wind velocity and the temperature). Trudy geofiz inst AN SSSR 24 (151): 43–111

Kretschmer, SI, Karpovitsch, JV (1973) Maloinercionnyj ultrafioletovyj vlagometer (Sensitive ultraviolet hygrometer). Izv AN SSSR, Fiz Atm Okeana 9: 642–645

Kristensen, L, Mann, J, Oncley, SP, Wyngaard, JC (1997) How close is close enough when measuring scalar fluxes with displaced sensors. J Atm Oceanic Techn 14: 814–821

Kristensen, L (1998) Cup anemometer behavior in turbulent environments. J Atm Oceanic Techn 15: 5–17

Kukharets, VP, Nalbandyan, HG, Foken, T (2000) Thermal Interactions between the underlying surface and a nonstationary radiation flux. Izv Atmos Ocenanic Phys 36: 318–325

Landau, LD, Lifschitz, EM (1987) Fluid Mechanics. Butterworth-Heinemann, Oxford, 539 pp.

Laubach, J (1996) Charakterisierung des turbulenten Austausches von Wärme, Wasserdampf und Kohlendioxid über niedriger Vegetation an Hand von Eddy-Korrelations-Messungen. Wiss Mitt Inst Meteorol Univ Leipzig und Inst Troposphärenforschung Leipzig 3: 139 pp.

Leclerc, MY, Thurtell, GW (1989) Footprint Predictions of Scalar Fluxes and Concentration Profiles using a Markovian Analysis 19th Conference of Agricultural and Forest Meteorology, Charleston, SC, March 7–10, 1989, American Meteorological Society

Leclerc, MY, Thurtell, GW (1990) Footprint prediction of scalar fluxes using a Markovian analysis. Boundary-Layer Meteorol 52: 247–258

Leclerc, MY, Shen, S, Lamb, B (1997) Observations and large-eddy simulation modeling of footprints in the lower convective boundary layer. J Geophys Res 102(D8): 9323–9334

Lee, R (1978) Forest Microclimatology. Columbia University Press, New York, 276 pp.

Lee, X (1998) On micrometeorological observations of surface-air exchange over tall vegetation. Agric Forest Meteorol 91: 39–49

Lee, X (2000) Air motion within and above forest vegetation in non-ideal conditions. For Ecol Managem 135: 3–18

Lee, X, Massman, WJ, Law, B (Editors) (2004) Handbook of Micrometeorology: A Guide for Surface Flux Measurement and Analysis. Kluwer, Dordrecht, 250 pp.

Lege, D (1981) Eine Betrachtung zur Bestimmung des Stroms fühlbarer Wärme und der Schubspannungsgeschwindigkeit aus Temperatur- und Windgeschwindigkeitsdifferenzen. Meteorol Rundschau 34: 1–4

Lehmann, A, Kalb, M (1993) 100 Jahre meteorologische Beobachtungen an der Säkularstation Potsdam 1893-1992. Deutscher Wetterdienst, Offenbach, 32 pp.

Lenschow, DH, Mann, J, Kristensen, L (1994) How long is long enough when measuring fluxes and other turbulence statistics? J Atm Oceanic Techn 11: 661–673

Lettau, H (1939) Atmosphärische Turbulenz. Akad. Verlagsges., Leipzig, 283 pp.

Lettau, H (1949) Isotropic and non-isitropic turbulence in the atmospheric surface layer. Geophys Res Pap 1: 86 pp.

Lettau, H (1969) Note on aerodynamic roughness-parameter estimation on the basis of roughness-element description. J Appl Meteorol 8: 828–832

Lettau, HH (1957) Windprofil, innere Reibung und Energieumsatz in den untersten 500 m über dem Meer. Beitr Phys Atm 30: 78–96

Lettau, HH, Davidson, B (Editors) (1957) Exploring the atmosphere's first mile, Vol.1. Pergamon Press, London, New York, 376 pp.

Leuning, R, Legg, BJ (1982) Comments on 'The influence of water vapor fluctuations on turbulent fluxes' by Brook. Boundary-Layer Meteorol 23: 255–258

Leuning, R, Judd, MJ (1996) The relative merits of open- and closed path analysers for measurements of eddy fluxes. Global Change Biology 2: 241–254

Lexikon (1998) der Physik in 6 Bänden. Spektrum Akademischer Verlag, Heidelberg.

Liebethal, C, Foken, T (2003) On the significance of the Webb correction to fluxes. Boundary-Layer Meteorol 109: 99–106

Liebethal, C, Foken, T (2004) On the significance of the Webb correction to fluxes, Corrigendum. Boundary-Layer Meteorol 113: 301

Liebethal, C, Huwe, B, Foken, T (2005) Sensitivity analysis for two ground heat flux calculation approaches. Agric Forest Meteorol 132: 253–262

Liebethal, C, Foken, T (2006) On the use of two repeatedly heated sensors in the determination of physical soil parameters. Meteorol Z 15: 293–299

Liebethal, C, Foken, T (2007) Evaluation of six parameterization approaches for the ground heat flux. Theor Appl Climat 88: 43–56

Lilly, DK (1967) The representation of small-scale turbulence in numerical simulation experiments. IBM Scientific Computing Symposium on Environmental Science, Yorktown Heights, N.Y., November 14–16, 1966, IBM Form No. 320–1951,195–210

Linacre, ET (1994) Estimating U.S. Class-A pan evaporation from climate data. Water Internat 19: 5–14

Liou, KN (1992) Radiation and cloud processes in the atmosphere. Oxford University Press, Oxford, 487 pp.

Liu, H, Foken, T (2001) A modified Bowen ratio method to determine sensible and latent heat fluxes. Meteorol Z 10: 71–80

Liu, H, Peters, G, Foken, T (2001) New equations for sonic temperature variance and buoyancy heat flux with an omnidirectional sonic anemometer. Boundary-Layer Meteorol 100: 459–468

Liu, H (2005) An alternative approach for CO_2 flux correction caused by heat and water vapour transfer. Boundary-Layer Meteorol 115: 151–168

Lloyd, J, Taylor, JA (1994) On the temperature dependence of soil respiration. Functional Ecology 8: 315–323

Logan, E, Fichtl, GH (1975) Rough to smooth transition of the equilibrium neutral constant stress layer. Boundary-Layer Meteorol 8: 525–528

Louis, JF (1979) A parametric model of vertical fluxes in the atmosphere. Boundary-Layer Meteorol 17: 187–202

Louis, JF, Tiedtke, M, Geleyn, JF (1982) A short history of the PBL parametrization at ECMWF. Workshop on Boundary Layer parametrization, Reading, ECMWF, 59–79

Lumley, JL, Panofsky, HA (1964) The structure of atmospheric turbulence. Interscience Publishers, New Yotk, 239 pp.

Lumley, JL, Yaglom, AM (2001) A century of turbulence. Flow, Turbulence and Combustion 66: 241–286

Mahrt, L (1991) Eddy asymmetry in the sheared heated boundary layer. J Atmos Sci 48: 472–492

Mahrt, L (1996) The bulk aerodynamic formulation over heterogeneous surfaces. Boundary-Layer Meteorol 78: 87–119

Mangarella, PA, Chambers, AJ, Street, RL, Hsu, EY (1972) Laboratory and field interfacial energy and mass flux and prediction equations. J Geophys Res 77: 5870–5875

Mangarella, PA, Chambers, AJ, Street, RL, Hsu, EY (1973) Laboratory studies of evaporation and energy transfer through a wavy air-water interface. J. Phys. Oceanogr. 3: 93–101

Manier, G (1975) Vergleich zwischen Ausbreitungsklassen und Temperaturgradient. Meteorol Rundschau 28: 6–11

Marquardt, D (1983) An algorithm for least-sqares estimation of nonlinear parameters. J Soc Indust Appl Math 11: 431–441

Martini, L, Stark, B, Hunsalz, G (1973) Elektronisches Lyman-Alpha-Feuchtigkeitsmessgerät. Z Meteorol 23: 313–322

Marunitsch, SV (1971) Charakteristiki turbulentnosti v yslovijach lesa po gradientnym i strukturnym nabljudenijam. Trudy Gos Gidrometeorol Inst 198: 154–165

Mason, PJ (1988) The formation of areally-averaged roughness length. Quart J Roy Meteorol Soc 114: 399–420

Massman, WJ (2000) A simple method for estimating frequency response corrections for eddy covariance systems. Agric Forest Meteorol 104: 185–198

Matzarakis, A, Mayer, H (1997) Heat stress in Greece. Int J Biometeorol 41: 34–39

Matzarakis, A, Mayer, H, Iziomon, M (1999) Applications of a universal thermal index: physiological equivalent temperature. Int J Biometeorol 43: 76–84

Mauder, M (2002) Auswertung von Turbulenzmessgerätevergleichen unter besonderer Berücksichtigung von EBEX-2000. Dipl.-Arbeit, Universität Bayreuth, Bayreuth, 86 pp.

Mauder, M, Foken, T (2006) Impact of post-field data processing on eddy covariance flux estimates and energy balance closure. Meteorol Z 15: 597–609

Mauder, M, Liebethal, C, Göckede, M, Leps, J-P, Beyrich, F, Foken, T (2006) Processing and quality control of flux data during LITFASS-2003. Boundary-Layer Meteorol 121: 67–88

Mauder, M, Jegede, OO, Okogbue, EC, Wimmer, F, Foken, T (2007a) Surface energy flux measurements at a tropical site in West-Africa during the transition from dry to wet season. Theor Appl Climat 89: 171–183

Mauder, M, Oncley, SP, Vogt, R, Weidinger, T, Ribeiro, L, Bernhofer, C, Foken, T, Kohsiek, W, DeBruin, HAR, Liu, H (2007b) The Energy Balance Experiment EBEX-2000. Part II: Intercomparison of eddy covariance sensors and post-field data processing methods. Boundary-Layer Meteorol 123: 29–54

Mauder, M, Foken, T, Clement, R, Elbers, J, Eugster, W, Grünwald, T, Heusinkveld, B, Kolle, O (2008) Quality control of CarboEurope flux data – Part 2: Inter-comparison of eddy-covariance software. Biogeosci 5: 451–462

McAllister, LG, Pollard, JR, Mahoney, AR, Shaw, PJR (1969) Acoustic sounding – A new approach to the study of atmospheric structure. Proc IEEE 57: 579–587

McBean, GA (1971) The variation of the statistics of wind, temperature and humidity fluctuations with stability. Boundary-Layer Meteorol 1: 438–457

McBean, GA, Bernhardt, K, Bodin, S, Litynska, Z, van Ulden, AP, Wyngaard, JC (1979) The planetary boundary layer. WMO, Note 530: 201 pp.

McMillen, RT (1988) An eddy correlation technique with extended applicability to non-simple terrain. Boundary-Layer Meteorol 43: 231–245

Meijninger, WML, Green, AE, Hartogensis, OK, Kohsiek, W, Hoedjes, JCB, Zuurbier, RM, DeBruin, HAR (2002) Determination of area-averaged water vapour fluxes with large aperture and radio wave scintillometers over a heterogeneous surface – Flevoland Field Experiment. Boundary-Layer Meteorol 105: 63–83

Meijninger, WML, Lüdi, A, Beyrich, F, Kohsiek, W, DeBruin, HAR (2006) Scintillometer-based turbulent surface fluxes of sensible and latent heat over heterogeneous a land surface – A contribution to LITFASS-2003. Boundary-Layer Meteorol 121: 89–110

Mengelkamp, H-T (1999) Wind climate simulation over complex terrain and wind turbine energy output estimation. Theor Appl Climat 63: 129–139

Mengelkamp, H-T, Beyrich, F, Heinemann, G, Ament, F, Bange, J, Berger, FH, Bösenberg, J, Foken, T, Hennemuth, B, Heret, C, Huneke, S, Johnsen, K-P, Kerschgens, M, Kohsiek, W, Leps, J-P, Liebethal, C, Lohse, H, Mauder, M, Meijninger, WML, Raasch, S, Simmer, C, Spieß, T, Tittebrand, A, Uhlenbrook, S, Zittel, P (2006) Evaporation over a heterogeneous land surface: The EVA_GRIPS project. Bull Amer Meteorol Soc 87: 775–786

Meyers, TP, Paw U, KT (1986) Testing a higher-order closure model for modelling airflow within and above plant canopies. Boundary-Layer Meteorol 37: 297–311

Meyers, TP, Paw U, KT (1987) Modelling the plant canopy microenvironment with higher-order closure principles. Agric Forest Meteorol 41: 143–163

Michaelis, L, Menton, ML (1913) Die Kinetik der Invertinwirkung. Biochem Z 49: 333

Mitsuta, Y (1966) Sonic anemometer-thermometer for general use. J Meteorol Soc Japan Ser. II, 44: 12–24

Mix, W, Goldberg, V, Bernhardt, K-H (1994) Numerical experiments with different approaches for boundary layer modelling under large-area forest canopy conditions. Meteorol Z 3: 187–192

Miyake, M, Stewart, RW, Burling, RW, Tsvang, LR, Kaprov, BM, Kuznecov, OA (1971) Comparison of acoustic instruments in an atmospheric flow over water. Boundary-Layer Meteorol 2: 228–245

Moeng, C-H, Wyngaard, JC (1989) Evaluation of turbulent transport and dissipation closure in second-order modelling. J Atmos Sci 46: 2311–2330

Moeng, C-H (1998) Large eddy simulation of atmospheric boundary layers. In: AAM Holtslag, PG Duynkerke (Editors), Clear and cloudy boundary layers. Royal Netherlands Academy of Arts and Science, Amsterdam, pp. 67–83

Moeng, C-H, Sullivan, PP, Stevens, B (2004) Large-eddy simulation of cloud-topped mixed layers. In: Fedorovich, E., R Rottunno, B Stevens (Editors), Atmospheric turbulence and mesoscale meteorology. Cambridge University Press, Cambridge, pp. 95–114

Mölders, N, Raabe, A, Tetzlaff, G (1996) A comparison of two strategies on land surface heterogeneity used in a mesoscale ß meteorological model. Tellus 48A: 733–749

Mölders, N (2001) Concepts for coupling hydrological and meteorological models. Wiss. Mitt. aus dem Inst. für Meteorol. der Univ. Leipzig und dem Institut für Troposphärenforschung e. V. Leipzig 22: 1–15

Molemaker, MJ, Vilà-Guerau de Arellano, J (1998) Control of chemical reactions by convective turbulence in the boundary layer. J Atmos Sci 55: 568–579

Moncrieff, JB, Massheder, JM, DeBruin, H, Elbers, J, Friborg, T, Heusinkveld, B, Kabat, P, Scott, S, Søgaard, H, Verhoef, A (1997) A system to measure surface fluxes of momentum, sensible heat, water vapor and carbon dioxide. J Hydrol 188-189: 589–611

Monin, AS, Obukhov, AM (1954) Osnovnye zakonomernosti turbulentnogo peremesivanija v prizemnom sloe atmosfery (Basic laws of turbulent mixing in the atmosphere near the ground). Trudy geofiz inst AN SSSR 24 (151): 163–187

Monin, AS, Yaglom, AM (1973) Statistical fluid mechanics: Mechanics of turbulence, Volume 1. MIT Press, Cambridge, London, 769 pp.

Monin, AS, Yaglom, AM (1975) Statistical fluid mechanics: Mechanics of turbulence, Volume 2. MIT Press, Cambridge, London, 874 pp.

Monteith, JL (1965) Evaporation and environment. Symp Soc Exp Biol 19: 205–234

Montgomery, RB (1940) Observations of vertical humidity distribution above the ocean surface and their relation to evaporation. Pap. Phys. Oceanogr. & Meteorol. 7: 1–30

Montgomery, RB (1948) Vertical eddy flux of heat in the atmosphere. Journal Meteorology 5: 265–274

Moore, CJ (1986) Frequency response corrections for eddy correlation systems. Boundary-Layer Meteorol 37: 17–35

Müller, C (1999) Modelling Soil-Biosphere Interaction. CABI Publishing, Wallingford, 354 pp.

Müller, H, Kramm, G, Meixner, FX, Fowler, D, Dollard, GJ, Possanzini, M (1993) Determination of HNO₃ dry deposition by modified Bowen ratio and aerodynamic profile techniques. Tellus 45B: 346–367

Neckel, T, Montenbruck, O (1999) Ahnerts Kalender für Sternfreunde 2000. Sterne und Weltraum, Heidelberg, 351 pp.

Neff, WD, Coulter, RL (1986) Acoustic remote sounding. In: DH Lenschow (Editor), Probing the Atmospheric Boundary Layer. American Meteorological Society, Boston, pp. 201–236.

Nicholls, S, Smith, FB (1982) On the definition of the flux of sensible heat. Boundary-Layer Meteorol 24: 121–127

Nieuwstadt, FTM (1978) The computation of the friction velocity u∗ and the temperature scale T∗ from temperature and wind velocity profiles by least-square method. Boundary-Layer Meteorol 14: 235–246

Nikuradse, J (1933) Strömungsgesetze an rauhen Rohren. VDI Forschungshefte 361

Nitzschke, A (1970) Zum Verhalten der Lufttemperatur in der Kontaktzone zwischen Land und Meer bei Zingst. Veröff Geophys Inst Univ Leipzig XIX: 339–445

Obukhov, AM (1946) Turbulentnost' v temperaturnoj - neodnorodnoj atmosfere (Turbulence in an atmosphere with a non-uniform temperature). Trudy Inst Theor Geofiz AN SSSR 1: 95–115

Obukhov, AM (1951) Charakteristiki mikrostruktury vetra v prizemnom sloje atmosfery (Characteristics of the micro-structure of the wind in the surface layer of the atmosphere). Izv AN SSSR, ser Geofiz 3: 49–68

Obukhov, AM (1960) O strukture temperaturnogo polja i polja skorostej v uslovijach konvekcii (Structure of the temperature and velocity fields under conditions of free convection). Izv AN SSSR, ser Geofiz: 1392–1396

Obukhov, AM (1971) Turbulence in an atmosphere with a non-uniform temperature. Boundary-Layer Meteorol 2: 7–29

Oertel, H (Editor), (2004) Prandtl's essentials of fluid mechanics. Springer, New York, VII, 723 pp.

Ohmura, A (1982) Objective criteria for rejecting data for Bowen ratio flux calculations. J Climate Appl Meteorol. 21: 595–598

Ohmura, A, Steffen, K, Blatter, H, Greuell, W, Rotach, M, Stober, M, Konzelmann, T, Forrer, J, Abe-Ouchi, A, Steiger, D, Niederbäumer, G (1992) Greenland Expedition, Progress Report No. 2, April 1991 to Oktober 1992, Swiss Federal Institute of Techology, Zürich.

Ohmura, A, Dutton, EG, Forgan, B, Fröhlich, C, Gilgen, H, Hegner, H, Heimo, A, König-Langlo, G, McArthur, B, Müller, G, Philipona, R, Pinker, R, Whitlock, CH, Dehne, K, Wild, M (1998) Baseline Surface Radiation Network (BSRN/WCRP): New precision radiometry for climate research. Bull Amer Meteorol Soc 79: 2115–2136

Oke, TR (1987) Boundary layer climates. Methuen, New York, 435 pp.

Oncley, SP, Businger, JA, Itsweire, EC, Friehe, CA, LaRue, JC, Chang, SS (1990) Surface layer profiles and turbulence measurements over uniform land under near-neutral conditions. 9th Symp on Boundary Layer and Turbulence, Roskilde, Denmark, April 30 – May 3, 1990, Am. Meteorol. Soc., 237–240

Oncley, SP, Delany, AC, Horst, TW, Tans, PP (1993) Verification of flux measurement using relaxed eddy accumulation. Atmos Environm 27A: 2417–2426

Oncley, SP, Friehe, CA, Larue, JC, Businger, JA, Itsweire, EC, Chang, SS (1996) Surface-layer fluxes, profiles, and turbulence measurements over uniform terrain under near-neutral conditions. J Atmos Sci 53: 1029–1054

Oncley, SP, Foken, T, Vogt, R, Kohsiek, W, DeBruin, HAR, Bernhofer, C, Christen, A, van Gorsel, E, Grantz, D, Feigenwinter, C, Lehner, I, Liebethal, C, Liu, H, Mauder, M, Pitacco,

A, Ribeiro, L, Weidinger, T (2007) The energy balance experiment EBEX-2000, Part I: Overview and energy balance. Boundary-Layer Meteorol 123: 1–28

Orlanski, I (1975) A rational subdivision of scales for atmospheric processes. Bull. Am. Meteorol. Soc. 56: 527–530

Osczevski, R, Bluestein, M (2005) The new wind chill equivalent temperature chart. Bull Amer Meteorol Soc 86: 1453–1458

Osczevski, RJ (2000) Windward cooling: An overlooked factor in the calculation of wind chill. Bull Amer Meteorol Soc 81: 2975–2978

Owen, PR, Thomson, WR (1963) Heat transfer across rough surfaces. J Fluid Mech 15: 321-334

Paeschke, W (1937) Experimentelle Untersuchungen zum Rauhigkeitsproblem in der bodennahen Luftschicht. Z Geophys 13: 14–21

Panin, GN (1983) Metodika rastscheta lokalnogo teplo- i vlagoobmena v sisteme vodoem – atmosfera (Method for calculation of local heat and humidity exchange of the system water – atmosphere). Vodnye resursy 4: 3–12

Panin, GN (1985) Teplo- i massomen meszdu vodoemom i atmospheroj v estestvennych uslovijach (Heat- and mass exchange between the water and the atmosphere in the nature). Nauka, Moscow, 206 pp.

Panin, GN, Nasonov, AE, Souchintsev, MG (1996a) Measurements and estimation of energy and mass exchange over a shallow see. In: M Donelan (Editor), The air-sea interface, Miami, 489–494

Panin, GN, Tetzlaff, G, Raabe, A, Schönfeld, H-J, Nasonov, AE (1996b) Inhomogeneity of the land surface and the parametrization of surface fluxes - a discussion. Wiss Mitt Inst Meteorol Univ Leipzig und Inst Troposphärenforschung Leipzig 4: 204–215

Panin, GN, Tetzlaff, G, Raabe, A (1998) Inhomogeneity of the land surface and problems in the parameterization of surface fluxes in natural conditions. Theor Appl Climat 60: 163–178

Panin, GN, Nasonov, AE, Foken, T, Lohse, H (2006) On the parameterization of evaporation and sensible heat exchange for shallow lakes. Theor Appl Climat 85: 123–129

Panofsky, HA (1963) Determination of stress from wind and temperature measurements. Quart J Roy Meteorol Soc 89: 85–94

Panofsky, HA, Tennekes, H, Lenschow, DH, Wyngaard, JC (1977) The characteristics of turbulent velocity components in the surface layer under convective conditions. Boundary-Layer Meteorol 11: 355–361

Panofsky, HA (1984) Vertical variation of roughness length at the Boulder Atmospheric Observatory. Boundary-Layer Meteorol 28: 305–308

Panofsky, HA, Dutton, JA (1984) Atmospheric Turbulence – Models and methods for engineering applications. John Wiley and Sons, New York, 397 pp.

Papale, D, Valentini, R (2003) A new assessment of European forests carbon exchanges by eddy fluxes and artificial neural network spatialization. Global Change Biology 9: 525–535

Pasquill, F (1961) Estimation of the dispersion of windborne material. Meteorol Mag 90: 33–49

Pasquill, F (1972) Some aspects of boundary layer description. Quart J Roy Meteorol Soc 98: 469–494

Pattey, E, Desjardins, RL, Rochette, P (1993) Accuracy of the relaxed eddy accumulation technique. Boundary-Layer Meteorol 66: 341–355

Paulson, CA (1970) The mathematical representation of wind speed and temperature profiles in the unstable atmospheric surface layer. J Climate Appl Meteorol. 9: 857–861

Paw U, KT, Qiu, J, Su, H-B, Watanabe, T, Brunet, Y (1995) Surface renewal analysis: a new method to obtain scalar fluxes. Agric Forest Meteorol 74: 119–137

Paw U, KT, Baldocchi, D, Meyers, TP, Wilson, KB (2000) Correction of eddy covariance measurements incorporating both advective effects and density fluxes. Boundary-Layer Meteorol 97: 487–511

Pearson jr., RJ, Oncley, SP, Delany, AC (1998) A scalar similarity study based on surface layer ozone measurements over cotton during the California Ozone Deposition Experiment. J Geophys Res 103 (D15): 18919–18926

Peltier, LJ, Wyngaard, JC, Khanna, S, Brasseur, JG (1996) Spectra in the unstable surface layer. J Atmos Sci 53: 49–61

Penman, HL (1948) Natural evaporation from open water, bare soil and grass. Proc R Soc London A193: 120–195

Perera, MD (1981) Shelder behind two dimensional solit and porous fences. J Wind Engin Industrial Aerodyn 8: 93–104

Petersen, EL, Troen, I (1990) Europäischer Windatlas. RISØ National Laboratory, Roskilde, 239 pp.

Peterson, EW (1969) Modification of mean flow and turbulent energy by a change in surface roughness under conditions of neutral stability. Quart J Roy Meteorol Soc 95: 561–575

Peterson, EW, Hennessey jr., JP (1978) On the use of power laws for estimates of wind power potential. J Climate Appl Meteorol. 17: 390–394

Philip, JR (1961) The theory of heat flux meters. J Geophys Res 66: 571–579

Philipona, R, Fröhlich, C, Betz, C (1995) Characterization of pyrgeometers and the accuracy of atmospheric long-wave radiation measurements. Applied Optics 34: 1598–1605

Poulos, GS, Blumen, W, Fritts, DC, Lundquist, JK, Sun, J, Burns, SP, Nappo, C, Banta, R, Newsom, R, Cuxart, J, Terradellas, E, Balsley, B, Jensen, M (2002) CASES-99: A comprehensive investigation of the stable nocturnal boundary layer. Bull Amer Meteorol Soc 83: 55–581

Prandtl, L (1925) Bericht über Untersuchungen zur ausgebildeten Turbulenz. Z angew Math Mech 5: 136–139

Priestley, CHB, Swinbank, WC (1947) Vertical transport of heat by turbulence in the atmosphere. Proc R Soc London A189: 543–561

Priestley, CHB, Taylor, JR (1972) On the assessment of surface heat flux and evaporation using large-scale parameters. Monthly Weather Review 100: 81–82

Profos, P, Pfeifer, T (Editors) (1993) Grundlagen der Meßtechnik. Oldenbourg, München, Wien, 367 pp.

Pruitt, WO, Morgan, DL, Lourence, FJ (1973) Momentum and mass transfer in the surface boundary layer. Quart J Roy Meteorol Soc 99: 370–386

Queitsch, P (2002) TA Luft, Technische Anleitung zur Reinhaltung der Luft; Systematische Einführung mit Text der TA Luft 2002. Bundesanzeiger-Verl.-Ges., Bonn, 180 pp.

Raabe, A (1983) On the relation between the drag coefficient and fetch above the sea in the case of off-shore wind in the near shore zone. Z Meteorol 33: 363–367

Raabe, A (1991) Die Höhe der internen Grenzschicht. Z Meteorol 41: 251–261

Raasch, S, Schröter, M (2001) PALM – A large-eddy simulation model performing on massively parallel computers. Meteorol Z 10: 363–372

Radikevitsch, VM (1971) Transformazija dinamitscheskich charakteristik vozduschnogo potoka pod vlijaniem izmenenija scherochovatosti postilajustschej poverchnosti (Transformation of the dynamical characteristics of the wind field under the influence of a changing surface roughness) Izv AN SSSR, Fiz Atm Okeana 7: 1241–1250

Rannik, U, Vesala, T (1999) Autoregressive filtering versus linear detrending in estimation of fluxes by the eddy covariance method. Boundary-Layer Meteorol 91: 259–280

Rannik, U, Markkanen, T, Raittila, T, Hari, P, Vesala, T (2003) Turbulence statistics inside and above forest: Influence on footprint prediction. Boundary-Layer Meteorol 109: 163–189

Rannik, Ü, Aubinet, M, Kurbanmuradov, O, Sabelfeld, KK, Markkanen, T, Vesala, T (2000) Footprint analysis for measurements over heterogeneous forest. Boundary-Layer Meteorol 97: 137–166

Rao, KS, Wyngaard, JC, Coté, OR (1974) The structure of the two-dimensional internal boundary layer over a sudden change of surface roughness. J Atmos Sci 31: 738–746

Raupach, MR, Thom, AS, Edwards, I (1980) A wind-tunnel study of turbulent flow close to regularly arrayed rough surface. Boundary-Layer Meteorol 18: 373–379

Raupach, MR (1992) Drag and drag partition on rough surfaces. Boundary-Layer Meteorol 60: 375–395

Raupach, MR (1994) Simplified expressions for vegetation roughness lenght and zero-plane displacement as functions of canopy height and area index. Boundary-Layer Meteorol 71: 211–216

Raupach, MR, Finnigan, JJ, Brunet, Y (1996) Coherent eddies and turbulence in vegetation canopies: the mixing-layer analogy. Boundary-Layer Meteorol 78: 351–382

Raynor, GS, Michael, P, Brown, RM, SethuRaman, S (1975) Studies of atmospheric diffusion from a nearshore oceanic site. J Climate Appl Meteorol. 14: 1080–1094

Rebmann, C, Göckede, M, Foken, T, Aubinet, M, Aurela, M, Berbigier, P, Bernhofer, C, Buchmann, N, Carrara, A, Cescatti, A, Ceulemans, R, Clement, R, Elbers, J, Granier, A, Grünwald, T, Guyon, D, Havránková, K, Heinesch, B, Knohl, A, Laurila, T, Longdoz, B, Marcolla, B, Markkanen, T, Miglietta, F, Moncrieff, H, Montagnani, L, Moors, E, Nardino, M, Ourcvial, J-M, Rambal, S, Rannik, U, Rotenberg, E, Sedlak, P, Unterhuber, G, Vesala, T, Yakir, D (2005) Quality analysis applied on eddy covariance measurements at complex forest sites using footprint modelling. Theor Appl Climat 80: 121–141

Reichardt, H (1951) Vollständige Darstellung der turbulenten Geschwindigkeitsverteilung in glatten Röhren. Z angew Math Mech 31: 208–219

Reynolds, O (1894) On the dynamical theory of turbulent incompressible viscous fluids and the determination of the criterion. Phil Trans R Soc London A 186: 123–161

Richardson, LF (1920) The supply of energy from and to atmospheric eddies. Proceedings Royal Society A 97: 354–373

Richardson, SJ, Brock, FV, Semmer, SR, Jirak, C (1999) Minimizing errors associated with multiplate radiation shields. J Atm Oceanic Techn 16: 1862–1872

Richter, D (1977) Zur einheitlichen Berechnung der Wassertemperatur und der Verdunstung von freien Wasserflächen auf statistischer Grundlage. Abh Meteorol Dienstes DDR 119: 35 pp.

Richter, D (1995) Ergebnisse methodischer Untersuchungen zur Korrektur des systematischen Meßfehlers des Hellmann-Niederschlagsmessers. Ber. d. Dt. Wetterdienstes 194: 93 pp.

Richter, SH, Skeib, G (1984) Anwendung eines Verfahrens zur Parametrisierung des turbulenten Energieaustausches in der atmosphärischen Bodenschicht. Geod Geophys Veröff, R II 26: 80–85

Richter, SH, Skeib, G (1991) Ein Verfahren zur Parametrisierung von Austauschprozessen in der bodennahen Luftschicht. Abh Meteorol Dienstes DDR 146: 15–22

Rink, J (1961) Thermistore und ihre Anwendung in der Meteorologie. Abh Meteorol Hydrol Dienstes DDR 63: 58 pp.

Rinne, HJI, Delany, AC, Greenberg, JP, Guenther, AB (2000) A true eddy accumulation system for trace gas fluxes using disjunct eddy sampling method. J Geophys Res 105(D20): 24791–24798

Roedel, W (2000) Physik unserer Umwelt, Die Atmosphäre. Springer, Berlin, Heidelberg, 498 pp.

Roll, HU (1948) Wassernahes Windprofil und Wellen auf dem Wattenmeer. Ann Meteorol 1: 139–151

Ross, J (1981) The radiation regime and architecture of plant stands. Dr. W. Junk Publishers, The Hague, 391 pp.

Rotach, M, Vogt, R, Bernhofer, C, Batchvarova, E, Christen, A, Clappier, A, Feddersen, B, Gryning, SE, Martucci, G, Mayer, H, Mitev, V, Oke, TR, Parlow, E, Richner, H, Roth, M, Roulet, Y-A, Ruffieux, D, Salmond, JA, Schatzmann, M, Voogt, JA (2005) BUBBLE - an urban boundary layer meteorology project. Theor Appl Climat 81: 231–261

Roth, R (1975) Der vertikale Transport von Luftbeimengungen in der Prandtl-Schicht und die Depositionsgeschwindigkeit. Meteorol Rundschau 28: 65–71

Rummel, U, Ammann, C, Gut, A, Meixner, FX, Andreae, MO (2002a) Eddy covariance measurements of nitric oxide flux within an Amazonian rainforest. J Geophys Res 107 (D20): 8050

Rummel, U, Ammann, C, Meixner, FX (2002b) Characterizing turbulent trace gas exchange above a dense tropical rain forest using wavelet and surface renewal analysis. 15th Symp on

Boundary Layer and Turbulence, Wageningen, NL, 15-19 July 2002, Am. Meteorol. Soc., 602–605

Ruppert, J, Wichura, B, Delany, AC, Foken, T (2002) Eddy sampling methods, A comparison using simulation results. 15th Symp on Boundary Layer and Turbulence, Wageningen, 15–19 July 2002, Am. Meteorol. Soc., 27–30

Ruppert, J, Mauder, M, Thomas, C, Lüers, J (2006a) Innovative gap-filling strategy for annual sums of CO_2 net ecosystem exchange. Agric Forest Meteorol 138: 5–18

Ruppert, J, Thomas, C, Foken, T (2006b) Scalar similarity for relaxed eddy accumulation methods. Boundary-Layer Meteorol 120: 39–63

Rutgersson, A, Sullivan, PP (2005) Investigating the effects of water waves on the turbulence structure in the atmosphere using direct numerical simulations. Dynamics Atm Oceans 38: 147–171

Salby, ML (1995) Fundamentals of atmospheric physics. Academic Press, San Diego, New York, 624 pp.

Santoso, E, Stull, RB (1998) Wind and temperature profiles in the Radix layer: The botton fifth of the convective boundary layer. J Appl Meteorol 37: 545–558

Sauer, TJ, Harris, AR, Ochsner, TE, Horton, R (2002) Errors in soil heat flux measurement: Effects of flux plate design and varying soil thermal properties. 25th Symp Agric & Forest Meteor: 11–12

Savelyev, SA, Taylor, PA (2001) Notes on an internal boundary-layer height formula. Boundary-Layer Meteorol 101: 293–301

Savelyev, SA, Taylor, PA (2005) Internal boundary layers: I. Height formulae for neutral and diabatic flow. Boundary-Layer Meteorol 115: 1–25

Schädler, G, Kalthoff, N, Fiedler, F (1990) Validation of a model for heat, mass and momentum exchange over vegetated surfaces using LOTREX-10E/HIBE88 data. Contr Atmosph Phys 63: 85–100

Schatzmann, M, König, G, Lohmeyer, A (1986) Physikalische Modellierung mikrometeorologischer Vorgänge im Windkanal. Meteorol Rundschau 39: 44–59

Schlichting, H, Gersten, K (2003) Boundary-layer theory. McGraw Hill, New York, XXIII, 799 pp.

Schmid, HP, Oke, TR (1988) Estimating the source area of a turbulent flux measurement over a patchy surface. 8th Symposium on Turbulence and Diffusion, San Diego, CA., April 26–29, 1988, Am. Meteorol. Soc., 123–126

Schmid, HP, Oke, TR (1990) A model to estimate the source area contributing to turbulent exchange in the surface layer over patchy terrain. Quart J Roy Meteorol Soc 116: 965–988

Schmid, HP (1994) Source areas for scalars and scalar fluxes. Boundary-Layer Meteorol 67: 293–318

Schmid, HP, Bünzli, D (1995a) Reply to comments by E. M. Blyth on 'The influence of surface texture on the effective roughness length'. Quart J Roy Meteorol Soc 121: 1173–1176

Schmid, HP, Bünzli, D (1995b) The influence of the surface texture on the effective roughness length. Quart J Roy Meteorol Soc 121: 1–21

Schmid, HP (1997) Experimental design for flux measurements: matching scales of observations and fluxes. Agric Forest Meteorol 87: 179–200

Schmid, HP (2002) Footprint modeling for vegetation atmosphere exchange studies: A review and perspective. Agric Forest Meteorol 113: 159–184

Schmidt, H, Schumann, U (1989) Coherent structures of the convective boundary layer derived from large eddy simulations. J Fluid Mech 200: 511–562

Schmidt, W (1925) Der Massenaustausch in freier Luft und verwandte Erscheinungen. Henri Grand Verlag, Hamburg, 118 pp.

Schmitz-Peiffer, A, Heinemann, D, Hasse, L (1987) The ageostrophic methode – an update. Boundary-Layer Meteorol 39: 269–281

Schoonmaker, PK (1998) Paleoecological perspectives on ecological scales. In: DL Peterson, VT Parker (Editors), Ecological Scale. Columbia University Press, New York, pp. 79–103

Schotanus, P, Nieuwstadt, FTM, DeBruin, HAR (1983) Temperature measurement with a sonic anemometer and its application to heat and moisture fluctuations. Boundary-Layer Meteorol 26: 81–93

Schotland, RM (1955) The measurement of wind velocity by sonic waves. J Meteorol 12: 386-390

Schrödter, H (1985) Verdunstung, Anwendungsorientierte Meßverfahren und Bestimmungsmethoden. Springer, Berlin, Heidelberg, 186 pp.

Schuepp, PH, Leclerc, MY, MacPherson, JI, Desjardins, RL (1990) Footprint prediction of scalar fluxes from analytical solutions of the diffusion equation. Boundary-Layer Meteorol 50: 355–373

Schumann, U (1989) Large-eddy simulation of turbulent diffusion with chemical reactions in the convective boundary layer. Atmos Environm 23: 1713–1727

Sedefian, L (1980) On the vertical extrapolation of mean wind power density. J Climate Appl Meteorol. 19: 488–493

Seibert, P, Beyrich, F, Gryning, S-E, Joffre, S, Rasmussen, A, Tercier, P (2000) Review and intercomparison of operational methods for the determination of the mixing height. Atmos Environm 34: 1001–1027

Seiler, W (1996) Results from the integrated research programme SANA, Phase I. Meteorol Z 5: 179–278

Seinfeld, JH, Pandis, SN (1998) Atmospheric chemistry and physics. Wiley, New York, 1326 pp.

Sellers, PJ, Dorman, JL (1987) Testing the simple biospere model (SiB) for use in general circulation models. J Climate Appl Meteorol. 26: 622–651

Sellers, PJ, Hall, FG, Asrar, G, Strebel, DE, Murphy, RE (1988) The first ISLSCP field experiment (FIFE). Bull Amer Meteorol Soc 69: 22–27

Sellers, PJ, Hall, FG, Kelly, RD, Black, A, Baldocchi, D, Berry, J, Ryan, M, Ranson, KJ, Crill, PM, Lettenmaier, DP, Margolis, H, Cihlar, J, Newcomer, J, Fitzjarrald, D, Jarvis, PG, Gower, ST, Halliwell, D, Williams, D, Goodison, B, Wickland, DE, Guertin, FE (1997) BOREAS in 1997: Experiment overview, scientific results, and future directions. J Geophys Res 102: 28 731–28 769

Sentelhas, PC, Folegatti, MV (2003) Class A pan coefficients (Kp) to estimate daily reference evapotranspiration (ETo). Revista Brasilleira de Engenharia Agricola e Ambiental 7: 111–115

Sevruk, B (1981) Methodische Untersuchungen des systematischen Messfehlers der Hellmann-Regenmesser im Sommerhalbjahr in der Schweiz. Mitt. d. Versuchsanstalt f. Wasserb., Hydrol. u. Glaziol. 52: 290 pp.

Shaw, RH (1977) Secondary wind speed maxima inside plant canopies. J Appl Meteorol 16: 514–521

Shaw, RH (1985) On diffusive and dispersive fluxes in forest canopies. In: BA Hutchinson, BB Hicks (Editors), The Forest-Atmosphere interaction. Reidel Publishing Company, pp. 407–419

Shearman, RJ (1992) Quality assurance in the observation area of the Meteorological Office. Meteorol Mag 121: 212–216

Shen, S, Leclerc, MY (1994) Large-eddy simulation of small scale surface effects on the convective boundary layer structure. Atmosphere-Ocean 32: 717–731

Shen, S, Leclerc, MY (1995) How large must surface inhomogeneous be before they influence the convective boundary layer structure? A case study. Quart J Roy Meteorol Soc 121: 1209–1228

Shir, CC (1972) A numerical computation of the air flow over a sudden change of surface roughness. J Atmos Sci 29: 304–310

Shukauskas, A, Schlantschiauskas, A (1973) Teploodatscha v turbulentnom potoke shidkosti (Heat exchange in the turbulent fluid). Izd. Mintis, Vil'njus, 327 pp.

Shuttleworth, WJ (1989) Micrometeorology of temperate and tropical forest. Phil Trans R Soc London B 324: 299–334

Siegel, S (1936) Messungen des nächtlichen thermischen Gefüges in der bodennahen Luftschicht. Gerl Beitr Geophys 47: 369–399

Sinclair, TR, Allen jr., LH, Lemon, ER (1975) An analysis of errors in the calculation of energy flux densities above vegetation by Bowen-ratio profile method. Boundary-Layer Meteorol 8: 129–139

Skeib, G (1980) Zur Definition universeller Funktionen für die Gradienten von Windgeschwindigkeit und Temperatur in der bodennahen Luftschicht. Z Meteorol 30: 23–32

Smagorinsky, J (1963) General circulation experiments with the primitive equations: I. The basic experiment. Monthly Weather Review 91: 99–164

Smajstrla, AG, Zazueta, FS, Clark, GA, Pitts, DJ (2000) Irrigation scheduling with evaporation pans. Univ of Florida, IFAS Ext Bul 254

Smedman, A-S (1991) Some turbulence characteristics in stable atmospheric boundary layer flow. J Atmos Sci 48: 856–868

Smith, SD, Fairall, CW, Geernaert, GL, Hasse, L (1996) Air-sea fluxes: 25 years of progress. Boundary-Layer Meteorol 78: 247–290

Snyder, RL, Spano, D, Paw U, KT (1996) Surface renewal analysis for sensible and latent heat flux density. Boundary-Layer Meteorol 77: 249–266

Sodemann, H, Foken, T (2004) Empirical evaluation of an extended similarity theory for the stably stratified atmospheric surface layer. Quart J Roy Meteorol Soc 130: 2665–2671

Sodemann, H, Foken, T (2005) Special characteristics of the temperature structure near the surface. Theor Appl Climat 80: 81–89

Sogachev, A, Lloyd, J (2004) Using a one-and-a-half order closure model of atmospheric boundary layer for surface flux footprint estimation. Boundary-Layer Meteorol 112: 467–502

Song, C, Woodcock, CE, Seto, KC, Lenney, MP, MacOmber, SA (2001) Classification and change detection using Landsat TM data: when and how to correct atmospheric effects? Remote Sensing of Environment 75: 230–244

Sonntag, D (1966-1968) Hygrometrie. Akademie-Verlag, Berlin, 1086 pp.

Sonntag, D (1990) Important new values of the physical constants of 1986, vapour pressure formulations based on the ITC-90, and psychrometer formulae. Z Meteorol 40: 340–344

Sonntag, D (1994) Advancements in the field of hygrometry. Meteorol Z 3: 51–66

Sorbjan, Z (1986) Characteristics in the stable-continuous boundary layer. Boundary-Layer Meteorol 35: 257–275

Sorbjan, Z (1987) An examination of local similarity theory in the stably stratified boundary layer. Boundary-Layer Meteorol 38: 63–71

Sorbjan, Z (1989) Structure of the atmospheric boundary layer. Prentice Hall, New York, 317 pp.

Sponagel, H (1980) Zur Bestimmung der realen Evapotranspiration landwirtschaftlicher Kulturpflanzen. Geologisches Jahrbuch F9: 87 pp.

Spronken-Smith, RA, Oke, TR (1999) Scale modelling of nocturnal cooling in urban parks. Boundary-Layer Meteorol 93: 287–312

Spronken-Smith, RA, Oke, TR, Lowry, WP (2000) Advection and the surface energy balance across an irrigated urban park. International Journal of Climatology 20: 1033–1047

Strong, C, Fuentes, JD, Baldocchi, D (2004) Reactive hydrocrbon flux footprints during canopy senescence. Agric Forest Meteorol 127: 159–173

Stull, R, Santoso, E (2000) Convective transport theory and counter-difference fluxes. 14th Symposium on Boundary Layer and Turbulence, Aspen, CO., 7.-11. August 2000, Am. Meteorol. Soc., Boston, 112–113

Stull, RB (1984) Transilient turbulence theorie, Part 1: The concept of eddy mixing across finite distances. J Atmos Sci 41: 3351–3367

Stull, RB (1988) An Introduction to Boundary Layer Meteorology. Kluwer Acad. Publ., Dordrecht, Boston, London, 666 pp.

Stull, RB (2000) Meteorology for Scientists and Engineers. Brooks / Cole, Pacific Grove, 502 pp.

Suomi, VE (1957) Sonic anemometer - University of Wisconsin. In: HH Lettau, B Davidson (Editors), Exploring the atmosphere's first mile. Pergamon Press, London, New York, pp. 256–266

Sutton, OG (1953) Micrometeorology. McGraw Hill, New York, 333 pp.

Sverdrup, HU (1937/38) On the evaporation from the ocean. J. Marine Res. 1: 3–14

Swinbank, WC (1951) The measurement of vertical transfer of heat and water vapor by eddies in the lower atmosphere. J Meteorol 8: 135–145

Swinbank, WC (1964) The exponential wind profile. Quart J Roy Meteorol Soc 90: 119–135

Swinbank, WC (1968) A comparison between prediction of the dimensional analysis for the constant-flux layer and observations in unstable conditions. Quart J Roy Meteorol Soc 94: 460–467

Swinbank, WC, Dyer, AJ (1968) An experimental study on mircrometeorology. Quart J Roy Meteorol Soc 93: 494–500

Tanner, CB, Thurtell, GW (1969) Anemoclinometer measurements of Reynolds stress and heat transport in the atmospheric surface layer. ECOM 66-G22-F, ECOM, United States Army Electronics Command, Research and Developement.

Tatarski, VI (1961) Wave propagation in a turbulent medium. McGraw-Hill, New York, 285 pp.

Taubenheim, J (1969) Statistische Auswertung geophysikalischer und meteorologischer Daten. Geest & Portig, Leipzig, 386 pp.

Taylor, GI (1915) Eddy motion in the atmosphere. Phil Trans R Soc London A 215: 1–26

Taylor, GI (1938) The spectrum of turbulence. Proc R Soc London A 164: 476–490

Taylor, PA (1969) On wind and shear stress profiles above a change in surface roughness. Quart J Roy Meteorol Soc 95: 77–91

Taylor, PA (1987) Comments and further analysis on the effective roughness length for use in numerical three-dimensional models: A research note. Boundary-Layer Meteorol 39: 403–418

Tennekes, H (1982) Similarity relations, scaling laws and spectral dynamics. In: FTM Nieuwstadt, H Van Dop (Editors), Atmospheric turbulence and air pollution modelling. D. Reidel Publ. Comp., Dordrecht, Boston, London, pp. 37–68

Thiermann, V, Grassl, H (1992) The measurement of turbulent surface layer fluxes by use of bichromatic scintillation. Boundary-Layer Meteorol 58: 367–391

Thomas, C, Foken, T (2002) Re-evaluation of integral turbulence characteristics and their parameterisations. 15th Conference on Turbulence and Boundary Layers, Wageningen, NL, 15–19 July 2002, Am. Meteorol. Soc., 129–132

Thomas, C, Foken, T (2007a) Organised motion in a tall spruce canopy: Temporal scales, structure spacing and terrain effects. Boundary-Layer Meteorol 122: 123–147

Thomas, C, Foken, T (2007b) Flux contribution of coherent structures and its implications for the exchange of energy and matter in a tall spruce canopy. Boundary-Layer Meteorol 123: 317–337

Tiesel, R, Foken, T (1987) Zur Entstehung des Seerauchs an der Ostseeküste vor Warnemünde. Z Meteorol 37: 173–176

Tikusis, P, Osczevski, RJ (2002) Dynamic model of facial cooling. J Appl Meteorol 41: 1241–1246

Tikusis, P, Osczevski, RJ (2003) Facial cooling during cold air exposure. Bull Amer Meteorol Soc 84: 927–933

Tillman, JE (1972) The indirect determination of stability, heat and momentum fluxes in the atmospheric boundary layer from simple scalar variables during dry unstable conditions. J Climate Appl Meteorol. 11: 783–792

Torrence, C, Compo, GP (1998) A practical guide to wavelet analysis. Bull Amer Meteorol Soc 79: 61–78

Troen, I, Peterson, EW (1989) European Wind Atlas, Risø National Laboratory, Roskilde, 656 pp.

Tromp, SW (1963) Medical Biometeorology. Elsevier, Amsterdam, XXVII, 991 pp.

Tschalikov, DV (1968) O profilja vetra i temperatury v prizemnom sloe atmosfery pri ustojtschivoj stratifikacii (About the wind and temperature profile in the surface layer for stable stratification). Trudy GGO 207: 170–173

Tsvang, LR (1960) Izmerenija tschastotnych spektrov temperaturnych pulsacij v prizemnom sloe atmosfery (Measurement of the spectra of the temperature fluctuations in the near surface layer of the atmosphere). Izv AN SSSR, ser Geofiz 10: 1252–1262

Tsvang, LR, Kaprov, BM, Zubkovskij, SL, Dyer, AJ, Hicks, BB, Miyake, M, Stewart, RW, McDonald, JW (1973) Comparison of turbulence measurements by different instuments; Tsimlyansk field experiment 1970. Boundary-Layer Meteorol 3: 499–521

Tsvang, LR, Zubkovskij, SL, Kader, BA, Kallistratova, MA, Foken, T, Gerstmann, W, Przandka, Z, Pretel, J, Zelený, J, Keder, J (1985) International turbulence comparison experiment (ITCE-81). Boundary-Layer Meteorol 31: 325–348

Tsvang, LR, Fedorov, MM, Kader, BA, Zubkovskii, SL, Foken, T, Richter, SH, Zelený, J (1991) Turbulent exchange over a surface with chessboard-type inhomogeneities. Boundary-Layer Meteorol 55: 141–160

Turc, L (1961) Évaluation des besoins en eau d'irrigation évapotranspiration potentielle. Ann Agron 12: 13–49

Twine, TE, Kustas, WP, Norman, JM, Cook, DR, Houser, PR, Meyers, TP, Prueger, JH, Starks, PJ, Wesely, ML (2000) Correcting eddy-covariance flux underestimates over a grassland. Agric Forest Meteorol 103: 279–300

UBA (1996) Entwicklung einer Methodik zur Abschätzung der Gesamtdeposition von Spurenstoffen. F/E-Nr. 10402818, Umweltbundesamt, Berlin.

Uttal, T, Curry, JA, Mcphee, MG, Perovich, DK, Moritz, RE, Maslanik, JA, Guest, PS, Stern, HL, Moore, JA, Turenne, R, Heiberg, A, Serreze, MC, Wylie, DP, Persson, OG, Paulson, CA, Halle, C, Morison, JH, Wheeler, PA, Makshtas, A, Welch, H, Shupe, MD, Intrieri, JM, Stamnes, K, Lindsey, RW, Pinkel, R, Pegau, WS, Stanton, TP, Grenfeld, TC (2002) Surface heat budget of the Arctic ocean. Bull Amer Meteorol Soc 83: 255–275

Valentini, R, Matteucci, G, Dolman, AJ, Schulze, E-D, Rebmann, C, Moors, EJ, Granier, A, Gross, P, Jensen, NO, Pilegaard, K, Lindroth, A, Grelle, A, Bernhofer, C, Grünwald, T, Aubinet, M, Ceulemans, R, Kowalski, AS, Vesala, T, Rannik, Ü, Bergigier, P, Loustau, D, Guomundsson, J, Thorgeirsson, H, Ibrom, A, Morgenstern, K, Clement, R, Moncrieff, J, Montagnani, L, Minerbi, S, Jarvis, PG (2000) Respiration as the main determinant of carbon balance in European forests. Nature 404: 861–865

van Bavel, CHM (1986) Potential evapotranspiration: The combination concept and its experimental verification. Water Resources Res 2: 455–467

van der Hegge Zijnen, BG (1956) Modified correlation formulae for heat transfer by natural and by forced convection from horizontal cylinders. Appl Sci Res A6: 129–140

van Loon, WKP, Bastings, HMH, Moors, EJ (1998) Calibration of soil heat flux sensors. Agric Forest Meteorol 92: 1–8

VDI (1988) Schallausbreitung im Freien. Beuth Verlag, Berlin, VDI 2714, 18 pp.

VDI (1998) Umweltmeteorologie: Methoden zur human-biometeorologischen Bewertung von Klima und Lufthygiene für die Stadt- und Regionalplanung – Teil 1: Klima. Beuth Verlag, Berlin, VDI 3787 Blatt 2, 29 pp.

VDI (2000) Umweltmeteorologie, Meteorologische Messungen für Fragen der Luftreinhaltung – Wind. Beuth Verlag, Berlin, VDI 3786, Blatt 2, 33 pp.

VDI (2006a) Umweltmeteorologie -Meteorologische Messungen – Messstation. Beuth Verlag, Berlin, VDI 3786, Blatt 13, 24 pp.

VDI (2006b) VDI-Richtlinien-Katalog. Beuth, Berlin, Wien, Zürich, 464 pp.

VDI (2008) Umweltmeteorologie – Meteorologische Messungen – Turbulenzmessungen mit Ultraschallanemometern (Environmental meteorology – Meteorological measurements – Turbulence measurements with sonic anemometers). Beuth-Verlag, Berlin, VDI 3786, Blatt 12, in print

Vesala, T, Rannik, U, Leclerc, MY, Foken, T, Sabelfeld, KK (2004) Foreword: Flux and concentration footprints. Agric Forest Meteorol 127: 111–116

Vesala, T, Kljun, N, Rannik, U, Rinne, J, Sogachev, A, Markkanen, T, Sabelfeld, K, Foken, T, Leclerc, MY (2008) Flux and concentration footprint modelling: State of the art. Environm Pollution 152: 653–666

Vickers, D, Mahrt, L (1997) Quality control and flux sampling problems for tower and aircraft data. J Atm Oceanic Techn 14: 512–526

Vogel, H-J, Roth, K (2003) Moving through scales of flow and transport in soil. J Hydrol 272: 95–106

von Driest, ER (1959) Convective heat transfer in gases. In: CC Lin (Editor), High speed aerodynamics and jet propulsion, Vol. V, Turbulent flow and heat transfer. Princeton University Press, Princeton, pp. 339–427

von Kármán, T (1934) Turbulence and skin friction. J. Aeronautic Sci. 1: 1–20

von Kármán, T, Howarth, L (1938) On the statistical theory of isotropic turbulence. Proc R Soc London A 164: 192–215

Waterhouse, FL (1955) Micrometeorological profiles in grass cover in relation to biological problems. Quart J Roy Meteorol Soc 81: 63–71

Webb, EK (1970) Profile relationships: the log-linear range, and extension to strong stability. Quart J Roy Meteorol Soc 96: 67–90

Webb, EK, Pearman, GI, Leuning, R (1980) Correction of the flux measurements for density effects due to heat and water vapour transfer. Quart J Roy Meteorol Soc 106: 85–100

Webb, EK (1982) On the correction of flux measurements for effects of heat and water vapour transfer. Boundary-Layer Meteorol 23: 251–254

Wendling, U, Schellin, H-G, Thomä, M (1991) Bereitstellung von täglichen Informationen zum Wasserhaushalt des Bodens für die Zwecke der agrarmeteorologischen Beratung. Z Meteorol 41: 468–475

Wendling, U, Fuchs, P, Müller-Westermeier, G (1997) Modellierung des Zusammenhangs von Globalstrahlung, Sonnenscheindauer und Bewölkungsgrad als Beitrag der Klimaüberwachung. Dt Wetterdienst, Forsch. Entwicklung, Arbeitsergebnisse 45: 29 pp.

Wichura, B, Buchmann, N, Foken, T (2000) Fluxes of the stable carbon isotope 13C above a spruce forest measured by hyperbolic relaxed eddy accumulation method. 14th Symposium on Boundary Layer and Turbulence, Aspen, CO., 7–11 August 2000, Am. Meteorol. Soc., Boston, 559–562

Wichura, B, Buchmann, N, Foken, T, Mangold, A, Heinz, G, Rebmann, C (2001) Pools und Flüsse des stabilen Kohlenstoffisotops ^{13}C zwischen Boden, Vegetation und Atmosphäre in verschiedenen Pflanzengemeinschaften des Fichtelgebirges. Bayreuther Forum Ökologie 84: 123–153

Wichura, B, Ruppert, J, Delany, AC, Buchmann, N, Foken, T (2004) Structure of carbon dioxide exchange processes above a spruce Forest. In: E Matzner (Editor), Biogeochemistry of Forested Catchments in a Changing Enivironment, A German Gase Study. Ecological Studies. Ecological Studies. Springer, Berlin, Heidelberg, pp. 161–176

Wieringa, J (1980) A revaluation of the Kansas mast influence on measurements of stress and cup anemometer overspeeding. Boundary-Layer Meteorol 18: 411–430

Wieringa, J (1989) Shapes of annual frequency distribution of wind speed observed on high meteorological masts. Boundary-Layer Meteorol 47: 85–110

Wieringa, J (1992) Updating the Davenport roughness classification. Journal of Wind Engineering and Industrial Aerodynamics 41: 357–368

Wilczak, JM, Oncley, SP, Stage, SA (2001) Sonic anemometer tilt correction algorithms. Boundary-Layer Meteorol 99: 127–150

WMO (1981) Meteorological aspects of the utilization of wind as an energy source. WMO, Techn Note 175: 180 pp.

WMO (1996) Guide to meteorological instruments and methods of observation. WMO, Note 8: 6th edition

Wrzesinsky, T, Klemm, O (2000) Summertime fog chemistry at a mountainous site in Central Europe. Atmos Environm 34: 1487–1496

Wyngaard, JC, Coté, OR (1971) The budgets of turbulent kinetic energy and temperature variance in the atmospheric surface layer. J Atmos Sci 28: 190–201

Wyngaard, JC, Coté, OR, Izumi, Y (1971a) Local free convection, similarity and the budgets of shear stree and heat flux. J Atmos Sci 28: 1171–1182

Wyngaard, JC, Izumi, Y, Collins, SA (1971b) Behavior of the refractive-index-structure parameter near the ground. J Opt Soc Am 61: 1646–1650

Wyngaard, JC, Businger, JA, Kaimal, JC, Larsen, SE (1982) Comments on 'A revaluation of the Kansas mast influence on measurements of stress and cup anemometer overspeeding'. Boundary-Layer Meteorol 22: 245–250

Wyngaard, JC, Moeng, C-H (1992) Parameterizing turbulent diffusion through the joint probability density. Boundary-Layer Meteorol 60: 1–13

Yaglom, AM (1977) Comments on wind and temperature flux-profile relationships. Boundary-Layer Meteorol 11: 89–102

Yaglom, AM (1979) Similarity laws for constant-pressure and pressure-gradient turbulent wall flow. Ann Rev Fluid Mech 11: 505–540

Zelený, J, Pretel, J (1986) Zur Problematik der Bestimmung der aerodynamischen Rauhigkeit der Erdoberfläche. Z Meteorol 36: 325

Zhang, G, Thomas, C, Leclerc, MY, Karipot, A, Gholz, HL, Foken, T (2007) On the effect of clearcuts on turbulence structure above a forest canopy. Theor Appl Climat 88: 133–137

Zilitinkevich, SS, Tschalikov, DV (1968) Opredelenie universalnych profilej skorosti vetra i temperatury v prizemnom sloe atmosfery (Determination of universal profiles of wind velocity and temperature in the surface layer of the atmosphere). Izv AN SSSR, Fiz Atm Okeana 4: 294–302

Zilitinkevich, SS (1969) On the computation of the basic parameters of the interaction between the atmosphere and the ocean. Tellus 21: 17–24

Zilitinkevich, SS, Mironov, DV (1996) A multi-limit formulation for the equilibrium depth of a stable stratified atmospheric surface layer. Boundary-Layer Meteorol 81: 325–351

Zilitinkevich, SS, Calanca, P (2000) An extended similarity theory for the stably stratified atmopheric surface layer. Quart J Roy Meteorol Soc 126: 1913–1923

Zilitinkevich, SS, Perov, VL, King, JC (2002) Near-surface turbulent fluxes in stable stratification: Calculation techniques for use in general circulation models. Quart J Roy Meteorol Soc 128: 1571–1587

Register of Sources

Many publishing houses, societies, authorities and other institutions very kindly agreed to publish figures of there publications in this book. These sources are in the following given:

American Meteorological Society, Boston MA, USA

figure	source	fig. in source	page in source
Fig. 2.3	Stull (1984)	Fig. 1	3352

Annual Review of Fluid Mechanics, Volume 32 ©2000 Annual Reviews
www.annualreviews.org

figure	source	fig. in source	page in source
Fig. 3.29	Finnigan (2000)	Fig. 12	549

B. G. Teubner GmbH, Wiesbaden, Germany

figure	source	fig. in source	page in source
Fig. 7.1	Hupfer (1996)	Fig. 7.1	246
Fig. 7.2	Hupfer (1996)	Fig. 7.14	269
Fig. 8.3	Hupfer (1996)	Fig. 8.4	285

Brooks / Cole, Pacific Grove, CA, USA

figure	source	fig. in source	page in source
Fig. 1.3	Stull (2000)	Fig. 4.8	69

Cambridge University Press, Cambridge, UK

figure	source	fig. in source	page in source
Fig. 1.11	Frisch (1995)	Fig. 7.2	104

German Meteorological Service, Offenbach, Germany

figure	source	fig. in source	page in source
Fig. 3.1	Foken (1990)	Fig. 4.8	96
Fig. 3.21	Baumgartner (1956)	Fig. 12	16
Fig. 6.19	Richter (1995)	Fig. 10	42

Elsevier Science, Oxford, UK

figure	source	fig. in source	page in source
Fig. 3.17	Schmid (1997)	Fig. 1	93

Friedrich Vieweg & Sohn, Wiesbaden, Germany

figure	source	fig. in source	page in source
Fig. 3.22	Geiger *et al.* (1995)	Fig. 14-2	82

Gebr. Bornträger Verlagsgesellschaft, Stuttgart, Germany

figure	source	fig. in source	page in source
Fig. 4.10– 4.12	Foken *et al.* (1995)	Fig. 8	103
Fig. 5.1	Foken (2002)	Fig. 3	704

Kluwer Academic Publisher B. V., Dordrecht, The Netherlands

figure	source	fig. in source	page in source
Fig. 1.5	Foken *et al.* (1978)	Fig. 1	290
Fig. 2.4	Stull (1988)	Fig. 5.4	155
Fig. 3.7	Stull (1988)	Fig. 14.11	598
Fig. 3.11	Hupfer *et al.* (1976)	Fig. 2	504
Fig. 3.16	Schmid (1994)	Fig. 1	296
Fig. 3.23	Amiro (1990)	Figs. 3 and 4	106, 107
Fig. 3.24	Denmead and Bradley (1985)	Fig. 5	434
Fig. 3.25	Gao *et al.* (1989)	Fig. 1a–c	353–355
Fig. 3.28	Raupach *et al.* (1996)	Fig. 6	366
Fig. 3.32	Holtslag and Nieuwstadt (1986)	Fig. 2	205
Fig. 4.2b	Wilczak *et al.* (2001)	Fig. 1	132
Fig. 4.4	Finnigan *et al.* (2003)	Fig. 1	3

Fig. 4.14	Snyder *et al.* (1996)	Fig. 1	251
Fig. 4.16	Hicks and Matt (1988)	Fig. 2	120
Fig. 6.12	Liu *et al.* (2001)	Fig. 1	461
Fig. 7.3	Stull (1988)	Fig. 14.2	589

Munkgaard Int. Publisher Ltd., Kopenhagen, Dänemark

figure	source	fig. in source	page in source
Figs. 5.6 and 5.7	Mölders *et al.* (1996)	Fig. 1	735

Österreichische Gesellschaft für Meteorologie, Wien, Austria

figure	source	fig. in source	page in source
Fig. 4.5	Foken *et al.* (1997b)	Fig. 5	72

Oxford University Press, New York, USA

figure	source	fig. in source	page in source
Fig. 2.8	Kaimal and Finnigan (1994)	Fig. 2.1	34
Fig. 2.9	Kaimal and Finnigan (1994)	Fig. 2.2	35
Fig. 3.6	Kaimal and Finnigan (1994)	Fig. 3.3	78

RISØ National Laboratory, Roskilde, Dänemark

figure	source	fig. in source	page in source
Fig. 3.15	Petersen and Troen (1990)	Fig. 3.2	45

Royal Meteorological Society, Reading, U.K.

figure	source	fig. in source	page in source
Fig. 2.10	Kaimal *et al.* (1972)	Fig. 17	580
Fig. 3.5	Waterhouse (1955)		
Fig. 3.10	Bradley (1968a)	Figs. 6 and 7	368, 369
Fig. 5.8	Schmid and Bünzli (1995b)	Fig. 1a	7

SPB Academic Publisher, The Hague, The Netherlands

figure	source	fig. in source	page in source
Fig. 4.15	Dlugi (1993)	Fig. 4	865

Umweltbundesamt, Berlin, Germany

figure	source	fig. in source	page in source
Fig. 4.9	(UBA 1996)		
Fig. 5.5	(UBA 1996)		

Verein Deutscher Ingenieure e. V., Düsseldorf, Germany

figure	source	fig. in source	page in source
Fig. 6.8	VDI 3786, sheet 2	Fig. 1	5
Fig. 8.1	VDI 2714	Fig. 8	9
Fig. 8.2	VDI 3787, sheet 2	Fig. 2	10

WILEY-VCH Verlag, Berlin, Germany

figure	source	fig. in source	page in source
Fig. 3.2	Hurtalová et al. (1983)	Fig. 1	370
Fig. 6.16	Foken (1979)	Fig. 2	302

World Meteorological Organisation, Geneva, Schwitzerland

figure	source	fig. in source	page in source
Fig. 3.13	WMO (1981)	Fig. 4.36	97
Fig. 3.14	WMO (1981)	Fig. 4.35	97

Some companies made figures available:

figure	company
Fig. 4.6	Campbell Scientific Inc., Logan UT, USA
Fig. 6.7	Th. Friedrichs & Co., Schenefeld near Hamburg, Germany
Fig. 6.15	

Fig. 6.10	R. M. Young Company, USA / GWU Umwelttechnik GmbH, Germany
Fig. 6.17	
Fig. 6.20	METEK GmbH, Elmshorn, Germany
Fig. 6.21	Scintec AG, Tübingen, Germany

Springer-Verlag Berlin, Heidelberg, Springer-Verlag Wien, or the author has the rights to all other figures.

Index

Printing: Mercedes-Druck, Berlin
Binding: Stein+Lehmann, Berlin